ENGINEERING ANALYSIS

An Introduction to Professional Method

ENGINEERING ANALYSIS

An Introduction to Professional Method

D. W. VER PLANCK
HEAD, DEPARTMENT OF MECHANICAL ENGINEERING
CARNEGIE INSTITUTE OF TECHNOLOGY

B. R. TEARE, Jr.
DEAN, COLLEGE OF ENGINEERING AND SCIENCE
CARNEGIE INSTITUTE OF TECHNOLOGY

NEW YORK · JOHN WILEY & SONS, INC.
London · Sydney

SEVENTH PRINTING, MAY, 1965

Library of Congress Catalog Card Number: 54-8420

Printed in the United States of America

Preface

This book is written as an aid in the teaching of courses in engineering analysis. We conceive such courses as having as a major objective the development of the student's capacity to deal with situations that are new to him, by the application of fundamental principles, on his own initiative, and with well-ordered analytical thought processes. The importance of this educational goal has been recognized by many engineers and educators, and needs no further justification here. Rather, the question is how existing courses may be modified or new ones devised to accomplish this purpose effectively.

In our experience the student's capacity to treat new situations with good professional style is rarely developed in the usual course in which the major objectives must include new subject matter, new principles, and new techniques. Another kind of course seems to be indicated in which the learning of new principles and techniques is made subsidiary to the use of old ones in unfamiliar situations, and in which the activity in the course is directed primarily toward using and learning to use familiar principles. Interest in such courses has developed at many engineering colleges, and some have added courses like the ones for which this book is intended.

In the hands of an enthusiastic teacher, who himself is familiar with advanced mathematics and the great generalizations of classic theoretical physics, an engineering analysis course often tends to be a course in engineering mathematics or applied theoretical physics or a combination of both. We know from experience that it is tempting to present new and more general theorems, and new and more elegant mathematical processes instead of putting the emphasis on making sound engineering use of principles and techniques as they already have been presented to the student in his sophomore, junior, or senior years. A course in advanced engineering mathematics or physics is not in itself undesirable—indeed, it may be excellent for many reasons—but it fails to accomplish the urgent need of learning to deal with new problems by means of familiar tools.

The professional engineer treats a new situation in the following stages: (1) defining the specific problem he will attempt to solve; (2) planning his attack, making such simplifications as may seem necessary, and deciding on what principle to base his attack; (3) executing his plan to the end that he reaches a decision or result; (4) checking his work thoroughly; and (5) taking stock to see what he has learned about the given situation as a whole, and what may be of future use, that is, learning and generalizing.

It is the purpose of the kind of engineering analysis courses that we are describing to give special emphasis to stages 1, 2, 4, and 5, the first two being in essence the translation of the engineering situation into mathematical language, the last two, what is done after the mathematical crank has been turned and a result obtained. Courses in advanced engineering mathematics and physics often put their major emphasis on stage 3, executing the solution after the problem has been formulated for the student and then considering the task finished when a mathematical result is first obtained. We feel that, whatever the merit of learning more powerful and more general techniques of mathematical manipulation, there is still urgent need to help the student learn how to do the whole process: how to take the often crude engineering situation and simplify it; then translate it into mathematics himself under conditions in which he does not know at the start just what kind of mathematics is needed; and after he gets a mathematical result carry the work toward an engineering conclusion.

In the courses for which this book is written the student learns how to deal with such whole problems by actually dealing with them. This is in accordance with the principle of learning which says that the student learns from what he himself does, from all that he does, and only from what he does. Hence we may judge the merit of the course in terms of what it permits and encourages the student to do by himself on his own initiative.

In essence the engineering analysis course gives the student an opportunity to work the kinds of problems we want him to be able to solve and which have the same kinds of pitfalls and require the same kinds of thinking that he will encounter later in his professional career. Equally, it is focused on the professional method of dealing with problems, the orderly mental processes the professional engineer uses in practice. The teacher's task is to provide suitable problem situations for the student to work with and to criticize constructively his methods and results. Above all, the teacher must not deprive the student of taking the initiative. The crux of the matter is the choice of problem material. The best problems are based on engineering situations in which the student must decide for himself what the real problem is, which require him to choose among alternative methods at various stages in the solution, and which lead to the making of a decision, rather than ending with a mere numerical result. Such problems are not necessarily extremely difficult; they may be set at almost any level. Level is important, for a problem that is so easy that it does not require hard thinking falls short of doing as much as it should for the student. And a problem that is too difficult fails in its purpose, either by stopping the student completely and leaving him discouraged, or by forcing the teacher to assume the initiative. The best problems offer the student real obstacles which he himself can overcome.

Thus the kind of engineering analysis course we are describing is based squarely on the student's own solution of whole problems that are new to him. Our practice has been to assign one problem each week (or every two weeks), return it promptly with written comments and criticisms, and then discuss it systematically with the class.

We often assign problems orally and somewhat informally in order to give the student an opportunity to define the problem for himself. Oral assignment, which is characteristic of engineering practice, adds further engineering reality to the situations used in the course which are chosen for their inherent interest. Indeed, interest and reality, together with a feeling of accomplishment when the problem is solved, provide abundant motivation. The discussion after the class has turned in its solutions is carried on as a joint task by teacher and class and is a step-by-step solution carried on as if the problem were being dealt with for the first time. Included are a consideration of the different ways of handling various parts of the problem, the consequences of making particular choices, ways of checking, ways of understanding and interpreting results, and finally the different decisions that may be made. Examinations, given every two weeks, are shorter complete problems, in which the student must work entirely without help and under some pressure of time. These are handled and discussed like the weekly problems. We find that after the engineering analysis course gets under way the recitations are devoted almost entirely to discussions of problems the students have done. Often the students take the initiative in discussions and their questions and comments take precedence over the instructor's plan. This procedure means that the choice of problems is the principal planning activity of the teacher. Once these are selected (and we find it best to repeat only sparingly from year to year), the course is defined.

Although such courses are directed primarily toward learning to deal with new situations in terms of fundamental principles they achieve a secondary purpose of great value, that of gaining a more thorough understanding of the principles themselves. There seems to be no better way of learning what a principle really means, its range of usefulness, and its limitations than by using it. And we have found that first-hand use of fundamental principles by the student himself adds immeasurably to the thoroughness of his mastery of them.

This book is an exposition of professional method as it applies to the analysis of engineering problems and of the philosophy that we seek to have our students learn. In the book as in the course we proceed by examples; thus most of the book comprises cases to the selection of which we have given much thought and which we have tried to develop in the same way that a good student might for himself with some guidance. Throughout we are trying to emphasize the thinking process; the form in which the analyses in the book are presented is that which an engineer might use in his own notebook. In the course itself we ask students to present their work in this form, although some teachers may prefer to have their students also perform the next step, the very vital one of communicating the findings by means of a formal report. Our point of view is that we want to be in a position to guide the thinking process, and we can do this better by seeing a notebook version of the student's work rather than a carefully rewritten report. One aspect of analytical power is the ability to make a neat and orderly first draft or notebook version even though portions may have to be crossed out or rewritten. In a few places in the text we have tried to recall and reproduce our own first floundering efforts in a given situation in order to show how mistakes may be uncovered and how after some experience and reflection the engineering novice can improve his treatment.

Also, at some cost of space, we have emphasized thoroughness of understanding in terms of physical pictures, together with appropriateness of the solution and its accuracy. Short cuts are deemphasized since speed does not have the same importance for the professional engineer studying a new situation that it does for the technician doing routine work. Indeed, we suspect that speed is overemphasized in much of engineering education.

The methods used in the course must be suited to the student's background, and in the book we have employed relatively simple and general principles and techniques in the forms in which they are given to engineering students, even though some more sophisticated statement of a principle might lead more quickly to a result. In several instances we have shown or indicated more direct approaches than we first used. In solving differential equations we have mostly chosen the simplest cut-and-try methods. In teaching the courses we encourage students to use any method they prefer and these are often more advanced than those of the text. We have made these choices in order to put greatest attention on the thinking process. In chapters 1 and 2 we have given further emphasis by showing exactly what the engineer, or student, might set down on paper as he works a problem.

The situations in which the problems arise usually concern a small simple device, often familiar to the student. Although the student should learn to deal with new situations, in the beginning the background material should not be *too* new and the facts of the situation should not be too hard to come by. After the student's analytical powers have begun to develop there is plenty of time to start dealing with really complicated situations.

The principles and background material dealt with in the book have been selected for engineering analysis courses to be taken by students in electrical and mechanical engineering and industrial management. The level may be adjusted as appropriate for juniors, seniors, or first-year graduate students. While different problems are used, the different courses are not as far apart as one might think.

For a one-semester beginning course in engineering analysis we have found it desirable to spend from four to six weeks on the problem-solving method itself. During this period we assign a problem each week, the solution to be done in something like the style illustrated in chapters 1 and 2, which we suggest as reference material. Suitable problems for this purpose may be chosen from the first part of the problem section at the end of the book; they should involve relatively little mathematical complexity but should be of a kind where it is not easy to see what the real problem is or how to attack it. We deliberately try to avoid types of problems which are likely to be treated in standard textbooks. Toward the end of this initial period we generally suggest chapter 3 as helpful reading. When we believe that the class is gaining confidence in the professional method we begin to increase the mathematical difficulty of the weekly problems, emphasizing in turn the translation into mathematics, mathematical solutions, checking, and finally the interpretation of the mathematical result. These in order are the topics treated in the remaining four chapters. The latter part of the problem section is useful for this part of the course.

For a more elementary course, parts of the book may be skipped without destroying the continuity of what comes later. For instance, the individual teacher may prefer to pass over some of the illustrative problems in chapter 2 and parts of chapter 3. Also, if it is desired not to become involved with differential equations, parts of chapter 4 and all of chapter 5 may be omitted without affecting the intelligibility of chapters 6 and 7. Indeed, these two chapters could be taken up quite early in the course, and we sometimes do so.

The problems we have included for the student to work are intended to help the teacher get started and as illustrative of the kinds of problems we have found most useful for the purposes of our courses. The teacher will surely wish to build up his own stock of problems based on his own professional experiences or suggested by current technical literature. He will be able to give his own problems a degree of life and reality which cannot be achieved in the ones we communicate by printed statements.

The book contains numerous exercises interspersed within the text material. It is not our intention that these be assigned to be handed in; their purpose is merely to help the student in the process of learning by himself as he reads.

The book is not planned for regular daily reading assignments; indeed, reading is a poor substitute for working problems. Yet we have found that it is desirable for students to have a reference book; that is, it is helpful to be able to read and reread examples of how problems are formulated, how the attack is planned, how results are checked, and so forth. A textbook provides an opportunity for repetition outside the classroom of the points we seek to put across and is particularly valuable to the students who find this point of view difficult to understand. In the classroom we want the students to be active participants in the discussion of problems and not to be diverted by the seeming necessity of taking notes. In giving individual guidance to the student time is saved by referring him to an appropriate chapter or example in the book. It is our experience that with the textbook students develop problem-solving ability more rapidly than without it.

Many of the ideas that we present in this book came to us from our association with the late Robert E. Doherty, both in industry and in education, and from Elliott Dunlap Smith, Provost at Carnegie Institute of Technology. We count it a rare privilege to have had their guidance in developing our philosophy. We gratefully acknowledge also the important contributions made by our many colleagues who have worked with us in the twenty years that we have been teaching such courses.

D. W. Ver Planck
B. R. Teare, Jr.

Pittsburgh, Pennsylvania
April, 1954

Contents

CHAPTER ONE

The Professional Method of Dealing with Engineering Problems

1–1 Problem Solving in Engineering

Problem solving will be one of your main tasks in engineering. You will find, however, a tremendous range in the kinds of problems. At one extreme will be problems of a routine nature solvable by substituting numerical values in familiar formulas. At the other extreme will be problems not solved before that will require creative thinking for their solution, but which nevertheless you probably will be able to solve if you deal with them in a professional manner even though you may have had little experience in the particular field. It is true that you will have many routine problems so well within your powers that you can handle them safely by short-cut methods. Indeed, these may comprise a considerable part of your work, but getting solutions to them is unlikely to bring either professional recognition or much financial reward; in fact, one whose work is entirely of this routine kind is hardly an engineer in the truly professional sense. On the other hand, recognition and rewards will result from your finding solutions to new problems or new solutions to old ones if these solutions have economic or scientific value.

Problems of the kind requiring creative thinking for their solution will not be given you in clearly worded written statements. Rather you will be confronted with situations out of which you yourself must formulate specific problems to be solved. Until you analyze the situation it may not be clear what the problem really is, nor is it likely that the important factors will be apparent at first. Often, in practice, it is harder to define a problem than to solve it.

Another way in which real engineering problems differ from those usually given in college is that, even after the problem has been clearly defined, the way in which the desired solution can be obtained may still not be evident. In a course designed to develop new subject matter or new techniques, you can be almost certain that the problems will be illustrative of the matter being studied. But in practice you will seldom know in advance that a problem will yield to a particular line of attack. Thus, it is important for you to be able to plan how to apply what you know in new situations. You must learn to combat your natural tendency to plunge into equations before you have considered carefully the facts which are to be expressed mathematically or perhaps even before you fully understand the real nature of your problem.

A characteristic that also distinguishes practical problem solving from the usual course work is that you must be more sure of your results. In industry much more than course grades are at stake; you cannot afford to make blunders. You must have sure means for checking your own work, and this does not mean comparing with the work of others because it is rare that two engineers, much less a whole group, are employed to do exactly the same job. Nor will you be able to check by comparison with the result in a book when you are working on the kinds of problems that lead to your advancement, for these will be new problems, never worked before and thus not in books. Methods of checking are, therefore, an essential part of practical problem solving.

In professional practice you will not be done with a problem when you have completed and checked the mathematical work. The end of a problem is rarely a single number or mathematical expression which you can lable *Answer* as you have so often been required to do. Frequently, the product of a mathematical analysis will be a formula which you will have to interpret carefully to yourself so as to discover its full meaning. Sometimes the desired end of an analysis may be numerical, as for example in a design, but it is more likely to be a decision or recommendation perhaps depending heavily on economic and human factors. Often you will need to study a situation from several angles and exercise judgment in selecting the best solution from among various possibilities. You may also have to persuade your associates or client that your recommendation is sound and the best under the circumstances.

Only by continual practice can you expect to develop your capacity to deal effectively with the real problems that will confront you in your professional life. The purpose of this book is to help you learn good professional method. As far as possible this is done by example, the examples being selected to illustrate the application of many of the fundamental principles most useful in engineering practice. This chapter is devoted to establishing in a broad way what the professional method of dealing with problems is; the following chapters develop the method further and present related matter important in engineering analysis.

This book can serve only as a guide to professional method; to learn the method you must apply it yourself to many problems which are unfamiliar and which tax your powers to the limit.

1-2 Example of Professional Method: An Accelerometer

What is meant by the professional method of dealing with an engineering problem is shown first by discussing the way that an engineer might deal with a particular problem. The problem is not difficult, and our discussion of it may therefore appear to be unduly long. Our purpose is to

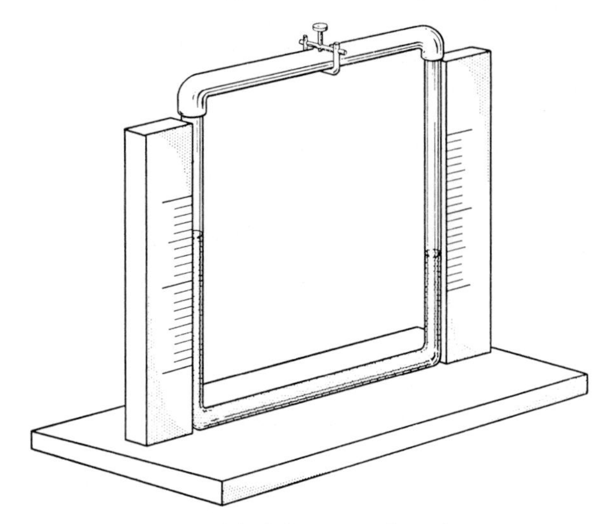

Fig. 1-1. Model of the proposed accelerometer.

make clear not only what an engineer might do at each stage of the solution, but why he might do it.

We will suppose that the engineer is faced with the task of analyzing a form of accelerometer which an inventor has proposed for use in an automobile and making a recommendation concerning its development. The inventor has submitted a model, illustrated in figure 1-1, to demonstrate his proposal. The device is a glass tube bent into a rectangular U shape and containing colored liquid. It is to be mounted in the automobile with the lower part of the tube fore and aft and the side parts vertical. The inventor says that, if the car moves at constant velocity, the liquid in each of the two vertical arms has the same height; if the car is accelerating, the liquid rises in one arm and falls in the other; if the brakes are applied and the car decelerates, there is again a difference of levels but the direction of the change is reversed. The change of level in one vertical tube as read against a scale gives the amount of acceleration or deceleration. Thus, the proposed instrument should be useful in testing the ability of a vehicle to accelerate or the capacity of its brakes to stop it. There is a rubber connection across the top of the U with a clamp to impede the air flow when the levels are changing, for the purpose of damping out the oscillations which occur when the acceleration changes suddenly and which interfere with the reading of the instrument. The inventor believes that making the bottom horizontal tube larger than the vertical ones makes the instrument more sensitive; that is, gives a greater change of level for a given acceleration.

The engineer wishes to determine whether the device is operable in the way that the inventor claims. We will suppose that he has neither seen nor heard of such a device before, but, in the basic science and engineering of his training, he has studied principles that will be useful in handling the problem. One characteristic of this problem that is typical is that numerical data are not given to the engineer, but he himself must supply what he needs.

To come to grips with the situation the engineer thinks about the physical implications of the inventor's claims, and as his thoughts develop he jots them down somewhat as follows:

Evidently the device is responsive to acceleration, for if it is accelerated horizontally the liquid certainly will tend to back up into the rear tube and out of the forward one, thus giving a difference in liquid levels which might be read against a scale. Such a difference in level, however, could also result from other influences as follows:

Tilting of the base.
Difference in capillarity in the two side tubes.
Difference in air pressures on the two free surfaces.

For a first analysis assume these influences are not present. That is, assume:

Base remains horizontal and the acceleration is horizontal. Vertical tubes identical so that surface-tension effects cancel.

Difference in height has been maintained constant long enough for air pressures above the two sides to equalize through the restricted passage, that is, the liquid comes to rest relative to the tube.

With these assumptions the problem is:

How is difference in the liquid levels related to acceleration?

Now let us pause to see what the engineer has accomplished so far. In the first place he has reduced vague wonderings such as "Is the device any good?" or "How does it work?" to a single definite question capable of a specific answer which will be just what he needs to know. Secondly, by making simplifying assumptions he has stripped away all but what he believes are the determining factors. This does not mean that he intends to ignore the other factors; rather he is reserving them for consideration after focusing his whole attention on what seems to him at this stage to be the essential problem.

Having defined his problem the engineer goes on organizing his thoughts and writing them down to keep them straight:

> Consider the horizontal portion of the liquid when the car is accelerating. If there were no net force on it there would be no acceleration and the car would move away from the liquid, which would back up into the rear vertical tube. Evidently this creates a greater pressure at the rear end of the tube, and provides the unbalanced force to keep the liquid accelerating with the car. To find how this force depends on acceleration, apply Newton's law $\Sigma f = ma$ to the liquid in the horizontal tube.

The engineer is mindful that Newton's law in this form applies not only to rigid bodies but also to any aggregation of particles, such as the liquid in this case, provided the acceleration is that of the center of mass.

Continuing his planning, the engineer writes:

> Since the liquid is at rest relative to the tube there will be no shearing forces between it and the glass surfaces. Also, the pressure acting across these surfaces gives no resultant horizontal force on the liquid because the axis of the tube is horizontal and its cross section is uniform. Thus the only horizontal forces acting on the liquid mass are those supplied by the pressures at the ends.

The engineer now begins to carry out his plan for solution by sketching the diagram of figure 1–2, showing the liquid in the horizontal tube as a free body and the horizontal forces acting on it. In doing this he is confronted with the desirability for another simplification, and he makes it:

> Neglect the effects of the corners of the tube by treating the liquid body as a cylinder extending between the center lines of the vertical tubes.

Now he is ready to state explicitly how Newton's law of motion applies, and he writes:

[The difference of the forces at the two ends of the liquid, which result from the columns of liquid in the vertical arms, is equal to the product of the mass of the liquid in the horizontal tube by its acceleration, which is the acceleration of the instrument.]

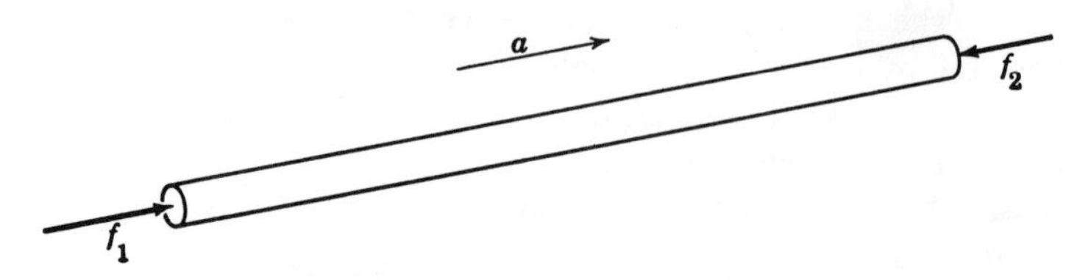

Fig. 1-2. Free-body diagram of the liquid in the horizontal tube.

Referring to this statement of Newton's law as applied to his particular case and aided by the free-body diagram, the engineer writes an equation in mathematical form:

$$f_1 - f_2 = ma \tag{1-1}$$

f_1 = force on left end of the liquid, positive when exerted to the right (pounds).
f_2 = force on right end, positive to the left.
m = mass of the liquid in the horizontal tube (slugs).
a = acceleration of the liquid (same as that of tube), positive to the right (ft sec^{-2}).

Notice that in starting to execute his plan for solving the problem the engineer did not try immediately to write a mathematical equation. Instead he proceeded in two distinct steps. First he applied the physical law by writing a clear statement in words using the free-body diagram to help him visualize. Then with the physical truth clearly set down before him he translated it into a mathematical equation, carefully defining the symbols and positive directions. A formulation in mathematical terms involves two different functions; applying a physical law and using arbitrarily defined symbols and coordinates. When both these functions are performed together, there is far greater chance for confusion and consequent error than when they are separated. Particularly in a problem that taxes your capacity to the limit, the complexity of doing these two things at once may make the difference between success and failure. Accordingly, good professional engineering method demands separation of the two functions. Application of physical law is accomplished by making a statement in precise English telling exactly what the chosen principle implies in the particular case at hand. Definition of symbols and arbitrary choice of coordinates is then made a separate matter, and the final translation into mathematics is relatively easy.

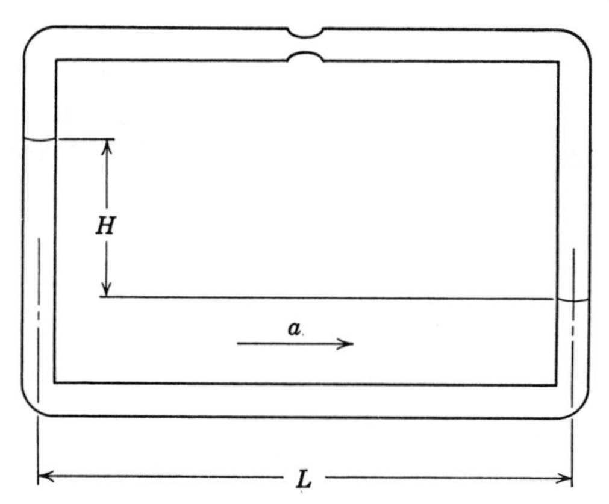

Fig. 1-3. Diagram showing the significant dimensions.

Having established equation 1–1 the engineer considers how he can put it in a useful form. He wishes to express the difference in forces $f_1 - f_2$ in terms of difference in height of the vertical columns of liquid. To do this he writes:

The force is average pressure times area, therefore

$$f_1 = p_1 A$$
$$f_2 = p_2 A \qquad (1\text{–}2)$$

A = cross-sectional area of the horizontal tube (ft²).
p_1 = average pressure over the cross section at the left end of the tube (lb ft⁻²).
p_2 = (similarly) the pressure at the right end.

Thus

$$f_1 - f_2 = (p_1 - p_2)A \qquad (1\text{–}3)$$

But the difference in pressures at the two ends $p_1 - p_2$ is proportional to the difference in heights of the two vertical liquid columns; that is

$$p_1 - p_2 = \rho g H \qquad (1\text{–}4)$$

ρ = mass density of the liquid (slugs ft^{-3}).
g = acceleration of gravity (ft sec^{-2}).
H = difference in height (ft) shown in the sketch figure 1–3, the difference being taken between corresponding points on the meniscuses.

Here the engineer has neglected, perfectly properly, the pressure differential associated with the air column of height H in the right-hand tube. Going on with his work he combines equations 1–3 and 1–4 with 1–1 and writes:

[

$$\rho g H A = ma \qquad (1\text{–}5)$$

but the mass m is density ρ times volume of the horizontal tube; thus

$$m = \rho A L \qquad (1\text{–}6)$$

L = length of the lower tube (feet).

]

The engineer also defines L on his sketch, figure 1–3, and going on he substitutes 1–6 into 1–5 and solves for H:

[

$$H = \frac{\rho A L}{\rho g A} a$$

$$= L \frac{a}{g} \qquad (1\text{–}7)$$

Since the vertical tubes have been assumed identical in section the rise of liquid in one will equal the fall in the other, and each will equal $H/2$. Then

$$h = \frac{La}{2g} \qquad (1\text{–}8)$$

where h is the rise of the liquid in the left-hand tube above its level when the acceleration is zero.

]

The left member of equation 1–8 represents a change of level on one of the scales of the instrument, and according to the equation is proportional to the acceleration as the inventor claims. The engineer, however, does not stop here, for in his analysis he deliberately ignored various factors which may affect the final conclusion significantly. Moreover, at this point he is not yet sure that there may not be errors in his analysis of even the simplified case which he has treated. So he goes on with his thinking and writing.

[The result checks dimensionally.]

By this he means that the terms of the equation each measure the same kind of physical quantity, in this case length. The term on the left is defined as a length; the right member is a length L multiplied by an acceleration a and divided by another acceleration g, and hence also has the dimensions of length. Thus the equation is dimensionally homogeneous as it should be.

The result 1–8 says that if a is zero h will be zero, and as a increases h also increases as it should.

If L is zero h is zero no matter what the value of acceleration. This is correct because with L = zero there would be no horizontal body of liquid to accelerate, hence no difference in heights would be brought about.

In each of these checks the result agrees in a special case with what is concluded by direct physical reasoning for the special case. These checks and the test for dimensional homogeneity are powerful means for detecting such errors as the misplacement or omission of factors. The engineer is aware, however, that he has not positively established that his result is correct. For example, is the numerical factor $\frac{1}{2}$ in equation 1–8 correct? To assure himself on this and similar points he reviews all of his analysis, carefully going over it step by step from the beginning.

As the engineer reexamines his work he collects the simplifying assumptions he has made and keeps alert to detect any additional assumptions that may be implied in his analysis. He jots down his assumptions with remarks about their effects.

1. Base and lower tube are horizontal. This is not an important limitation because the instrument can be mounted to permit leveling, or if necessary the analysis can be redone for the case where the instrument is not horizontal.

2. Vertical tubes identical. This assumption was introduced so that the effects of capillarity would balance, but, since the reading of the instrument can be made on one scale as the difference in two positions of the meniscus, the assumption need not have been made. If, however, the tubes are of unequal area the rise of liquid in one would not equal the fall in the other, but the analysis could be modified to take this into account if necessary.

3. Transient effects have been neglected. The analysis assumes the acceleration constant or changing very slowly. Actually there would be transient effects such as there are in many measuring devices. Whether or not these effects would be serious in this case would require further study.

4. The horizontal tube is assumed uniform in section and straight throughout its length, but its section need not be the same as that of the vertical tubes. This analysis does not hold if the horizontal tube is of non-uniform section between the planes where the vertical tubes join it.

By this systematic examination of what he has done the engineer not only has gained further assurance that his analysis is sound but he has also checked to see what limitations are imposed by his assumptions. Further, he has learned a good deal about the significance of his result in relation to what he is trying to find out. He is now in a position to draw his analysis to a conclusion, and does so by writing the following general statements.

Change in height of either vertical liquid column is directly proportional to acceleration, as the inventor claims.

In addition to the change in height only two other quantities need be known to determine the acceleration: the horizontal distance between the vertical tubes, and g.

The result is independent of density of the fluid although density may influence the transient response.

Contrary to the inventor's opinion the sensitivity is independent of the diameter of the horizontal tube.

With this generalization by the engineer let us turn from the study of the accelerometer to a study of the main features of his method of analysis. In the first place it should be pointed out that there are other more sophisticated ways of analyzing the U-tube accelerometer that may seem to arrive at a solution more quickly or precisely. Judgment, taste, and the engineer's background will determine which he chooses. We have chosen the particular approach here because it employs very elementary principles. This possibility of alternate schemes of approach is typical of most engineering problems.

An important characteristic of the work is the completeness with which the engineer wrote out his thoughts about the problem and his analysis of it. This writing is not a report to communicate the findings to someone else; it is a careful record of the analysis which the engineer made for himself as he went along to keep his thinking consistent and orderly.

Now let us review the engineer's written work to see what he was doing at each stage. The first writing begins on page 4. Here the engineer was evidently collecting the facts of the problem and deciding which were important and which less so. This led him to the question "How is difference in liquid levels related to acceleration?" Thus the engineer defined a specific problem for himself.

Next the engineer turned to a consideration of the way to answer his question. His reasoning toward a plan for doing this is summarized in the next two bits of writing (on page 5). The plan which he adopted was to apply Newton's law $\Sigma f = ma$ to the liquid in the horizontal tube.

It is important to observe that the principle the engineer planned to use is a very fundamental one; it has broad applicability, and he thoroughly understands its uses and limitations.

Then the engineer carried out his plan. First he wrote out clearly and exactly how the chosen principle applied to his case; then he translated this statement into the mathematical form equation 1–1. As he developed his mathematical analysis he found it necessary to introduce additional physical definitions and concepts as he went along, including the definition of pressure, the relation of pressure to height of a column of fluid, and the concept of mass density. The execution of his plan culminated with equation 1–8.

Next we find the engineer checking his work in various ways, first in detail and then more broadly as he considered the effects of the simplifications which he thought desirable to make.

Checking the effects of the simplifications led naturally to a consideration of what was to be learned from the analysis that would be useful in relation to the real problem. This led to the last of the writing which we have reproduced here; generalizations which the engineer was able to deduce from his analysis.

Thus we can summarize by saying that the analysis of this problem proceeded in five distinct stages: defining the problem, planning a way to solve it, executing the plan, checking the work, and finally learning and generalizing from what was done. These five stages characterize the professional method. They are illustrated further in section 1–3 by another example of problem solving, and in section 1–4 they are defined in general terms.

1–3 A Second Example: Magnetic Force on the Iron Core of a Solenoid

Now that you have seen how the engineer's analysis of the accelerometer proceeded in five distinct stages—the five stages of the professional method—let us investigate the method further by considering an entirely different problem, this time identifying the stages as we go along.

Defining the Problem

Suppose that we are working on a scheme for controlling the alternating current that is supplied to a heating element. We are using a variable inductor in series with the heater and a constant voltage a-c source as shown in figure 1–4. The variable inductor consists of a solenoid with movable iron core; when the core is fully in the solenoid, the inductance is high and the current a minimum; as the core is moved out the inductance becomes less and the current increases. The core is laminated to minimize eddy currents that are induced in it. In the present state of our development we have an inductor built which is crude mechanically but satisfactory as far as current-carrying capacity and range of inductance are concerned. Now we want to incorporate in the design a remotely controlled motor for moving the core. One of the first questions that

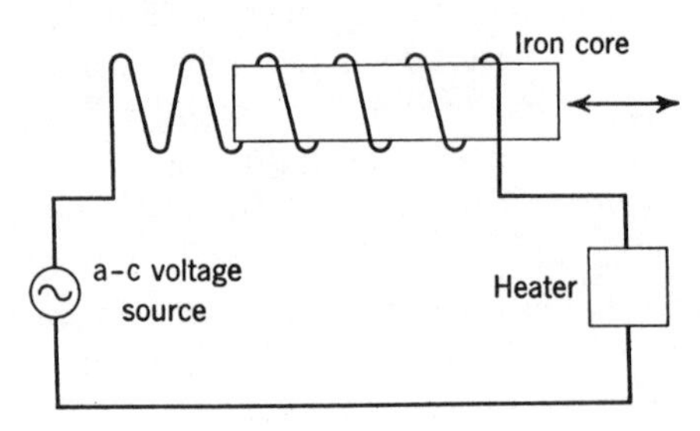

Fig. 1-4. Variable inductance method of control.

arises, and the one we want to study now, is that of the amount of force required to move the core.

To keep our objective clearly before us we write it down thus:

> Find the magnetic force exerted on the core of the inductor when it carries alternating current.

Observe that we have not included in this definition of the problem numerical data or statements about what is supposed to be known, as is usual with problems assigned in college courses. This is because at this early stage of the analysis we do not know exactly what data will be needed. Later as we plan the treatment of the problem we must consider what information either is at hand or can be obtained, and we may find it necessary then to shape our plan according to the availability of data.

Planning the Treatment

At the outset in thinking about our attack we have to choose between determining the force analytically or measuring it in the laboratory. In spite of the approximations that may be entailed we decide in this case on analysis rather than experiment for two reasons: analysis often tells more about how the main factors enter a problem than simple experiments would, and it is often less expensive. In our case the existing inductor is mechanically crude and would have to be rebuilt with considerable care and precision and also means would have to be provided to measure force.

The problem of determining the force analytically is complicated by the fact that the current is alternating, varying sinusoidally in time. We think that the way to deal with the alternating current may be first to find a way of calculating force when the current is constant and then replace the constant current by one that varies sinusoidally. As a check on this idea we suppose that the current has a particular value and consider whether anything makes the force on the iron core depend on whether or not the current is changing. We think the instantaneous state of magnetization of the iron depends only on the instantaneous value of the current in the solenoid if the effects of eddy currents and hysteresis can be neglected. To neglect these effects appears reasonable since the core is designed for low power loss, being both laminated to reduce eddy currents and made of a steel having low hysteresis. To fix in mind the part of the plan developed so far we set it down:

> First find the force due to a constant current and then afterward extend the result to the case where the current alternates, and in so doing neglect eddy currents and hysteresis.

In thinking about the relation of magnetic force to current we realize that a current in the solenoid will produce a magnetic field at every point within and on the surface of the core. If this field could be calculated we should be able to determine the magnetic force on each part of the core and then sum to get the total force. But it is rather difficult to find the field in a case such as this; it probably would be necessary to map it graphically. As we would like to avoid this, we look for some other approach.

We recall that there is energy stored in the magnetic field, and it can be expressed in terms of inductance ($\frac{1}{2}Li^2$). In our case we happen to have measured the inductance at various positions of the core in connection with our previous work, but even if we had not it still might be reasonable to think in terms of inductance.

We could relate energy stored to mechanical force if the core happened to move in the direction of the force because then the force would do work and we could apply the law of conservation of energy. So we suppose the core is at first in mechanical equilibrium; the magnetic force on it is balanced by an equal and opposite restraining force as shown in figure 1–5 applied mechanically by some external means, the mechanism for holding or positioning the core. Next we imagine that the restraining force is very slightly increased so that the magnetic force is overcome and the core is pulled out a little bit and then the restraining force is reduced enough to restore equilibrium. In this process the restraining force does work which we should be able to relate to change in stored energy, and through this relationship find the restraining force (equal and opposite to the magnetic force) in terms of inductance.

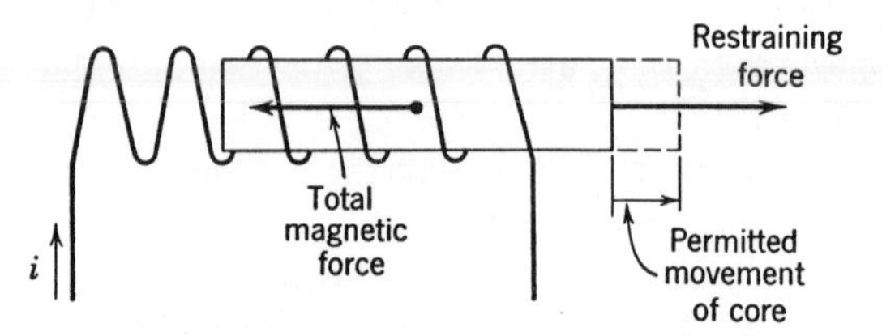

Fig. 1-5. Equilibrium and motion of core.

This method seems easier than the one that involved field mapping, so we decide to try it; and to consolidate our ideas we write the following plan:

> To find the force on the core for any particular value of current regarded for the moment as constant, suppose the core is at rest, the magnetic force being balanced by an external restraining force. Then let the restraining force be increased by an infinitesimal amount so that the core moves very slightly, after which the restraining force is modified so that there is equilibrium again. Neglect any frictional forces which might be present. During the motion there are energy changes that involve mechanical work and energy stored in the magnetic field of the solenoid. Use the law of conservation of energy to relate these energy changes, and try to solve for the restraining force in terms of inductance.

We are now ready to carry out our plan.

Execution of the Plan

First we state clearly how the physical principle we have selected, the law of conservation of energy, applies to our problem.

> During the displacement of the core as shown in figure 1–5 the work done by the externally applied restraining force is equal to the gain of energy stored in the magnetic field of the inductance.

To check this statement before going on we think carefully about what is happening. As the core is moved the energy stored in the magnetic field certainly changes and the restraining force certainly does work. Is there any other way in which energy is put into or taken out of the system comprising the solenoid, the core, and the magnetic field? What happens to the current as the core moves? We reason that, if the current were supposed to be constant before the core moves, the field is also constant; if we neglect the resistance of the solenoid there is no voltage across the solenoid and hence no electric power put in or taken out. But during the motion of the core if current were constant the field probably changes, causing a voltage to be induced in the solenoid, and thus during the motion energy *is* put in or taken out electrically. So our written statement is wrong; we have to modify it to include energy that may be exchanged with the electrical source.

In fact our planning now seems to be incomplete; we should have considered the electric circuit more carefully. If the current is to stay constant during the motion of the core there will have to be a source of voltage to maintain it so.

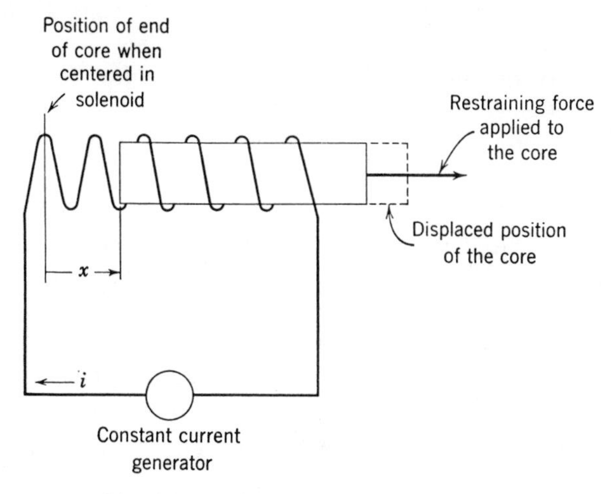

Fig. 1-6. Inclusion of the current source.

We think of two different ways to complete the circuit: either with a battery and resistance or with an automatically controlled source that supplies just enough voltage always to maintain the current constant at its initial value. The first introduces the complexity of another place that energy can go, into i^2R heating; the second is simpler in that resistance need not be included, and in this day of electronic gadgets is not too hard to visualize. So choosing the second we add to our plan.

[Neglect the resistance of the solenoid and suppose constant current to be supplied from a generator that develops just the right voltage during motion of the core to balance the magnetically induced voltage of the solenoid. See figure 1–6.]

It is not unusual to find as one progresses that a plan has to be modified, and so we cross out the incorrect statement of the application of conservation of energy to our problem and instead write:

[During the displacement of the core as shown in figure 1–6 the work done by the externally applied force plus the energy put in by the generator is equal to the gain of energy stored in the magnetic field of the inductance.]

Next we proceed to translate this statement in words into mathematics, carefully defining symbols so that their algebraic signs have meaning and their units are clear.

[

W_M = mechanical work put in by the externally applied restraining force as the core is moved a small amount (joules).

W_G = energy supplied by the generator during the movement (joules).

$\Delta(\frac{1}{2}Li^2)$ = increase in stored energy (joules). (L is the inductance in henries and i the current in amperes.)

]

At this point we need not define the positive direction of current; experience tells us that the core will tend to be drawn into the solenoid no matter which way the current flows. (When electromagnets pick up iron they appear to do so equally well with the current in either direction.) Clearly also the energy $\frac{1}{2}Li^2$ stored in the magnetic field does not depend on the direction of current. Proceeding, then, we write the statement of conservation of energy in mathematical symbols:

$$W_M + W_G = \Delta(\tfrac{1}{2}Li^2) \quad (1\text{–}9)$$

Now we turn to the expression of each of these quantities in terms which will be useful. To express W_M we need to define the coordinate locating the core and the force which acts. Thus:

[

x = distance in meters, positive to the right, locating the core with respect to its position of symmetry in the solenoid. This is also the position in which the inductance is a maximum if the solenoid is wound uniformly. See figure 1–6.

Δx = small displacement of the core in meters, positive when it is in the $+x$ direction.

f_R = externally applied restraining force, in newtons directed along the axis, positive when it has the direction of $+x$. (f_R is equal and opposite to the magnetic force on the core.)

]

In the MKS system of units we are using, force is measured by the newton equal to 10^5 dynes, and the work done by a force of 1 newton acting through a distance of 1 meter is 1 joule. Going on with our writing:

During the movement of the core the force f_R differs only infinitesimally from its equilibrium value, so it is essentially constant; hence the work it does is

$$W_M = f_R \, \Delta x \tag{1-10}$$

The energy put into the system by the generator is equal to the time integral of the product of current by voltage drop through the solenoid. The voltage drop is

$$e = N \frac{d\phi}{dt} \tag{1-11}$$

where N is the number of turns and ϕ is the flux in webers.

Use of this relationship implies that the same flux links each turn; a more general expression might have been written in terms of the summation of flux linkages with each turn.

The flux linkage $N\phi$ depends linearly on the current i and on the position x of the core. (See figure 1-7.) The slope of the line for each constant value of x is the inductance L of the solenoid for the core in that position. Thus $N\phi = Li$, and so

$$\phi = \frac{Li}{N} \tag{1-12}$$

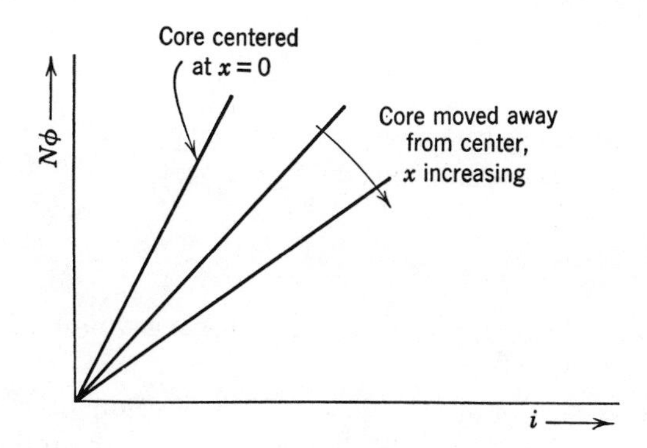

Fig. 1-7. Flux linkage as a function of current for various positions of the core.

In sketching the graphs of figure 1-7 as straight lines we are assuming that since the flux path is largely in air the iron does not saturate. Thus we assume inductance L is a function of position of the core but not of current.

Substituting 1–12 in 1–11,

$$e = N\frac{d}{dt}\left(\frac{Li}{N}\right) = \frac{d}{dt}(Li) \tag{1–13}$$

since N is a constant. Thus the energy W_G put in by the generator is

$$W_G = \int_0^{\Delta t} ei\,dt = \int_{(Li)_0}^{(Li)_0+\Delta(Li)} id(Li) \tag{1–14}$$

The limits of the second integral are the values of linkage at the beginning of the motion $t = 0$ and at the end when $t = \Delta t$. Thus since i is being held constant

$$W_G = i^2\,\Delta L \tag{1–15}$$

where ΔL is the change in inductance when the core moves from x to $x + \Delta x$.

The energy stored in the inductance expressed in joules is $\frac{1}{2}Li^2$; thus the gain of stored energy is

$$\Delta(\tfrac{1}{2}Li^2) = \tfrac{1}{2}i^2\,\Delta L \tag{1–16}$$

since i is constant.

Now putting each of the energy terms 1–10, 1–15, and 1–16 into 1–9,

$$f_R\,\Delta x + i^2\,\Delta L = \tfrac{1}{2}i^2\,\Delta L \tag{1–17}$$

or

$$f_R = -\tfrac{1}{2}i^2\frac{\Delta L}{\Delta x} \tag{1–18}$$

and, if Δx is small, approaching zero,

$$f_R = -\tfrac{1}{2}i^2\frac{dL}{dx} \tag{1–19}$$

The force f of the magnetic field on the iron core is equal and opposite to the externally applied restraining force f_R, hence

$$f = \tfrac{1}{2}i^2\frac{dL}{dx} \tag{1–20}$$

where f like f_R is in newtons and is positive when it has the $+x$ direction.

Thus the first part of our plan, to find the force due to a constant current, has been accomplished. Going on to extend the result to the case of alternating current we write:

Equation 1–20 has been derived assuming i fixed at a constant value. Neglecting eddy currents and hysteresis, 1–20 should give the instantaneous value of force corresponding to the instantaneous value i of a varying current. If the alternating current is sinusoidal

$$i = \sqrt{2}\, I \cos \omega t \tag{1-21}$$

where I is the rms value of the current in amperes and ω is the angular frequency. Then substituting 1-21 in 1-20,

$$f = \tfrac{1}{2}(2I^2 \cos^2 \omega t)\,\frac{dL}{dx} \tag{1-22}$$

$$= \tfrac{1}{2}I^2\,\frac{dL}{dx}\,(1 + \cos 2\omega t) \tag{1-23}$$

This force is made up of a constant term

$$f_0 = \tfrac{1}{2}I^2\,\frac{dL}{dx} \tag{1-24}$$

and a sinusoidal term whose peak value is the same quantity and which has a frequency double that of the current.

To clarify this to ourselves we make a sketched plot of 1-23, figure 1-8, showing how the force varies with time. Evidently 1-24 gives the time average of the force.

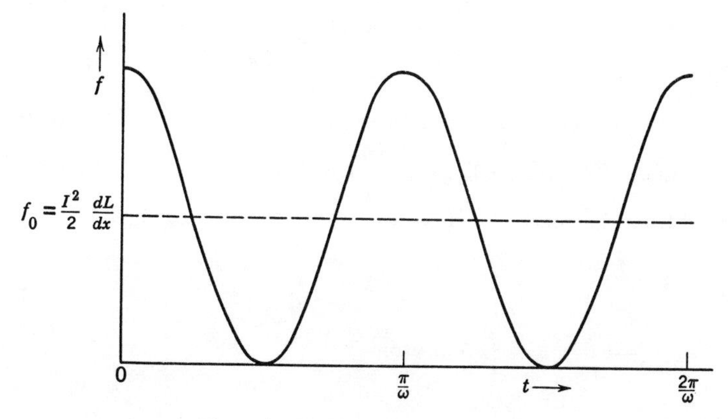

Fig. 1-8. Sketch of equation 1-23.

For finding the numerical magnitude of f_0 we turn to the experimental determination of the inductance for different core positions and from these data plot the curve of L versus x of figure 1-9. Then by graphical means we find its slope, the quantity needed. A curve of $-\dfrac{dL}{dx}$ is included in the figure. In testing the use of 1-24 we write

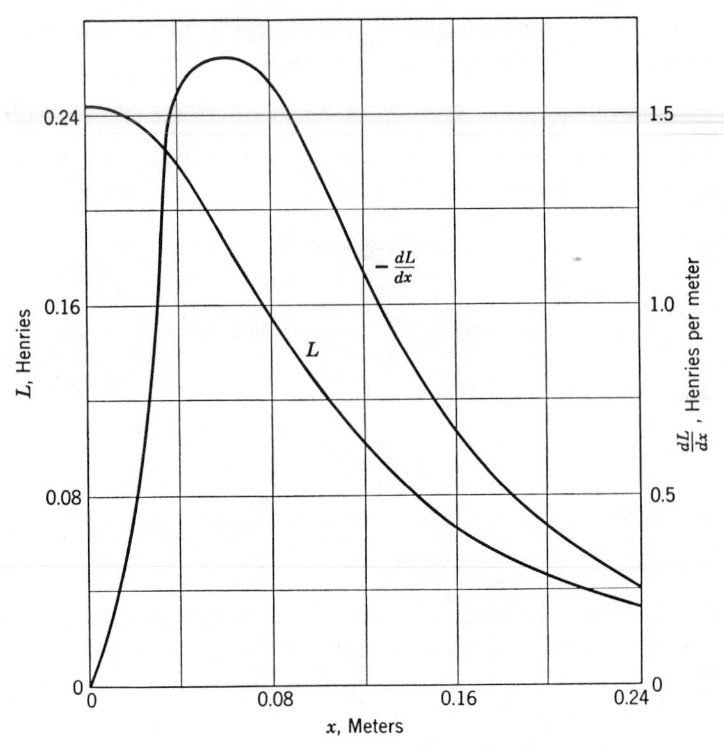

Fig. 1-9. Inductance and its derivative.

Suppose the current I is 10 amperes and we wish to find the greatest value of f_0. This will occur where $\dfrac{dL}{dx}$ has its greatest magnitude, and this is read from the curve as -1.65 henries per meter. Then

$$
\begin{aligned}
f_0 &= \tfrac{1}{2}I^2 \frac{dL}{dx} \\
&= 50(-1.65) = -82.5 \text{ newtons} \\
&= -82.5 \times 10^5 \text{ dynes} \\
&= \frac{-82.5 \times 10^5}{981} \text{ grams force} \\
&= \frac{-82.5 \times 10^5}{981 \times 454} = -18.5 \text{ lb}
\end{aligned}
$$

The minus means that the magnetic force is opposite to $+x$, that is, toward the center of the solenoid.

Thus we have done what we set out to do: found an analytical means for determining the force on the core. But the job is not finished, for we are by no means sure that our analysis is correct; we must check.

Checking the Result

First we check dimensionally and write

The left member of 1–20 is a force; the right member is

$$\frac{[\text{Current}]^2[\text{Inductance}]}{[\text{Length}]} = \frac{[\text{Energy}]}{[\text{Length}]} = [\text{Force}]$$

That is, inductance multiplied by current squared has the dimensions of energy; energy has the dimensions of force times distance. Thus the equation is dimensionally homogeneous, as it should be.

Now we seek limit checks.

According to 1–20 the force is zero when the current is zero and increases with the current. These deductions from 1–20 agree with expectation.

Moreover, 1–20 says that the direction of force is independent of the direction of current and this agrees with experience.

The fact that 1–20 shows force to be proportional to the square of current, rather than some other power, seems to agree with the fact that the force on iron in a magnetic field is proportional to the square of flux density and flux density is proportional to current.

Now check the signs of 1–19 and 1–20. If the core is actually in the position shown in figure 1–6 the inductance decreases as x increases, hence $\frac{dL}{dx}$ is negative. Equation 1–19 indicates that f_R is then positive; 1–20 that f is negative. This agrees with experience; the solenoid tends to draw the core into it and the mechanically applied force f_R has such a direction as to pull the core out, the $+x$ direction in fact.

If x is zero and the core is symmetrically placed in the solenoid, $\frac{dL}{dx}$ must be zero (L is a maximum and decreases if the core is moved either way). According to 1–20 the force then is zero and this seems correct, that is, $x = 0$ is a position of equilibrium toward which the solenoid draws the core.

Although the foregoing checks are quite gratifying, the result could be dead wrong and still withstand these tests. To check the analysis as a whole a graphical interpretation may be helpful and we proceed to try it.

During the small displacement the inductance of the solenoid decreases. The inductance at any position of the core is the ratio of flux linkages $N\phi$ to current; or the slope of the $N\phi$-i curve in figure 1–10. This is shown for the first position x as the line with greater slope, for the second, $x + \Delta x$, as that with lesser slope. The energy stored in an inductance $\frac{1}{2}Li^2 = (\frac{1}{2}N\phi)i$ is the triangular area oa_1b_1 in the first case and oa_2b_2 in the second.

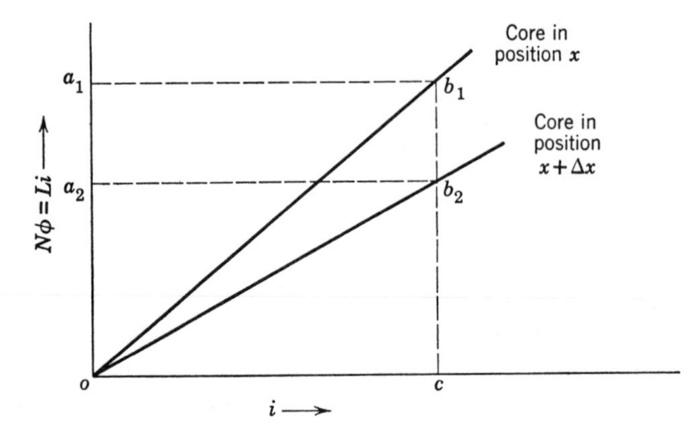

Fig. 1-10. Graphical interpretation.

That the stored energy is represented by these triangles and not by the equal ones ob_1c and ob_2c we know from the fundamental derivation of the expression for the stored energy. This is found as the energy that must be supplied at the terminals to build up the current against the voltage of self-induction. Thus the stored energy E_s is

$$E_s = \int_0^t ei\,dt = \int_0^t N\frac{d\phi}{dt}i\,dt$$

$$= \int_0^{N\phi} i\,d(N\phi) = \int_0^{Li} i\,d(Li) = \tfrac{1}{2}Li^2$$

and from the second line it is clear that the area of the triangle with base on the $Li = N\phi$ axis represents the stored energy. Going on with our graphical checking:

The two triangles representing the stored energy before and after have the common altitude $a_1b_1 = a_2b_2$ but different bases oa_1 and oa_2. The difference in their areas is the triangle ob_1b_2; hence this area represents the decrease of energy stored when the inductance decreases, the core being moved from x to $x + \Delta x$.

The energy put into the system by the constant current generator is, from 1–14,

$$W_G = i\,\Delta(Li).$$

Since the flux linkage Li decreases as the core moves from x to $x + \Delta x$, $\Delta(Li)$ is negative and thus W_G is negative, that is, the generator actually *receives* energy. In figure 1–10 this energy is represented by the area $a_1b_1b_2a_2$.

By conservation of energy,

Work done by f_R + Decrease of stored energy
= Energy received by the generator

or in terms of areas on figure 1–10

$$\text{Work done by } f_R + \text{Area } ob_1b_2 = \text{Area } a_1b_1b_2a_2$$

But the area of the rectangle $a_1b_1b_2a_2$ is just twice the area of the triangle ob_1b_2, therefore

$$\text{Work done by } f_R = \text{Area } ob_1b_2$$

That is, during the motion of the core a certain amount of energy is transferred into the generator, and half of this comes from the stored energy and half is supplied by the external mechanical agent which moves the core out through the distance Δx.

The area ob_1b_2 representing the work done is $-\frac{1}{2}i\,\Delta(Li) = -\frac{1}{2}i^2\,\Delta L$, the negative sign being inserted to account for the fact that the work done is positive when ΔL is negative, that is, when the inductance decreases as in the case of figure 1–10. If the displacement is small enough the work done by f_R is $f_R\,\Delta x$, and so

$$f_R\,\Delta x = -\tfrac{1}{2}i^2\,\Delta L$$

$$f_R = -\tfrac{1}{2}i^2\frac{dL}{dx}$$

when $\Delta x \to 0$.

This confirmation of the former result does not, of course, constitute an independent check because we have merely rederived the relationship by a graphical treatment that parallels the analytical one. It is a check none the less since we have taken a fresh view of the process and found nothing wrong. Indeed, we have increased our understanding and have gained an inkling that the analysis may be made more general. To follow up this possibility we write:

Redo the graphical analysis without making the assumption that the current is held constant during the motion of the core.

During the motion from x to $x + \Delta x$ suppose that the current increases so that the relation between linkage and current is along the curve b_1b_2 (see figure 1-11).

$$\text{The increase of stored energy} = \text{Area } oa_2b_2 - \text{Area } oa_1b_1$$
$$= \text{Area } opb_2 - \text{Area } a_1b_1pa_2$$

By 1-14, which is general and not restricted to constant current, the energy supplied by the source is $\int id(Li)$ taken over the change of linkage from c_1b_1 to c_2b_2 along the path b_1b_2. It is represented by the area $a_1b_1b_2a_2$ and is negative since the linkage Li is decreasing. That is, this area taken as positive represents energy transferred *to* the generator.

By conservation of energy the work done by the restraining force f_R during core motion Δx

$$= \text{Increase of stored energy} + \text{Energy transferred to the generator}$$
$$= \text{Area } opb_2 - \text{Area } a_1b_1pa_2 + \text{Area } a_1b_1b_2a_2$$
$$= \text{Area } opb_2 + \text{Area } pb_1b_2$$
$$= \text{Area } ob_1b_2$$

To express this last area in terms of linkages and current, divide it into the three triangles which meet at q. Thus:

$$\text{Area } ob_1b_2 = \text{Area } ob_1q + \text{Area } oqb_2 + \text{Area } qb_1b_2$$
$$= -\tfrac{1}{2}i\,\Delta(Li) + \tfrac{1}{2}[Li - \Delta(Li)]\,\Delta i - \tfrac{1}{2}\,\Delta i\,\Delta(Li)$$

where the last term assumes b_1b_2 to be a straight line. The minus signs are each associated with $\Delta(Li)$ being negative in the case represented in figure 1-11. Expanding this expression and dropping terms containing products of increments by increments results in

$$\text{Area } ob_1b_2 = -\tfrac{1}{2}i^2\,\Delta L$$

and thus

$$f_R\,\Delta x = -\tfrac{1}{2}i^2\,\Delta L$$

which is equivalent to 1-18, and therefore the original result does not depend on the assumption of a constant-current generator. That the force turns out to depend on current, but to be entirely independent of the source which supplies the current, seems as it should be.

Observe that incidental to checking the result of the original analysis we have reworked the problem twice. In the last solution we were able to remove a restriction because of insight gained during the earlier work.

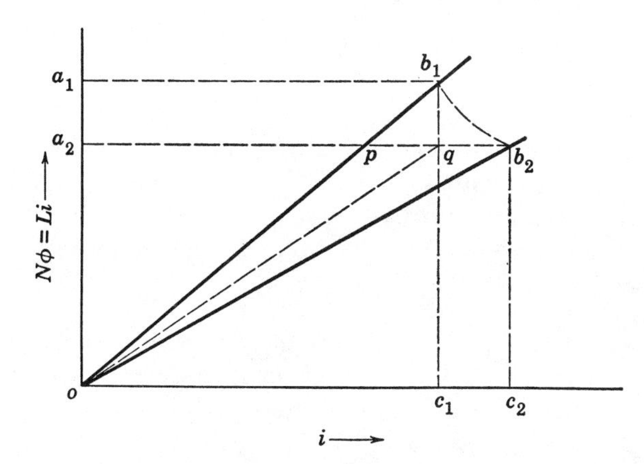

Fig. 1-11. Extension of the graphical analysis.

Going on with our checking we summarize the simplifying assumptions that remain and try to assess their effect on the accuracy of our solution. We write

Consider the assumptions and their effect on the validity of the solution:

1. Eddy currents in the iron core were neglected. This seems reasonable because the core is laminated in order to minimize such currents and their effects of reducing the inductance and heating the core.

2. Resistance of the solenoid was neglected. Had we included its effect in the application of conservation of energy two additional terms would have entered: one representing electric energy input to the resistance and a term precisely equal but opposite in sign representing outflow of heat from the resistance. Thus this assumption has no effect on our result.

3. Mechanical friction was ignored. This makes no difference whatever in this analysis since the restraining force f_R could be the resultant of all forces acting on the core in the x direction except the magnetic force f. Friction must, of course, be considered in the design of the mechanism.

4. Inductance was assumed independent of current, which means that saturation and hysteresis are neglected. While the iron core might be expected to show these effects, much of the flux paths are air, and even a small air gap usually takes the non-linearity out of a reactor. Hence this assumption seems amply justified.

Learning and Generalizing

Having checked our result and satisfied ourselves as to the validity of our assumptions we look for what can be learned from the analysis. With respect to the immediate problem we have established a means for calculating numerically the magnetic force on the core in terms of data readily available. Indeed, the curve of $\frac{dL}{dx}$ in figure 1-9 with a suitable change of scale gives the force as a function of core position for any particular rms current. But there is more to be learned from the analysis than the result for the specific case of immediate interest. To consolidate what may be of value in future problems we might write somewhat as follows:

An expression for the force on the iron core of a solenoid has been derived in terms of inductance and current. Equation 1-20 gives only the axial or x component of force. Similar expressions for other Cartesian components such as f_y and f_z would be

$$f_y = \frac{1}{2}i^2 \frac{\partial L}{\partial y}$$

and

$$f_z = \frac{1}{2}i^2 \frac{\partial L}{\partial z}$$

partial derivatives being used since L is now a function of x, y, and z coordinates of position.

The expression is very general in that nowhere were any dimensions or shapes specified, except as they appear in L. Thus it applies to the force exerted on a piece of iron of any shape that is anywhere in the field of a solenoid.

Moreover, a similar method could be used to find torques. It would involve derivatives of L with respect to one or more angles of rotation.

The method was based upon the assumed slight displacement of the core. However, in finding the limit at the close of the derivation the displacement approached zero. It is clear that even if the core were rigidly fastened the force on it still could be found by this method; the displacement takes place only in the mind and need not even be able to take place physically.

This method has many applications and is commonly called "virtual displacement" or "virtual work." In summary it amounts to giving an imaginary displacement to a system and relating by means of the law of conservation of energy the various transfers of energy that occur. Its distinctive feature is that force or torque may be calculated in terms of an externally (with respect to terminals) measurable quantity such as inductance and the internal details such as shapes of parts and distribution of magnetic field need not be known.

Let us turn now from the purely technical aspects of this problem to a consideration of what more may be learned about the professional method by which it and the problem of section 1–2 were analyzed.

1–4 Professional Method: Summary

In section 1–3 we inserted headings to call attention to the beginnings of the stages in the treatment of the problem. These were the same five stages of defining, planning, executing, checking, and learning which we identified in section 1–2 in studying the analysis of the accelerometer problem. But observe that in the solenoid problem we did not always proceed smoothly through one stage and on to the next. For instance, in starting the execution stage we decided to check and discovered an error in the plan, making it necessary to return to the planning stage before proceeding. Again within the stage labeled Checking we started a graphical check which developed into a new execution of the plan under a less restrictive assumption than the original one. Thus while there is general progress through the five stages it is not necessarily uninterrupted. At any stage there may be reason to return to an earlier stage for correction or modification; moreover, at any stage what is done must be in the light of what is anticipated will make for success in a later stage.

Looking for other features which distinguish the analyses of the foregoing problems, notice the part played by simplification. In the accelerometer problem the engineer began by eliminating all but what he thought were the really essential factors. In the solenoid problem we decided to postpone the complication of alternating current until a simpler situation was treated; likewise consideration of friction, eddy currents, and hysteresis were left out of the initial treatment. Such use of simplification is a characteristic of professional method.

But how, you may wonder, do you know what things to leave out in the simplification? Indeed, how do you know how to choose one plan rather than another? Such questions can be answered in part by such logical reasoning as we have endeavored to show in the study of these problems; but insight also is very important, and this kind of insight grows with experience in treating many problems.

Another feature of professional method is the frequent use of checking throughout the work, not merely near the end. And associated with this is readiness to learn from the check things which may make for improvement in the whole treatment.

Thus there are characteristics of professional method in addition to the five stages. These are use of simplification, willingness to be guided by insight, critical checking at every appropriate opportunity, and using what is learned as the work progresses.

The professional way of dealing with concrete problems may be formulated as follows:*

Stage 1. *Define the problem*

Collect and analyze the facts in relation to the original question in order to fully discover and define the problem.

Stage 2. *Plan its treatment*

Determine what values, principles, attitudes, and basic practices are applicable to the problem. Plan the means of dealing with the facts in the light of these ways of approach.

Stage 3. *Execute the plan*

Carry through the plan so as to reach a decision or result. (Often the decision does not end the problem but clarifies or changes the issue so that the problem is started over in a new aspect.)

Stage 4. *Check the work as a whole before using the solution*

Go over the results, first systematically, then realistically in terms of use, and at last with reference to the general knowledge and experience of that field.

Stage 5. *Learn and generalize if possible*

Take thought to find what can be learned that is of value in the situation at hand and what can be learned that may be of use in future problems.

However, these stages cannot be considered a rigid sequence to be followed in handling any problem. Often, you must first go through the problem roughly in order to discover its essentials before attempting a thorough treatment. Often also, you may have to return to prior stages, or even repeat them all, as enlightenment comes in the course of developing the problem. And whenever appropriate, one or more of the following functions must be performed in connection with any stage, or in connection with the problem as a whole.

Function 1. *Simplification*

a. By restricting assumptions.
b. By condensed, exact statement in words or in symbols.

Function 2. *Alternation between analysis and insight*

* This formulation is by a group of our associates at Carnegie Institute of Technology which included not only engineers and physical scientists but social scientists as well.

Function 3. *Checking validity,* both systematically and realistically in terms of use.

Function 4. *Using all that can be learned by experience* as the solution proceeds as a basis for correction and as a guide to future steps, even if this involves radical change in the problem or its treatment.

The professional method formulated above is not the inductive process commonly called the scientific method by which scientific knowledge is extended. In the scientific method, many individual instances are studied with the object of constructing a hypothesis. The hypothesis is then tested by carefully designed experiments, and, if it successfully sustains all the tests that can be devised, it is dignified by being called a principle. The professional method for dealing with concrete problems is by contrast a deductive process. It is used to deduce from scientific principles or other generalizations what must be true under a particular set of circumstances.

To show further the nature of the stages and functions of the professional method, chapter 2 is devoted to additional illustrative examples.

CHAPTER TWO

Application of Professional Method

2–1 Further Illustration

This chapter is devoted to further illustration of professional method as applied to six cases in a variety of fields. The intention is to show professional method applied in some situations which are less straightforward than those of the preceding chapter. The first three problems are chosen to illustrate particularly the early stages of the method: defining and planning. The fourth case depends heavily on human factors, and mathematics is not involved at all. In the last two cases, the emphasis is particularly on the learning and generalizing stage and on the repetitive functions.

In applying professional method to the analysis of an engineering problem, what you write as your thoughts develop is very important. Just as you would almost never try to do a long algebra or calculus exercise in your head without using pencil and paper, so you should not try to collect the facts and go through the complexities of analyzing an entirely unfamiliar engineering problem without the aid of a carefully written record of your steps. What you write down as your thinking progresses will depend on your experience in relation to the problem, but good professional method, in any case, calls for writing out clearly the significant facts and thoughts pertaining to the analysis as they develop. Four of the illustrations in this chapter, sections 2–2, 2–3, 2–4, and 2–6, are arranged like the ones in chapter 1 to show this kind of writing.

Careful writing in precise English serves not only to clarify and fix the ideas but is essential for purposes of checking the validity of the conclusions and for retracing the steps of the analysis if modifications must be made. This writing is part of the analytical process and is not to be confused with the report which the engineer may prepare afterwards in order to communicate his findings to others.

2–2 An Electric Accelerometer

An engineer is frequently responsible for working out a means for attaining some broad objective but is given no specific instruction as to what the means shall be, and there may be many possibilities. For him to cope with the situation, he must somehow formulate for himself a problem which is sufficiently definite and simple enough for him to be able to solve, and whose solution either will be useful directly in attaining the desired objective or will serve to eliminate one of the possibilities for attaining it. In solving the problem he sets himself, the engineer almost inevitably uncovers a series of additional problems, at least some of which must be solved, before the desired end can be reached. In the illustration of this process which follows, it is important to note that, as in many cases, nobody gives the engineer the problem, except perhaps in the sense of assigning him a broad responsibility to see that certain things are accomplished.

Let us suppose that the engineer who was studying the liquid-level accelerometer of section 1–2 is seeking a direct-reading accelerometer for permanent mounting in an automobile which is being used in development work. He decides that the liquid type is unsuitable for this purpose because of its sensitivity to level of the vehicle, and that he would prefer a device indicating acceleration on an electric instrument such as a voltmeter or ammeter.

It occurs to him that a d-c generator, geared directly to the drive shaft of the car, would give a voltage accurately proportional to car speed provided the magnetic flux remained constant, as it would with a permanent magnet field. Now if he could connect the generator terminals to something that would draw a current proportional to rate of change of voltage, that is, proportional to rate of change of speed, acceleration, he would have a solution. He considers the common circuit elements: resistance, inductance, and capacitance, and their effects on the flow of current. He chooses a condenser for his purpose because its charge will be proportional to its terminal voltage, $q = Cv$; hence rate of change of charge, or current $i = dq/dt$, will be proportional to rate of change of voltage.

Thus the engineer decides to try a condenser in series with an ammeter, and to fix ideas he sketches and labels the diagram figure 2–1 and writes:

> The d-c generator is geared directly to the drive shaft of the automobile and is connected in series with a condenser and ammeter. How will the current vary with car speed?

It is significant that, although this statement of the problem is much like those which fill common engineering textbooks, it was not given to the engineer in this clearly defined form. Rather, he made up the problem statement himself in an effort to answer the broader question, also put by himself, "How can I make an instrument that will measure acceleration electrically?" He hopes that solution of the definite problem which he has stated will supply the answer to his question, but he fully realizes

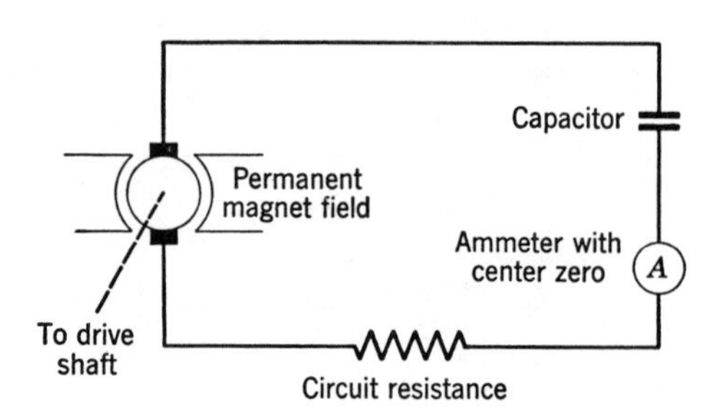

Fig. 2-1. Circuit for accelerometer.

that it may not, in which case he must try again, and he realizes also that there may be many answers.

Having formulated for himself a definite problem, the engineer next seeks a plan for its solution. Observing that he has a simple series circuit, he sees that Kirchhoff's voltage law will yield a relationship involving the current. So, to fix ideas, he writes as his plan:

[Apply Kirchhoff's voltage law to the circuit and solve for the current. Assume that the voltage generated will be exactly proportional to car speed and independent of current (that is, neglect armature reaction) and assume that inductance is negligibly small.]

In execution of this plan, he writes:

[Voltage generated equals the sum of the voltage drops through the condenser and the circuit resistance.]

Translating this statement into mathematical form, he writes:

$$Ku = \frac{q}{C} + Ri \qquad (2\text{–}1)$$

where K = constant of proportionality between generated voltage and car speed (volts sec ft^{-1}). This involves generator constants, gear ratios, and wheel diameter.

u = car speed (ft sec^{-1}).
q = charge on the condenser at any instant (coulombs).
C = capacitance of the condenser (farads).
R = total resistance of the circuit (ohms).
$i = \frac{dq}{dt}$ = current in the circuit (amp).

This gives a relationship between car speed, condenser charge, ammeter current, and the system constants. Does it show that current is in proportion to rate of change of car speed? To try to answer this question, the engineer proceeds and writes:

Differentiate with respect to time

$$K\frac{du}{dt} = \frac{1}{C}\frac{dq}{dt} + R\frac{di}{dt} \tag{2–2}$$

But

$$i = \frac{dq}{dt}$$

so

$$K\frac{du}{dt} = \frac{i}{C} + R\frac{di}{dt} \tag{2–3}$$

Inspection of this equation indicates that i will be proportional to $\frac{du}{dt}$ as is desired if the last term is negligibly small compared to the other two, or, as may be seen from 2–1, if the Ri drop can be neglected. The engineer continues:

Assume Ri negligible in comparison with $\frac{q}{C}$ at all times. Then

$$K\frac{du}{dt} = \frac{1}{C}\frac{dq}{dt} \tag{2–4}$$

or

$$i = \frac{dq}{dt} = CK\frac{du}{dt} \tag{2–5}$$

that is, the current is proportional to $\frac{du}{dt}$ which is the acceleration of the car.

Before putting any faith in this very promising result, the engineer checks it in various ways.

[Dimensional check:

$$i \text{ (amperes)} = C \text{ (farads) } K \left(\frac{\text{volt sec}}{\text{feet}}\right) \frac{du}{dt} \left(\frac{\text{feet}}{\text{sec}^2}\right)$$

$$= \text{(farads)} \left(\frac{\text{volts}}{\text{sec}}\right)$$

$$= \text{(amperes)} \quad \text{check} \tag{2-6}$$

]

The basis for the last step is that capacitance in farads times voltage is charge in coulombs ($q = Cv$) and rate of change of charge in coulombs per second is current in amperes.

[The result appears reasonable because, if the acceleration is zero, so is the current, and, if the acceleration increases, so does the current. Having K and C in the numerator appears reasonable because the larger C, the greater the current; also the larger K, the greater the voltage and hence the greater the current.]

Satisfied with these checks, the engineer turns to a consideration of what he has learned about the answer to his original question, which was how to make an electric accelerometer. He writes the following conclusion:

[A permanent-magnet d-c generator geared directly to the wheels and connected to a condenser will give a current proportional to acceleration if resistance drop, armature inductance, and armature reaction can be neglected. If in addition the mechanical transient in the ammeter may be ignored the reading will be proportional to acceleration.]

The engineer also adds to his general stock of learning by noting that

[Measuring the current through a condenser constitutes a way of obtaining the time derivative of a voltage, provided the voltage, undiminished by resistance, can be applied directly to the condenser.]

Apparently the engineer had a glimmering of this idea or he would not have thought of trying the circuit he did. On the other hand, had he been perfectly familiar with differentiating circuits, he might have used this knowledge as a fundamental basis for a solution instead of proceeding as he did.

The seemingly favorable result obtained above immediately raises a number of questions which also must be answered before the engineer would be justified in considering his original question answered. For example: Are the sizes of the components necessary in the generator-condenser scheme practically realizable? Is it possible for the resistance drop to be negligibly small? Is it safe to ignore the effect of inductance of the circuit? Is it reasonable to neglect armature reaction in a small permanent-magnet generator? Can the mechanical transient in the ammeter be neglected? Are there wholly different electrical schemes which might be better?

Let us see how the engineer might answer the question, "Will the scheme work in an automobile using reasonable values of the parameters?" To answer a question such as this requires some experience with apparatus of the kinds involved or perhaps some searching through sources of data such as handbooks and manufacturers' catalogues. It is remarkable, though, how far one can go on a very modest fund of experience. For example, let us see how the engineer might proceed in answering this question if he had no more practical experience with the quantities involved than he had been able to pick up as an undergraduate. Thinking over the components which comprise his electric accelerometer, the generator, the ammeter, and the condenser, the engineer plans to solve the new problem he has set himself in this way:

[Choose a generator and ammeter arbitrarily and see how big a condenser will give a suitable meter deflection.]

In carrying out this plan, the engineer recalls his undergraduate laboratory experience and decides on a generator which he remembers using as a component of an electric tachometer. He writes:

[Use a tachometer generator which develops 200 volts at a maximum speed and gear it so that this voltage is reached at 100 miles per hour (2 volts per mph).]

The voltage output of commercially available tachometer generators varies widely among manufacturers. A common rating is 6 volts per 1000 rpm, but models are available where this figure is as high as 75 volts per 1000 rpm. It is probably one of these latter that our friend has in mind. Then, thinking of a small portable meter he once used, which was quite sensitive and still fairly rugged, he decides to:

[Use an ammeter with center zero and a full scale reading of 1 milliampere (0.001 amp).]

Then he writes further:

[It would be desirable to be able to read a moderate rate of acceleration, say from 10 to 60 mph in 20 sec, at half scale on the meter.

Thus the change of voltage will be 100 volts in 20 sec or

$$K \frac{du}{dt} = 5 \text{ volts/sec} \tag{2-7}$$

The current corresponding to this should be 0.5 milliampere or

$$i = 0.0005 \text{ amp} \tag{2-8}$$

Then from the formula developed

$$C = \frac{i}{K \dfrac{du}{dt}} = \frac{0.0005}{5} = 10^{-4} \text{ farads}$$

$$= 100 \text{ microfarads} \tag{2-9}$$

If this value turns out to be uneconomically high, it could be reduced by using a more sensitive ammeter or a generator which gives a still higher voltage.]

So far the scheme looks feasible. In considering a generator of higher voltage and a more sensitive meter, the engineer would realize at once that these probably would involve increased resistance. Thus as the next step, he might set himself the problem of determining how much resistance can be tolerated. In attacking this problem, he writes:

[What is the effect of circuit resistance? Return to equation 2–1, in which resistance was included, and attempt to solve it without neglecting Ri.

$$Ku = \frac{q}{C} + Ri$$

Differentiate and rearrange the terms

$$\frac{di}{dt} + \frac{i}{RC} = \frac{K}{R}\frac{du}{dt} \tag{2-10}$$

For simplicity, assume a constant rate of acceleration suddenly applied, that is, let

$$\frac{du}{dt} = A \tag{2-11}$$

Then separate variables in 2–10 and integrate between limits:

$$\int_0^i \frac{di}{\frac{KA}{R} - \frac{i}{RC}} = \int_0^t dt \tag{2-12}$$

In choosing the lower limits as he did, the engineer reasoned as follows: Assuming the acceleration applied with the vehicle initially at rest, the voltage generated at the first instant will be zero. Also, the charge on the condenser, and hence its voltage, will be zero. Therefore at the first instant, there is no voltage available to cause current to flow through the resistance of the circuit; consequently $i = 0$ at $t = 0$. Integrating 2–12, he proceeds:

$$-RC \ln \frac{\frac{KA}{R} - \frac{i}{RC}}{\frac{KA}{R}} = t \tag{2-13}$$

$$1 - \frac{i}{KCA} = \epsilon^{-t/RC} \tag{2-14}$$

and finally

$$i = KCA(1 - \epsilon^{-t/RC}) \tag{2-15}$$

Dimensional check of the coefficient:

$$i \text{ (amp)} = K\left(\frac{\text{volt sec}}{\text{ft}}\right) C \text{ (farads) } A\left(\frac{\text{ft}}{\text{sec}^2}\right)$$

$$= \frac{\text{volts}}{\text{sec}} \times \text{farads} = \text{amp} \qquad \text{check}$$

The exponent should be dimensionless:

$$\frac{t \text{ (sec)}}{R \text{ (ohms) } C \text{ (farads)}} = \frac{\text{sec}}{\frac{\text{volts}}{\text{amp}} \times \text{farads}} = \frac{\text{amp}}{\frac{\text{volts}}{\text{sec}} \times \text{farads}}$$

$$= \frac{\text{amp}}{\text{amp}} \qquad \text{check}$$

With $R = 0$ the result 2–15 becomes $i = KCA$. This agrees with 2–5 derived directly for the case of zero resistance if the acceleration du/dt has the constant value A. With $R = \infty$, 2–15 gives $i = 0$, which evidently is correct.

Thus the engineer has derived a formula showing the effect of resistance in the circuit for the case of a suddenly applied constant acceleration. To interpret his formula 2–15 to himself, he sketches roughly a graph, figure 2–2, showing as functions of time the acceleration and the currents

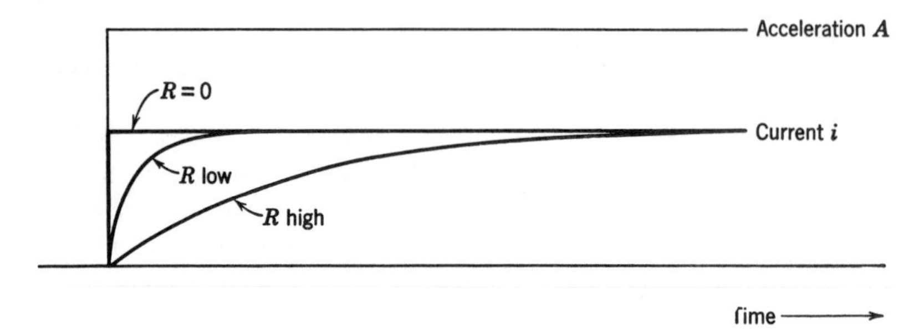

Fig. 2-2. Sketch to interpret equation 2-15.

that would result with different values of resistance. As further interpretation of the result, he writes:

The current approaches the value corresponding to the steady acceleration exponentially with a time constant equal to RC. At a time equal to RC, the current will reach about 63 per cent of its final value. In three time constants, the current will be within 5 per cent of its final value.

EXERCISE. Verify the preceding two sentences.

In addition to this time lag, there also will be a mechanical transient dependent on the characteristics of the meter. The part of the transient associated with the circuit would be imperceptible to the eye if the time constant RC is made $\frac{1}{20}$ second.

$$RC = 0.05$$

$$R = \frac{0.05}{C} = \frac{0.05}{10^{-4}}$$

$$= 500 \text{ ohms}$$

using the value of $C = 100$ microfarads found in the earlier calculation. This is the maximum tolerable resistance, and is probably higher than the resistances likely for the assumed generator and meter.

Thus far it appears that this scheme for an electric accelerometer is practically feasible. Before embarking on an experimental installation, however, the engineer certainly would analyze carefully the effects of the mechanical behavior of the meter element, for these effects might well determine the success of the scheme. Moreover, he might consider the influence of other factors and possibly also entirely different schemes.

EXERCISE. Investigate the possibility of using a similar permanent-magnet generator but instead of a condenser some arrangement of inductance and resistance.

In this illustration we have seen how a single question can give rise to a series of engineering problems each of which must be solved before the original question can be answered satisfactorily. It is evident that the engineer's judgment in formulating the questions to ask himself, that is, his skill in defining his problems, will have an important bearing on his success in obtaining useful solutions.

2–3 A Non-Fogging Mirror

This illustration is somewhat like the preceding one in that it shows that what may appear at first like one problem is resolved by the engineer into a number of separate problems. During the analysis of one of these problems, the engineer learns something which, as frequently happens, indicates the desirability of replanning the solution and making a fresh start.

A young electrical engineer newly employed by a small manufactuer of electric appliances is told to investigate the feasibility of a bathroom mirror heated electrically from behind so as to prevent fogging. The proposed heater is to be a sheet of electrically conductive rubber covered with suitable electrical insulation and cemented to the back of the mirror.

Many questions occur at once to the engineer. Why does fogging occur? How much power must the heater supply? Will the heater get too hot? Should the heater be turned on and off manually or automatically? Will the costs be reasonable? He cannot answer all these at once, so, to focus his attention on what seems to be the heart of the problem, he writes:

The mirror is to be made non-fogging by heating it from behind. How much heat will this require?

After further cogitation, he begins to see how to break into a solution. To organize his ideas he writes out the following chain of reasoning:

Fog on the mirror is moisture which condenses there because the temperature of the surface is below the dew point, that is, below the temperature at which air becomes saturated with water vapor. If the mirror surface can be maintained at a temperature above the dew point, the immediately adjacent air will remain unsaturated and there will be no condensation of water, hence no fogging. But how much above the dew point must the surface temperature be?

The air in the bathroom approaches saturation because of the flow of hot water in the tub or shower. The water may be as hot as 130°F. (Some automatic water heaters are set for 135°F.) The air in the room may rise to 80°F or even 90°F fairly quickly and be close to saturation with moisture at this temperature (a bathroom often feels like a muggy summer day.) The solid surfaces, however, will tend to remain at their original temperatures, which after a cold night might be as low as 60 or 65°F. (The temperatures of the surfaces will rise but slowly, because of the thermal capacity behind them.) Under these conditions, condensation on the surfaces would occur quickly because the air soon would become saturated at temperatures above 60 or 65°F, that is, the dew point would become higher than the surface temperatures.

Assume that the surface and air temperatures of the bathroom are originally at 65°F, and that the dew point quickly rises to 85°F; then, if the mirror surface is originally maintained at 85°F, it will not fog and some margin of safety will be provided because the temperature of the mirror will tend to increase as the room warms.

Thus, having considered carefully the phenomena which must occur and having drawn heavily on his common experience for data, the engineer is ready to revise the statement of his problem. He writes:

The problem now is: What heat flow from the mirror surface to the surroundings is necessary to maintain a 20°F temperature rise of the surface above the air?

In planning how to answer this question, he goes on:

Heat transfer from the surface of the mirror will be by free convection to the air and by radiation to the walls of the room. The heat transferred can be calculated as the sum of the parts by convection and by radiation.*

* This is discussed in section 3–10.

This calculation will be made for a square foot of mirror surface assuming that the surface is at a uniform and constant temperature of 85°F and that the air and walls are at 65°F.

For convection, the heat flow per unit area is proportional to the temperature difference and an empirical coefficient:

$$\frac{q_c}{A} = h(T_s - T_a) \tag{2–16}$$

where q_c = heat power transferred to the air by convection (Btu hr^{-1}).
A = surface area (ft^2).
h = natural convection heat transfer coefficient (Btu hr^{-1} ft^{-2} $°\text{F}^{-1}$).
T_s = surface temperature (°F).
T_a = air temperature (°F).

For vertical surfaces in air

$$h = 0.27(T_s - T_a)^{0.25} \tag{2–17}$$

The engineer does not burden his memory with formulas such as this, but refers to an appropriate source* when the need arises. Proceeding,

$$\frac{q_c}{A} = 0.27(T_s - T_a)^{1.25}$$

$$= 0.27(85 - 65)^{1.25} = 11.4 \text{ Btu hr}^{-1} \text{ ft}^{-2} \text{ by convection}$$

Turning to the consideration of the radiation from the mirror, our young electrical engineer happens to find himself in unfamiliar territory. Nevertheless, he recalls what he *does* know of the subject and writes it down as it applies to his problem:

Radiation heat transfer between bodies is a process of exchange, each body radiating to all other bodies which it can "see" and receiving radiation back from them. In this exchange of radiation, the geometric arrangement is important. The mirror surface radiates to the walls of the room and the walls radiate back to the mirror. The intervening air has little effect.

A body radiates energy in proportion to its absolute temperature raised to the fourth power (Stefan-Boltzmann law). Thus, if the mirror

* In this case, Marks, *Mechanical Engineers' Handbook*, Fifth Edition, McGraw-Hill New York, 1951, p. 374.

is at a higher temperature than the walls, the net radiation of heat will be from the mirror.

The amount of radiation emitted from a body and the fraction absorbed of that arriving from other bodies depend on the nature of the surface of the body and on the wave length of the radiation.

And this is as far as our young engineer can go without help, so he turns to a reference,* which, having collected his thoughts in relation to his problem, he is prepared to read intelligently. After a little study, he finds his prior understanding confirmed, and he obtains the detailed information he requires to make a calculation. Again, he writes:

Assume that the mirror and walls radiate as gray bodies (emissivity same for all wave lengths involved), and that the emissivities all are equal to 0.9.

The geometry is such that the mirror can be taken as a convex body wholly enclosed by the walls which constitute a concave body. Then, for this case, the net loss of heat by radiation from the front surface of the mirror is

$$\frac{q_r}{A} = \sigma\epsilon[(T_s + 460)^4 - (T_w + 460)^4] \tag{2-18}$$

where q_r = net heat power lost from mirror by radiation (Btu hr^{-1}).
σ = Stefan-Boltzmann constant (Btu hr^{-1} ft^{-2} °F^{-4}) (= 0.174×10^{-8}).
ϵ = emissivity of the glass surface (dimensionless) (= 0.9).
T_w = wall temperature (°F).

Exercise. Make a dimensional check of 2–18.

$$\frac{q_r}{A} = 0.174 \times 0.9[(5.45)^4 - (5.25)^4]$$

$$= 0.174 \times 0.9(5.25)^4 \left[\left(1 + \frac{0.20}{5.25}\right)^4 - 1\right]$$

$$= 0.174 \times 0.9 \times 762 \times 0.161$$

$$= 19.2 \text{ Btu hr}^{-1} \text{ ft}^{-2} \quad \text{by radiation}$$

* *Elements of Heat Transfer and Insulation*, Second Edition, by Max Jakob and George A. Hawkins, Wiley, New York, 1950, pp. 168–186.

Thus the total heat transfer necessary to maintain the 20°F temperature rise of the glass surface is

$$\frac{q_c}{A} + \frac{q_r}{A} = 11.4 + 19.2$$

$$= 30.6 \text{ Btu hr}^{-1} \text{ ft}^{-2}$$

The engineer is rather surprised to find that radiation plays such a large part in this process. As a check on his calculation, he consults another reference* and finds a table of combined heat transfer coefficients for natural convection and radiation where the surrounding surfaces and the air are at about 70°F. For vertical plane surfaces, he interpolates to a surface temperature of 85°F and finds a combined coefficient of 1.48 Btu hr^{-1} ft^{-2} $°\text{F}^{-1}$. This multiplied by the temperature difference of 20°F gives 29.6 Btu hr^{-1} ft^{-2}, a reasonable check on the earlier calculation.

Being in an unfamiliar field, however, the engineer still is not satisfied that somehow he may not have made a gross error, so he converts the unfamiliar Btu per hour to watts and gets for the total heat transfer about 9 watts per square foot. Then, to get an independent and overall check on the order of magnitude of his result, he thinks of familiar cases where heat is transferred with small temperature drop into a room. A small electric clock in his house consumes 2 watts and has a surface area that he estimates to be 0.8 ft^2, thus dissipating about 2.5 watts per square foot. It runs just barely perceptibly warmer than nearby objects. Then he thinks of an enclosed motor of ¼ horsepower whose losses he estimates to be 25 per cent of its output and exposed surface area to be about 2 square feet. This gives a dissipation of about 23 watts per square foot. The motor, when running fully loaded, gets very hot to the touch; its temperature rise above room air he estimates at perhaps 40 or 50°F (he remembers it as about as uncomfortable to touch as water pipes carrying water which he knows to be at 130°F). With these checks based on experience with familiar things, the engineer feels quite confident in using his calculated value of heat dissipation as a basis for further work. So he sets down the following new problem:

To maintain a surface temperature of 85°F in a room at 65°F requires 9 watts ft^{-2}. How much will this cost if the heater is run continuously?

The heater must supply not only the heat to be dissipated from the front surface of the glass, but also heat losses from the back of the heater.

* Kent's *Mechanical Engineers' Handbook*, Power volume, Twelfth Edition, Wiley, New York, pp. 3–30.

Assume that the losses to the back can be kept down by suitable insulation to 20 per cent of the useful heat. (This figure can be investigated and revised later if necessary).

A bathroom mirror of the type on a medicine-cabinet door has an area of about 2 square feet.

Therefore, the heater must supply

$$9 \times 1.2 \times 2 = 22 \text{ watts}$$

For a residential customer, the cost of electric energy will be between, say, 3 and 1.5 cents per kilowatt hour depending on consumption.

Using the lower figure, the monthly cost of operating the mirror continuously would be

$$\frac{22}{1000} \times 24 \times 30 \times \frac{1.5}{100} = \$0.24 \text{ per month}$$

Having found that the heated mirror appears to be economically feasible even if operated continuously, the engineer decides to look into the possibilities of operation only as needed. To do this, he starts by making a rough estimate of the time that would be required to heat the mirror to the desired temperature. He writes:

How long will it take the mirror to heat up?

For an estimate, find the energy stored in the glass in the steady state and then see how long it would take to build this up at the steady-state power input assuming no dissipation during the storage process. The actual time to reach steady state may be several times the value so calculated in view of the fact that during the storing process dissipation also will be taking place.

The energy stored above the starting temperature is equal to the heat capacity of the glass times its final average temperature. The latter is the surface temperature plus half the drop through the glass.

Calculate the steady-state temperature drop through the glass using Fourier's law of heat conduction.

$$\Delta T = \frac{qL}{kA} \tag{2-19}$$

ΔT = temperature drop through glass from heater side to front surface (°F).

q = heat flow in direction of the temperature drop (Btu hr^{-1}).

L = thickness of glass (ft).

k = thermal conductivity of glass (Btu hr^{-1} ft^{-1} °F^{-1}).

A = cross-sectional area of path (= mirror surface area) (ft^2).

Assume $L = \frac{1}{8}$ in. $= \frac{1}{96}$ ft

$q/A = 30.6$ Btu hr^{-1} ft^{-2} (= total heat dissipated from surface, already calculated).

$k = 0.3$ Btu hr^{-1} ft^{-1} °F^{-1} (Marks *Handbook*).

$$\Delta T = \frac{30.6}{0.3 \times 96} = 1.06\text{°F} \qquad (2\text{–}20)$$

The small amount of this temperature drop through the glass in comparison to the drop of 20°F from the surface to the air and walls suggests a change in the plan for solution, so the engineer writes:

Revised plan: neglect the temperature drop through the glass and, instead of estimating the transient time roughly, calculate it more accurately. Do this by applying the law of conservation of energy as follows:

Heat flow into the glass from the heater equals the heat flow out of the glass into the room plus the time rate of increase of energy stored in the glass.

Assume that the temperature throughout the glass is uniform.

Assume that the heat transfer from the surface is proportional simply to the first power of the temperature rise above the surroundings (Newton's law of cooling).*

Assume that the heat flow from the heater into the glass is constant and equal to its final steady-state value (already calculated).

Then, translating to a mathematical statement, he writes:

$$q = hAT + C\rho AL\frac{dT}{dt} \qquad (2\text{–}21)$$

q = heat flow from heater into glass (Btu hr^{-1}).
T = temperature rise of glass above surroundings (°F).
t = time (hr).
h = combined heat-transfer coefficient surface to surroundings (Btu hr^{-1} ft^{-2} °F^{-1}).
C = specific heat of glass (Btu lb^{-1} °F^{-1}).
ρ = density of glass (lb ft^{-3}).

* See section 3–10 for a discussion of the accuracy of Newton's law in this case.

Checking dimensionally:

$$q\left(\frac{\text{Btu}}{\text{hr}}\right) = h\left(\frac{\text{Btu}}{\text{hr ft}^2\ ^\circ\text{F}}\right) A\ (\text{ft}^2)\ T\ (^\circ\text{F})$$
$$+ C\left(\frac{\text{Btu}}{\text{lb}\ ^\circ\text{F}}\right) \rho \left(\frac{\text{lb}}{\text{ft}^3}\right) A\ (\text{ft}^2)\ L\ (\text{ft})\ \frac{dT}{dt}\left(\frac{^\circ\text{F}}{\text{hr}}\right)$$
$$= hAT\left(\frac{\text{Btu}}{\text{hr}}\right) + C\rho AL\frac{dT}{dt}\left(\frac{\text{Btu}}{\text{hr}}\right)$$

The equation also checks in the steady state when $\frac{dT}{dt}$ becomes zero and all the heat from the heater goes right through and is dissipated from the surface. It checks at the first instant when $T = 0$ and all the heat from the heater goes into storage.

Solve by separating variables and integrating between limits:

$$-\frac{C\rho L}{h}\int_0^T \frac{-hA\ dT}{q - hAT} = \int_0^t dt \tag{2-22}$$

$$\ln\ (q - hAT) - \ln q = -\frac{ht}{C\rho L} \tag{2-23}$$

$$\ln\left(1 - \frac{hAT}{q}\right) = -\frac{ht}{C\rho L} \tag{2-24}$$

$$T = \frac{q}{hA}\left(1 - \epsilon^{-\frac{ht}{C\rho L}}\right) \tag{2-25}$$

Dimensional check of the coefficient:

$$T\ (^\circ\text{F}) = \frac{q\ (\text{Btu hr}^{-1})}{h\ (\text{Btu hr}^{-1}\ \text{ft}^{-2}\ ^\circ\text{F}^{-1})\ A\ (\text{ft}^2)} = (^\circ\text{F}) \qquad \text{check}$$

Dimensional check of the exponent:

$$\frac{h\ (\text{Btu hr}^{-1}\ \text{ft}^{-2}\ ^\circ\text{F}^{-1})\ t\ (\text{hr})}{C\ (\text{Btu lb}^{-1}\ ^\circ\text{F}^{-1})\ \rho\ (\text{lb ft}^{-3})\ L\ (\text{ft})} = (\text{dimensionless}) \qquad \text{check}$$

The result says that T finally approaches q/hA. But this is the steady-state temperature rise, which is as it should be.

The time constant of the exponential is $\frac{C\rho L}{h}$. Calculate its numerical value.

$$\left.\begin{aligned} C &= 0.19 \\ \rho &= 162 \end{aligned}\right\} \text{(Marks } \textit{Handbook}\text{)}$$

$$h = \frac{30.6}{20} = 1.5 \text{ (from previous calculations)}$$

$$\frac{C\rho L}{h} = \frac{0.19 \times 162}{1.5 \times 96} = 0.21 \text{ hr } (= 13 \text{ minutes})$$

Thus, it will take the temperature several times 13 minutes, say half to three-quarters of an hour, to come within a degree or two of the desired 85°F. This is probably too long for operation only as needed to be practical under the conditions assumed.

But is it really too long—are there perhaps other factors, a consideration of which may lead to a different conclusion? To try to answer this question, the engineer goes back over his work to reconsider his analysis and to see what he can learn that may be useful. Going back to the beginning, he makes the following observation:

The dew point initially is probably well below the room temperature. Fogging occurs only when the dew point increases and passes the mirror temperature. Thus, to prevent fogging, it is necessary only that the mirror temperature increase just fast enough to keep ahead of the dew point.

Going along through his work, he notes that:

The mirror temperature can be made to rise faster by decreasing the thickness of glass, but it may not be practical to make the glass thinner than ⅛″ already assumed.

The rate of rise also can be increased by raising the power input from the heater. This will entail a higher steady-state temperature which may not be objectionable if the heater is to be turned off when not in use.

The engineer sees now that he has a new problem to solve:

How does the dew point vary with time after the hot water is turned on in a bathroom?

After thinking about this awhile, he concludes that a calculation is not feasible but that an answer for a typical case might be obtained very simply by direct measurement. Accordingly, he plans to run an experiment in his own bathroom the next morning. To get advice on how best to measure dew point, he calls on a friend in the air-conditioning business. The friend lends him a sling psychrometer, shows him how to use it, and explains how to get the dew point from a psychrometric chart,* using the temperatures indicated by the wet- and dry-bulb thermometers of the instrument.

During the evening, he prepares for his experiment by fastening a thermometer to the front surface of the mirror with adhesive tape so that it will be in thermal equilibrium in the morning. Also before going to bed, he practices using the sling psychrometer in the confined space of the bathroom.

In the morning, first he observes the temperature of the mirror and then, using the sling psychrometer, the wet- and dry-bulb temperatures in the room. Next, he closes the door and turns on the hot water in the shower. He then observes the three temperatures at definite times. This is his data sheet:†

Time, Minutes	*Mirror Temperature, °F*	*Dry Bulb, °F*	*Wet Bulb, °F*
0	61	58	52
2	65	75	72
5	69	80	77
10	74	81	82

Mirror fogged at 1 min
Size of room 8′ × 9′ × 8′ ceiling
Mirror 5′ from shower head
Water temperature 122°F
Rate of water flow 6.5 gal/min

The dew point he reads from a psychrometric chart by entering with the wet- and dry-bulb temperatures. He then plots the curves in figure 2-3. He observes that the intersection of the mirror temperature and dew-point curves is very close to the time at which fogging occurs, thus confirming his theory of what happens.

Next, he concludes that, if he had the heater and it were turned on simultaneously with the shower, the mirror would not fog if its temperature

* Marks *Handbook*, p. 358.

† These data are those actually obtained by a senior student who planned and executed this experiment on his own initiative.

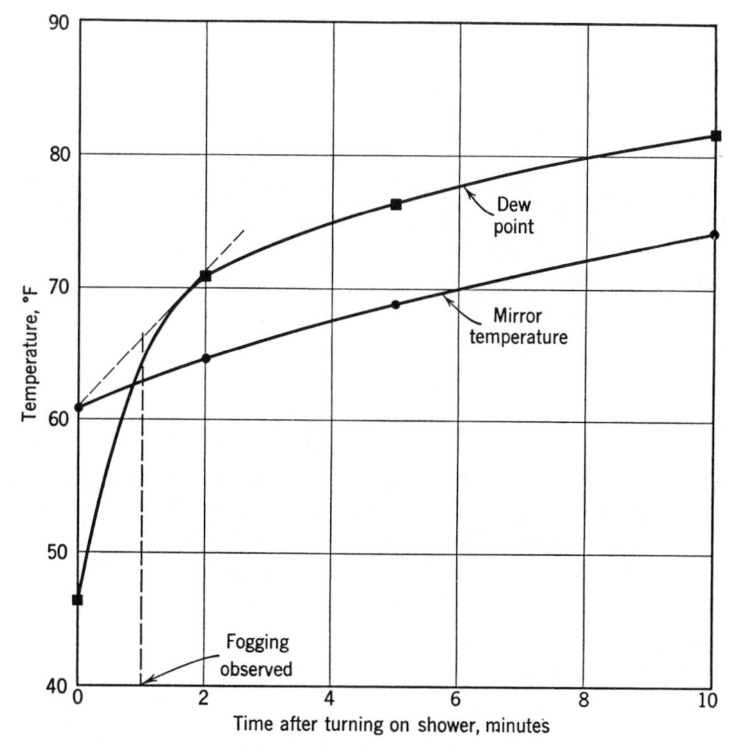

Fig. 2-3. Fogging test of bathroom mirror.

were to rise initially at a rate of not less than 5.5°F per minute (dashed line). This allows him to make the following calculation.

From differential equation 2–21, the initial rate of rise is related to the input by

$$\frac{q}{A} = C\rho L\left(\frac{dT}{dt}\right)_{t=0}$$

$$= 0.19 \times 162 \times \tfrac{1}{96} \times 5.5 \times 60$$

$= 106$ Btu hr^{-1} ft^{-2} heat input necessary to prevent fogging with operation only as needed.

If the heater is left on with this input, the final steady-state temperature rise also can be calculated from the differential equation, this time with

$$\frac{dT}{dt} = 0.$$

$$(T)_{t=\infty} = \frac{q}{Ah}$$

$$= \frac{106}{1.5} = 71°\text{F}$$

This temperature rise added to a room temperature of 80°F or higher could hardly be tolerated, and it might be necessary to include a thermostat with the heater to turn it off if the user forgets to do so, as certainly would often be the case.

Thus the engineer has carried the analysis of his problem to the point where it is apparent that the proposed scheme for making a non-fogging mirror is technically sound. Moreover, either continuous operation of the heater or operation only as needed probably is practicable. The first is attractive because of its simplicity and freedom from possibility of human error; and the cost of operation would not be high (the experiment suggests that the power necessary might be even less than at first calculated.) With operation only as needed, the cost for energy would be substantially less, but the first cost might be considerably more if a thermostat were needed.

Among the additional things the engineer would need to do before he could consider his job finished would be:

1. Investigate the question of suitable thermal insulation to prevent the escape of heat from the back of the heater and the related question of what temperature the contents of a medicine cabinet behind the mirror could be permitted to attain.
2. Investigate the electrical design of the heating element, including its electrical insulation, and consider whether the material of which the heater is to be made would have a suitably long life under the temperature conditions to be imposed.
3. See that the proposed design not only would comply with existing safety codes, but that the device would indeed be safe in the hands of the general public.

Also, to the engineer's lot might fall the job of investigating whether or not there was a potential market for the device.

In following the young engineer through his study of the mirror problem, we see that first he had to collect his ideas so he could define a problem whose solution would be both possible and useful. In carrying out his solution, he had to study the subject of radiation, and again later it was necessary for him to extend his learning in order to make the dew-point measurements. Indeed, it is rare for an engineer not to have to study in unfamiliar fields as he goes along in his professional work. In several instances, as his work on the mirror progressed, our engineer learned things about his problem which enabled him either to redefine an objective or replan the work so as to come closer to the truth being sought. Another way in which this illustration is typical of professional engineering is the way in which an experiment was integrated with the analysis. Lastly, it is important to note that a final decision as to whether to operate the mirror continuously or only as needed, or indeed whether to proceed with the development at all, depends as heavily on economic and human factors as it does on technological ones, and this too is typical of engineering problems.

2–4 Life Expectancy of a Bearing

Problems in engineering practice are not always difficult to define. Sometimes the problem is crystal clear, but a plan for solving it may be discouragingly elusive. In trying to solve such a problem, it may seem at first that, because of lack of knowledge, it is hopeless to proceed. Then sometimes consideration of a parallel, but greatly simplified, situation, may serve to clarify your thoughts, so that you are led to a plan for successfully attacking the actual problem. An example of such a case follows.

An engineer is faced with the necessity for estimating the life expectancy of a certain roller bearing under conditions of varying load, and he has had no experience whatever with such matters. He has gathered some facts about the bearing with which he is concerned, and to get started he writes them down:

According to the manufacturer's catalogue, the life expectancy at rated load and rated speed, if both are constant, is 9×10^7 revolutions, and the life expectancies at other constant loads at rated speed can be calculated using the empirical formula

$$\frac{N}{N_R} = \left(\frac{P_R}{P}\right)^{3.3} \tag{2–26}$$

where N = life expectancy in revolutions with a constant load P.
N_R = life expectancy with rated load P_R ($= 9 \times 10^7$ revolutions).

The life expectancy is the number of revolutions that will be exceeded without failure by 9 out of 10 bearings.

The load cycle for which the life expectancy is desired is:
150 per cent of rated load for ¼ of the time
90 per cent of rated load for ¾ of the time all at rated speed.

The engineer now tries to find his way to a quick solution. He senses that he will need to know the life expectancies for each of the two loads acting alone, and then he can find the desired result as the time-weighted average of these two life expectancies. Proceeding to carry out this plan, he writes:

For 150 per cent load:

$$N = \frac{9 \times 10^7}{(1.5)^{3.3}}$$

$$= 2.36 \times 10^7 \text{ revolutions}$$

For 90 per cent load:

$$N = \frac{9 \times 10^7}{(0.9)^{3.3}}$$

$$= 12.7 \times 10^7 \text{ revolutions}$$

The time-weighted average gives the life expectancy for the given cycle:

$$N_0 = (2.36 \times \tfrac{1}{4} + 12.7 \times \tfrac{3}{4}) \times 10^7$$

$$= (0.59 + 9.56) \times 10^7$$

$$= 10.1 \times 10^7 \text{ revolutions} \qquad (2\text{–}27)$$

The result, of course, lies between the values calculated separately for each load, and is greater than that for rated load. The engineer decides to check further by seeing what would happen to his result 2–27 if the heavy load were increased greatly, and the light load made still lighter. Evidently, this would make the first term in 2–27 become much smaller and the second larger. Thus in the limit, the lighter load would be the controlling factor in determining the life. But this is not reasonable! It should be the other way around, for the heavier load must be the one that causes failure and thus determines the life. The trouble comes from the use of a time-weighted average in a situation where it has no rational basis; indeed, the engineer now realizes that what he has done is absurd. He has been looking for an easy short cut, rather than analyzing the problem professionally. He decides to start over and this time to work carefully, to search for a more rational basis for a solution. After some reflection, it becomes apparent to him that knowing the mechanism by which roller bearings fail is the key to the problem. However, he has not studied in this field and so does not have the special knowledge which seems to be required. Not to be deterred by this seeming lack, he decides to see if he can simplify the situation enough to bring a solution within his grasp. He writes:

Suppose the roller bearing wears in the same ways as a sleeve bearing might be thought to wear. Assume that the surface wears away a constant amount per revolution, the amount depending on the load, and that failure occurs when the total amount worn away reaches some fixed limit which is independent of the load; i.e., when the clearance reaches some critical value. Even if the progress toward failure is not linear with number of revolutions turned, the assumption of linearity should be a good first approximation to the truth. Then, for 100% load, a certain fraction of the total allowable wear, $1/N_R = (9 \times 10^7)^{-1}$, occurs in each revolution. If the load were doubled, the amount worn away per revolution according to the formula then would be $2^{3.3}$ (= about 10) times as much.

It now becomes clear to the engineer that he has an assumption on which to base a solution, and he writes as his working principle:

The wear under the one condition of load plus the wear under the other is equal to the wear to cause failure.

Let the wear to cause failure be designated by K. Then, if the rated life is N_R revolutions under rated load P_R, the wear per revolution under this load is K/N_R. Similarly for another load P_1, under which the rated life is N_{R1} revolutions, the wear per revolution is K/N_{R1}, and under load P_2, with life N_{R2} revolutions, the wear per revolution is K/N_{R2}.

Hence under the given varying load if, up to failure, N_1 revolutions are turned under load P_1, and N_2 are turned under load P_2,

$$N_1 \frac{K}{N_{R1}} + N_2 \frac{K}{N_{R2}} = K$$

or

$$\frac{N_1}{N_{R1}} + \frac{N_2}{N_{R2}} = 1 \qquad (2\text{–}28)$$

The engineer realizes the vague, and hence unsatisfactory, character of the quantity K which he has used in his argument. In the sleeve-bearing analogy with which he started K is clear enough. There it would be simply the amount by which the clearance could be allowed to increase before it was decided that the bearing was no longer useful. But, in the roller bearing, he has no clear concept of the real nature of the failure, and hence the vagueness in K. It turns out, however, that it makes no difference, for as has been seen K cancels out. To explain this to himself, the engineer interprets 2–28 by writing out its meaning in words:

N_1/N_{R1} is the fraction of the total useful life expended under one load, and similarly N_2/N_{R2} is the fraction expended under the other load. The sum of these fractions equals the whole, that is, 1.

Having first derived this principle rather laboriously, the engineer sees now that he might have written it down directly. This realization is in fact a check on the soundness of his reasoning so far. He proceeds with his writing.

The fractions of time turned under loads P_1 and P_2 are known; designate them by T_1 and T_2 respectively, where $T_1 + T_2 = 1$.

Since the speed is constant,

$$\left.\begin{aligned} N_1 &= T_1 N_0 \\ N_2 &= T_2 N_0 \end{aligned}\right\} \tag{2–29}$$

where N_0 is the life expectancy which is sought.

Substituting 2–29 into 2–28,

$$\frac{T_1 N_0}{N_{R1}} + \frac{T_2 N_0}{N_{R2}} = 1 \tag{2–30}$$

By 2–26,

$$\frac{N_{R1}}{N_R} = \left(\frac{P_R}{P_1}\right)^{3.3}$$

or

$$\frac{1}{N_{R1}} = \frac{1}{N_R}\left(\frac{P_1}{P_R}\right)^{3.3} \tag{2–31}$$

Similarly,

$$\frac{1}{N_{R2}} = \frac{1}{N_R}\left(\frac{P_2}{P_R}\right)^{3.3} \tag{2–32}$$

whence, using 2–31 and 2–32 in 2–30,

$$T_1 \frac{N_0}{N_R}\left(\frac{P_1}{P_R}\right)^{3.3} + T_2 \frac{N_0}{N_R}\left(\frac{P_2}{P_R}\right)^{3.3} = 1$$

or

$$\frac{N_0}{N_R} = \frac{1}{T_1\left(\frac{P_1}{P_R}\right)^{3.3} + T_2\left(\frac{P_2}{P_R}\right)^{3.3}} \tag{2–33}$$

Checks

The result obviously is correct dimensionally.

Suppose only one load, say P_1, acts throughout, and $T_1 = 1$, $T_2 = 0$. Then N_0 should equal N_{R1}. The result checks for this case, as it does also if only the other load P_2 acts.

Suppose $P_2 = P_1$, that is, the load is constant throughout. Then $N_{R2} = N_{R1}$ and N_0 should equal N_{R2}. The result also checks for this case.

If one of the loads P_1 becomes very large, N_0 should become zero, and the result checks this.

With one of the loads P_2 reduced to zero, the bearing will turn idle for T_2/T_1 revolutions for every one while P_1 acts, and the life N_0 should then be $N_{R1}\ (1 + T_2/T_1)$. The result obtained checks.

The result having stood these tests satisfactorily, the engineer is ready to make a numerical substitution for the case of interest to him:

$$T_1 = \tfrac{1}{4} \qquad P_1 = 1.5P_R$$

$$T_2 = \tfrac{3}{4} \qquad P_2 = 0.9P_R$$

Then

$$N_0 = \frac{9 \times 10^7}{\frac{1}{4}(1.5)^{3.3} + \frac{3}{4}(0.9)^{3.3}}$$

$$= 6.1 \times 10^7 \text{ revolutions}$$

To check on the reasonableness of this result, the engineer uses the life expectancy of 2.4×10^7 revolutions for the heavier load (150 per cent) acting constantly. Now if the light load did not act at all, but the bearing just turned idly for ¾ of the time, he would expect a life of $4 \times 2.4 \times 10^7 = 9.6 \times 10^7$ revolutions. This is an upper limit for the life expectancy and so the value of 6.1×10^7 revolutions which has been calculated appears reasonable. Moreover, it is substantially greater than the value 2.4×10^7 which would obtain if the heavier load acted continuously. On the basis of this and the other checks, the engineer feels justified in using a value of 6×10^7 revolutions as an estimate of the life expectancy, although he realizes that it may be only a rough estimate because of the assumption which he has made that the fraction of life expended in each revolution is a constant depending only on the load.

Thus the engineer, working in good professional style except for his initial fumble, has planned and carried out a reasonable solution to a problem in a field where he has had no training or experience. As a matter of fact, the formula which he has derived is exactly equivalent to the one used in current practice by machine designers and bearing manufactuers, and is apparently the best answer there is at the present time.

2-5 Relay Heater Production Test

From the preceding illustrations, it should be clear that the more difficult stages of a problem solution may occur before the mathematical part is reached. Once a mathematical formulation is made, there are definite rules for proceeding toward a solution, and this of course is the reason for trying to reduce a situation to mathematical terms. It may come as a surprise to you, therefore, to find that many, perhaps most, of the problems you will meet in your professional career will not involve mathematics. The professional method, however, is applicable to *all* problems whether or not mathematics is appropriate in their solution. The following instance of an engineering problem, not involving mathematics, is based on an actual experience of a young engineer working as a trainee in a large manufacturing plant.

The engineer was told that there was a production bottleneck in the testing of relay heaters, the output of which had recently been stepped up, and that he should investigate. After some inquiry, he found, a relay heater to be an electrical resistance element molded into a plastic block (figure 2-4) and intended for mounting close to a thermally sensitive relay

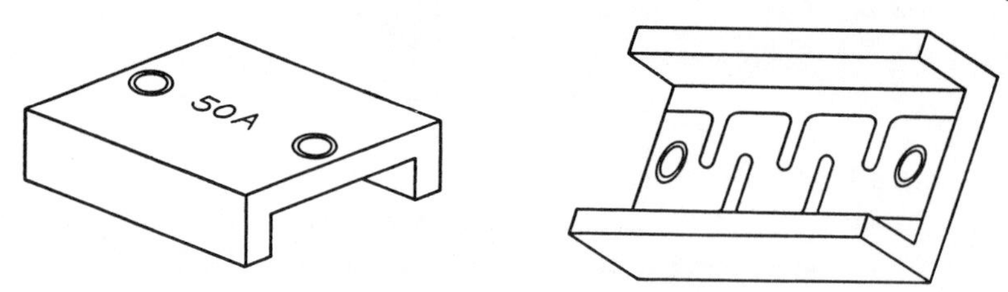

Fig. 2-4. A relay heater. The resistance element shows in the right-hand view.

in a motor controller. The motor line current flows through the heater, and, if the motor is overloaded for a sufficient length of time, the heater makes the thermal relay hot enough to trip, causing the motor to be shut down before it is damaged. The heaters installed in a particular controller are dependent on the size of the motor to be controlled and are mechanically interchangeable and of identical appearance except for the line-current rating marked on the top. Each current rating corresponds to a particular value of resistance of the metallic element which is fastened into the under surface of the block.

The engineer found the test to be a troublesome spot indeed. It was surrounded by piles of trays holding hundreds and hundreds of heaters waiting to be tested. The operator of the test, a girl apparently in a nearly frantic state, was being badgered by a couple of production expediters who were carrying the heaters away, a handful at a time, as she finished with them. The engineer felt that this was no time to intrude with questions, so he stayed quietly in the background to see what facts he could gather about the problem by just watching.

The girl put each heater over a fixture mounted on the bench, so that a couple of pins came through the mounting holes. Then she pushed the heater down, apparently with considerable effort, while the light spot of a wall-mounted galvanometer moved slowly and jerkily up a graduated scale. When the spot stopped, she took the heater off and either passed it to one of the production men or, at very rare intervals, dropped it into a bin marked "scrap." Having made these observations, the engineer decided to go away and come back after the shop quitting time when he could study the test more closely.

The closer inspection revealed that the test was simply a measurement of resistance by the voltmeter-ammeter method, the wall galvanometer being the voltmeter. The fixture (see figure 2–5) on which the heaters

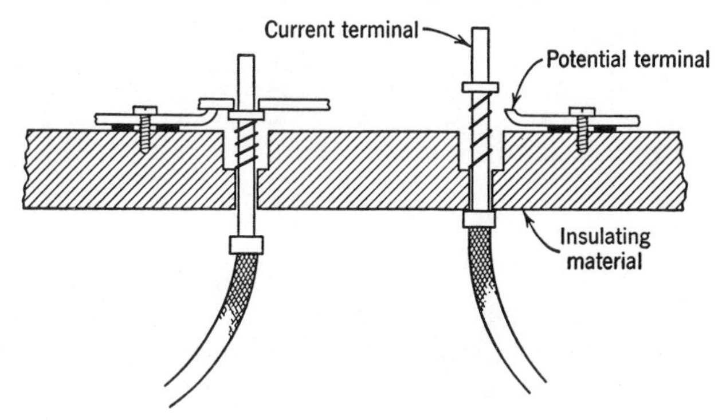

Fig. 2-5. Text fixture. Left-hand part shown with terminal of heater element in contact.

were held consisted of the two pins to engage the mounting holes in the heater, and to act as current terminals, and two leaf springs to act as potential terminals. The latter could not touch the heater element unless the heater was pushed down, so that springs under the current terminals were deflected enough to give very firm contact of the current terminals with the heater, thus insuring that the galvanometer could not be subjected to excessive voltage. The circuit proved to be as shown in figure 2–6. An instruction card fastened to the bench indicated that, with a jumper connecting the current terminal pins together, the rheostat was to be set so that the ammeter would indicate a current equal to the rating of the heater to be tested. The galvanometer shunt setting and the range of acceptable galvanometer readings for each heater rating were specified in a table, the tolerance on resistance being ± 5 per cent. Evidently it was intended that the resistance of the rheostat would be large enough so that the presence of the heater in the circuit would have no effect on the current setting, and this seemed reasonable to the engineer, for he estimated the heater resistance at not more than a few thousandths of an ohm.

Once he understood the circuit arrangements, the engineer decided to try his hand at testing some heaters. He found everything as expected,

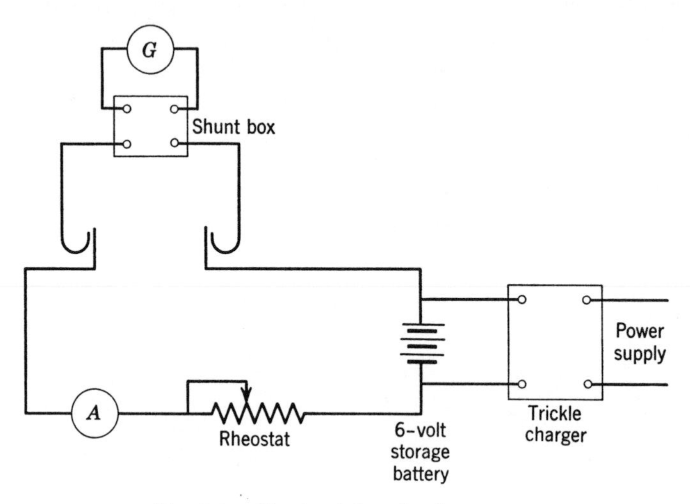

Fig. 2-6. Circuit of the relay-heater test.

except that he had to push the heaters down exceedingly hard to make good contact. If they were not held down very firmly, the potential contacts apparently did not "make" properly and any galvanometer deflection would start to go back toward zero. After testing a dozen or so units, his wrists were tired and his finger tips sore. Clearly it was unreasonable to expect anyone to perform such an awkward and painful operation hour after hour. Obviously the way to get speed was to modify the test rig so that less physical exertion would be needed.

Having thus gathered the facts and defined the problem, the young engineer considered plans for a solution. He eliminated at once the possibility of making changes in the electrical arrangements; considering the low resistances being measured, the method employed seemed the only feasible one, and in fact appeared to be well conceived and reasonably foolproof. Moreover, changes of this character would be time consuming since they undoubtedly would involve not only getting other apparatus together but also obtaining approval for altering the basic method of the test. Likewise it seemed best not to try reducing the stiffness of the springs under the current terminal pins because there might be good reason for the heavy contact pressure which was thus necessitated. Accordingly he decided to leave the electrical characteristics of the test entirely unaltered and to seek a means for pushing the heater down with less effort. His plan was to use a lever which would give a mechanical advantage of three or so and have a handle more comfortable to push on than the heaters themselves.

In execution of his plan, he made a lever (figure 2–7) out of parts from a large knife switch, which he appropriated from a neighboring department where they were made.

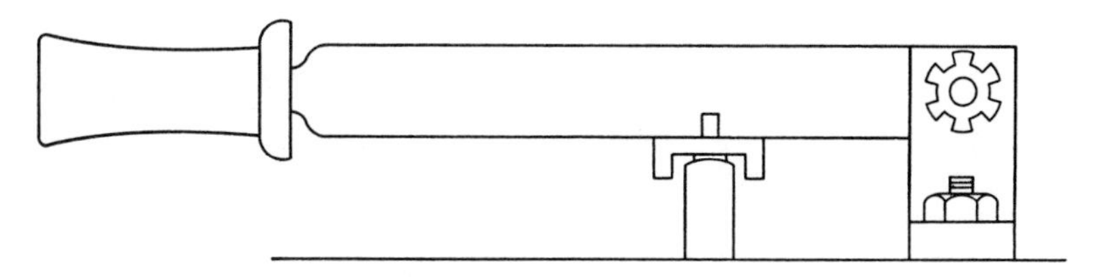

Fig. 2-7. Improvised lever with a heater in position on the test fixture.

To check his solution to the problem, he tried out his lever by testing several dozen heaters. It was easy, and, because he no longer had to wait between heaters for circulation to recommence in his finger tips, it was faster than before. But inherently it was a slow process since for each heater it took the galvanometer nearly 10 seconds to reach its final deflection. He found he could gain a little speed by changing heaters very fast, so that the galvanometer had not returned all the way to zero before he got another unit in the circuit. This gave him the germ of an idea for a further improvement, but he decided to sleep on it as it was now long after midnight, and anyway he had already improved the situation so greatly that there should be a decided spurt in the next day's production.

In the morning, he returned to the test to be sure the girl understood the lever he had rigged. She did and already had made a substantial dent in the backlog of untested heaters; so he went away to think about the new problem he had set himself: how to reduce the delay inherent in the long-period galvanometer?

In considering plans for a solution, he decided to assume again that the electrical features of the test were good and would be left basically unchanged. His attempts to gain speed by trying to change heaters so fast that the galvanometer deflection would not return to zero between measurements suggested the use of a dummy resistor to be switched into the circuit by action of the lever while the tested heater was being replaced by a new one. To keep the galvanometer at nearly constant deflection, the dummy would need to have the same resistance as the heaters being tested and should therefore be easily replaceable by ones of different values for batches of heaters of different ratings. Clearly, suitable dummies would be heater elements themselves, ones known to be good.

In thinking of ways to switch the dummy in and out of the circuit automatically and to make it easily replaceable, it came as an inspiration to use two fixtures connected electrically in parallel and arranged so that heaters would be tested alternately on the two fixtures, the connections

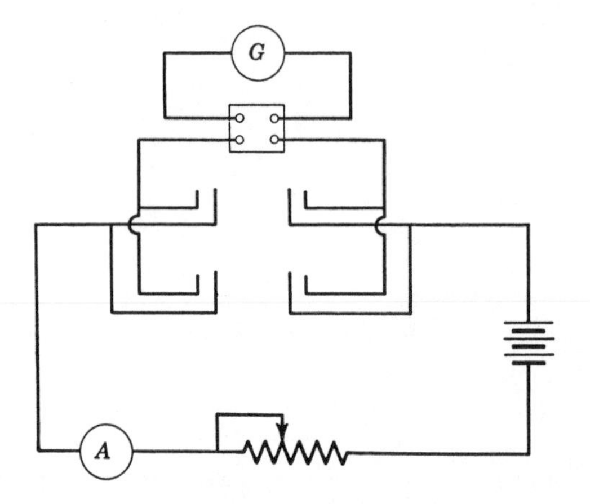

Fig. 2-8. Electrical connections to the double test rig.

being changed quickly from one to the other. (Inspirations such as this come frequently when you analyze thoroughly in good professional style.) The planned modification of the circuit is shown in figure 2–8. Switching from one set of connections to the other would be done by having but one heater in contact at a time. This would be accomplished by using test fixtures like figure 2–5, but modified as follows:

1. Insulate the upward-projecting parts of the current terminal pins with thin tape.
2. Provide a light spring in the center of the fixture to keep the heater out of electrical contact unless it is pushed down.

In planning the mechanical means for pressing the heaters down on the fixtures, and the placing of the two fixtures on the bench, the engineer gave careful thought to the comfort of the operator and to economy of motion, for his own attempts at fast testing had taught him the importance of these matters. His plan was to have a foot-operated lever system for pushing the heaters down so as to leave both the operator's hands free to move heaters on and off the fixtures. The two fixtures would be parallel to the front of the bench and one behind the other. Heaters would be set on with the left hand and removed with the right; this left-to-right flow conforming to the direction of movement to and from the test bench already established by the shop layout. Figure 2–9 shows the arrangement planned. To test the front heater, the pedal is depressed; for the

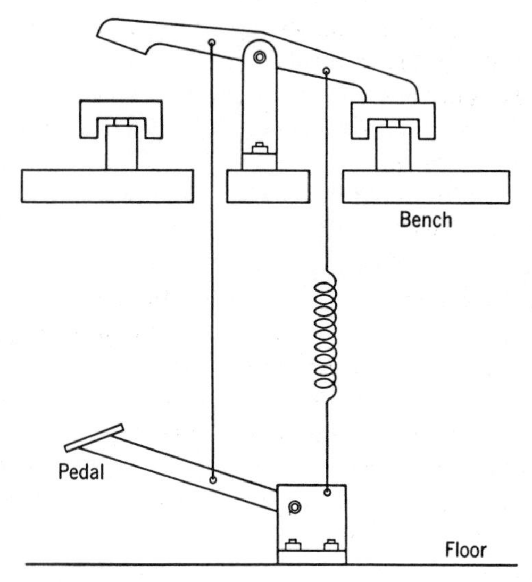

Fig. 2-9. Double test rig. Front of bench is to the left.

rear one it is released, allowing the tension spring to supply the necessary pressure.

In executing his plan, the engineer made working sketches in proper form of the parts needed, and to obtain an order for their manufacture went to the shop foreman. This gentleman was a testy old curmudgeon, but the young engineer had thought out his plan so carefully and had made his sketches so clear that he was able to counter all arguments diplomatically and convincingly and get what he wanted. The parts were made quickly, and the engineer himself installed them that night so he could check the operation personally and work out any difficulties that might develop.

There were no difficulties. He found that there were long runs of heaters so nearly alike that the galvanometer spot never moved at all from a standard deflection, and that the only limitation on his speed was the necessary muscular coordination between his two hands and his foot. When the galvanometer spot started to move, as it did occasionally, it was a simple matter to wait with the pedal up or down until the spot came to rest and he could see whether or not to reject the heater.

As a result of the improved test set-up, the backlog of untested heaters was disposed of in a few days and thereafter it was necessary to operate the test for only an hour or two a day to keep up with production.

An unfortunate consequence of the improvement resulted from the fact that the operator was paid on a piece-rate basis. The rate had been set to give a fair wage with the old set-up, and, since no one informed the Rate and Time Study Department that a change had been made in the operation, the operator was able to make a small fortune the first week as her output was some ten times normal. The piece rate then was cut, and this caused the girl to quit in anger and disillusionment. Our young engineer learned too late that it was his responsibility to have seen that a new time study was made immediately, so that the disturbing effects on the person concerned would be as little as possible.

In reviewing the analysis of this problem, the use of good professional method is not difficult to discern. It is clear that our young friend spent a good deal of effort gathering the pertinent facts before attempting to form conclusions as to a solution. Proceeding to do this deliberately and systematically, he decided in due course that the real problem was to reduce the physical exertion required of the operator, and that, if this could be accomplished, the desired speed would result. In planning a way to reduce the effort, two alternatives appeared possible to him: one, to accept the electrical features of the test and provide some means for mechanical advantage; and the other, to try to devise a system wholly new in all respects. He saw that the existing electrical arrangements were sound, and so he chose to reduce the physical effort by mechanical means. After executing this plan, he checked to see if the result was what he anticipated. He found that in part it was not; although the effort was reduced suitably and the speed increased considerably, the testing still was too slow. This finding led to a new problem: how to overcome the inherent delay. Again in planning, he was faced with the choice of altering the present system or starting from scratch, and again he decided that the existing electrical arrangements should be retained basically intact. The idea for a plan to remove the delay came to him after thinking systematically about the causes. In executing his plan, he was forced by circumstances to enlist the cooperation of other people to make necessary parts. In this he was successful principally because his careful analysis had given him such a thorough understanding of the problem that he was able to counter objections to his solution in an immediately convincing manner. After getting the parts and checking their operation both personally and later in the hands of the regular operator, he found that he had successfully solved the problem in all respects but one. He had failed to take proper steps to cushion the impact on the operator of the necessary reduction in the piece rate. It is to be hoped he learned from this experience that the professional responsibility of an engineer extends beyond the merely technological results of his work.

2-6 Phonograph Record Changer

In contrast to the preceding case where the analysis involved no mathematics at all, let us consider an example of professional method in which the logic of mathematics is essential in arriving at useful conclusions.

The problem concerns improvement of the design of an automatic record changer for a phonograph with a view to reducing the time for changing records. In this changer, the new records are supported above the turntable on a stationary spindle. When the record has been played, a mechanism driven from the turntable moves the tone arm out of the way, releases the next record, which drops down onto the one already on the table, and finally causes the needle to come down into the beginning of the groove. The turntable is driven by a synchronous motor through a friction drive consisting of a small rubber-tired idler wheel which bears against the rim of the table and against a small metal wheel on the motor shaft. This friction drive may slip, allowing the turntable speed to decrease below its normal value if the torque required exceeds a definite value. The record is also driven through friction, between it and the next lower record, or between it and the table if it is at the bottom of the stack.

The time duration of a change is compounded of the following parts, some of which may overlap one another:

a. Interval for the tone arm to move out from the center.
b. Interval for the new record to fall from its support to the table.
c. Interval for the record to accelerate from rest to playing speed.
d. Interval for the tone arm to move in and drop onto the record.

Part *b* may overlap part *a*. Also, parts *c* and *d* may overlap or be concurrent, but to avoid unpleasant sounds it is important that the record be at normal playing speed before the needle enters the groove.

For purposes of the present illustration, let us confine our attention to interval *c* during which the record is accelerated from rest to normal playing speed. The object is to determine the length of this interval analytically so as to see how it depends on factors which might be influenced by design. Although we shall deal in this section only with interval *c*, the whole problem, of course, would require study of the other intervals, too; and indeed some of the others may be more important, but we cannot tell this until all have been investigated. To break into our problem, we

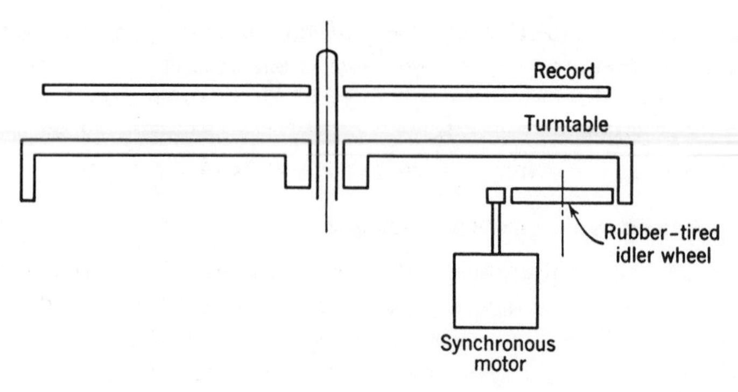

Fig. 2-10. Phonograph turntable and drive.

make a sketch (figure 2–10) and define a situation somewhat restricted by simplifying assumptions.

> The synchronous motor drives the turntable through a friction drive. With the table turning at normal speed corresponding to synchronous speed of the motor, a single record, initially not turning, drops freely onto the table.
>
> How will the speed of the record vary with time?

Note that the more general case where several records are already on the table is deliberately postponed until after the simpler situation involving only one record is thoroughly understood. In seeking a principle to apply, we write:

> We need a relation between angular velocity of the record and time. The record experiences accelerated angular motion, therefore application of the rotation law $T = I\alpha$ to the record is indicated. If the drive between motor and table slips, that is, if the table at first slows down and then comes up to speed again, the table will undergo accelerated motion too, and therefore, $T = I\alpha$ may be applied to it as well.
>
> The plan, then, is to apply the law $T = I\alpha$ for rotation of a rigid body separately to:
>
> 1. The record.
> 2. The turntable.
>
> Assume in doing this that the motor speed does not change, and that the inertia of the rubber-tired wheel is negligibly small.
>
> To carry out this plan, consider first the record. The only torques

which can act on the record (assuming that the tone arm does not come down until after synchronous speed is reached) are:

a. Torque of friction with the turntable
b. Drag of the surrounding air
c. Friction with the central pivot

The drag of the air and of the central pivot are probably small and for the present will be neglected. Thus:

The torque of friction of the turntable on the record must equal the moment of inertia of the record times its angular acceleration, or

$$T_r = I_r \frac{d\omega_r}{dt} \tag{2-34}$$

where T_r = torque of turntable friction on the record (in. lb).
I_r = mass moment of inertia of the record (lb in. sec^2).
ω_r = instantaneous angular velocity of record, positive in direction of turntable rotation (rad sec^{-1}).
t = time (sec).

T_r is written as positive in the equation because it acts on the record in the direction assumed positive for rotation. This will be defined as the positive direction for all angular quantities.

Now consider the turntable. Acting on it are:

a. The torque exerted by the friction drive.
b. The torque of the record on the turntable.
c. The torque of air friction on the table.
d. The torque of bearing friction.
e. The torque required to drive the mechanism which moves the tone arm.

Assume the windage and bearing-friction torques to be negligibly small, and for simplicity in a first analysis ignore *e*, the torque required to actuate the tone-arm mechanism. Then applying $T = I\alpha$ to the turntable:

The sum of the torque of the drive on the table and the torque of the record on the table must equal the moment of inertia of the table times its angular acceleration; thus

$$T_d - T_r = I_t \frac{d\omega_t}{dt} \tag{2-35}$$

where T_d = torque of the drive on the table (in. lb).
I_t = mass moment of inertia of the table (lb in. sec^2).
ω_t = instantaneous angular velocity of the table, positive in the normal direction of turning (rad sec^{-1}).
t = time (sec).

T_r is the same as defined for the record, but here is taken as negative, because it acts on the table oppositely to positive rotation.

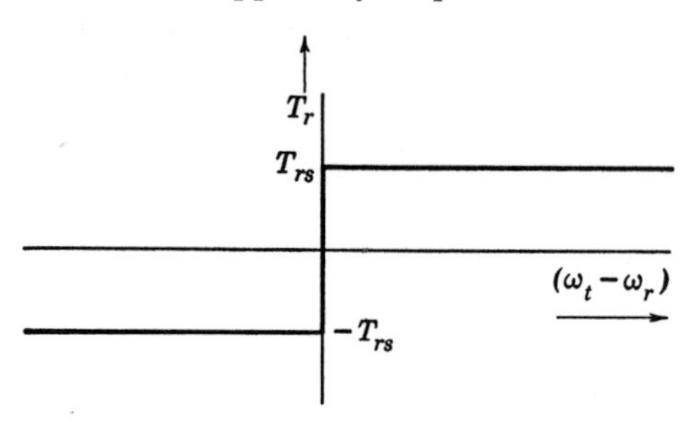

Fig. 2-11. Friction torque exerted on the record by the turntable.

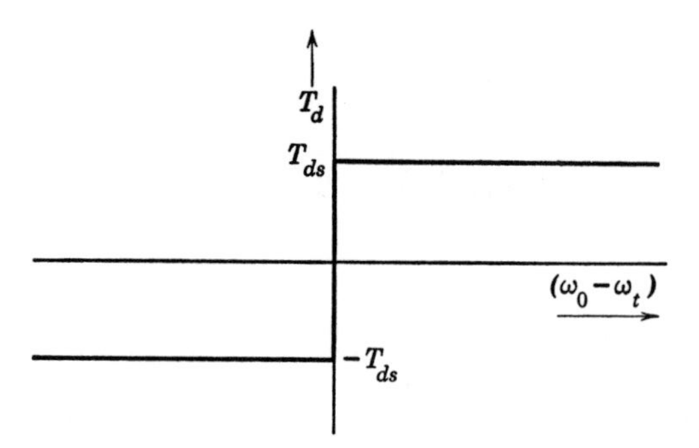

Fig. 2-12. Torque of the drive exerted on the turntable.

To find ω_t in terms of t, it is necessary to know how the torques T_r and T_d depend on time or speed.

These torques are due to dry rubbing friction. Assume that, while there is slipping, their magnitudes are constant and independent of the rates of slip. Then T_r varies as in figure 2–11. Thus, while there is slipping and the turntable is moving faster than the record, the torque T_r acting on the record has a constant value T_{rs}, but, when $\omega_r = \omega_t$, it can have any value between $+T_{rs}$ and $-T_{rs}$. If the record were to turn faster than the table, then there would be a constant torque T_{rs} acting backward on the record. Similarly, T_d would vary as in figure 2–12, where ω_0 = normal speed of the turntable.

Since the record initially is not turning, it must slip at first. The turntable may or may not slip with respect to the driving motor. Two cases, then, are possible.

CASE 1. The turntable remains at normal speed throughout. Then, since the record necessarily slips, equation 2–34 becomes

$$T_{rs} = I_r \frac{d\omega_r}{dt} \tag{2-36}$$

and, since the turntable remains at normal speed and the drive does not slip, equation 2–35 becomes

$$T_d - T_{rs} = 0 \tag{2-37}$$

Since always $T_{ds} \geq T_d$, the condition for occurrence of case 1 is

$$T_{ds} > T_{rs} \tag{2-38}$$

That is, the maximum friction torque that can be applied to the table by the drive is greater than the maximum friction torque between record and table.

From 2–36 the time t_0 for the record to reach normal speed is

$$t_0 = \int_0^{t_0} dt = \frac{I_r}{T_{rs}} \int_0^{\omega_0} d\omega_r$$

$$= \frac{I_r\omega_0}{T_{rs}} \tag{2-39}$$

CASE 2. The turntable slips with respect to the motor. Then, since the record must slip at first, equation 2–36 applies just as in case 1. Equation 2–35 for the turntable becomes

$$T_{ds} - T_{rs} = I_t \frac{d\omega_t}{dt} \tag{2-40}$$

But, if the turntable begins to slip with respect to the motor, it begins to lose speed, $\frac{d\omega_t}{dt}$ must be negative and

$$T_{ds} < T_{rs} \tag{2-41}$$

This is the condition for case 2.

During the period that the record slips with respect to the table, T_r is constant at the value T_{rs}, hence $\frac{d\omega_t}{dt}$ continues to be negative and the slip velocity increases or the table continues to lose speed.

During the first part of case 2, the record accelerates at a constant rate and the turntable decelerates at a constant rate as shown to the left of t_e in the sketch figure 2–13.

This continues until the speeds become equal at some time $t = t_e$ when slipping of the record ceases and torque T_r decreases from the slipping value T_{rs} to whatever value is necessary to keep the record turning with the table. From then on $\omega_r = \omega_t$, and record and table can be treated as one rigid body, so that

$$T_{ds} = (I_r + I_t)\frac{d\omega_t}{dt} \tag{2-42}$$

until $\omega_t = \omega_0$, that is, until normal speed is reached by record and table. During this latter period, the torque required to accelerate the record, and which must be exerted between table and record, is

$$T_r = I_r\frac{d\omega_t}{dt} = \frac{I_r}{I_r + I_t}T_{ds} \tag{2-43}$$

This shows that the required T_r is less than T_{ds}; therefore the record will not slip during this period because for this case T_{rs}, the slipping value of T_r, must be greater than T_{ds}; and this is a check on the reasoning so far.

To find the total time t_0 to reach synchronism, first find t_e, the time to reach equal speeds ω_e, and then the time $t_0 - t_e$ to go from speed ω_e to ω_0.

From 2–36 and 2–40, we have for $t < t_e$,

$$\omega_r = \frac{T_{rs}}{I_r}t \tag{2-44}$$

$$\omega_t = \omega_0 - \frac{T_{rs} - T_{ds}}{I_t}t \tag{2-45}$$

Setting these equal in order to find t_e,

$$\frac{T_{rs}}{I_r}t_e = \omega_0 - \frac{T_{rs} - T_{ds}}{I_t}t_e$$

$$t_e = \frac{I_r I_t \omega_0}{T_{rs}I_t + (T_{rs} - T_{ds})I_r} \tag{2-46}$$

Now substitute this into 2–44 to find ω_e.

$$\omega_e = \frac{T_{rs}I_t\omega_0}{T_{rs}I_t + (T_{rs} - T_{ds})I_r} \tag{2-47}$$

The time t_0 is then t_e plus the time to go from ω_e to ω_0 (found by inspection of 2–42), or

$$t_0 = t_e + \frac{\omega_0 - \omega_e}{T_{ds}}(I_r + I_t) \tag{2-48}$$

Substituting for t_e from 2–46 and for ω_e from 2–47,

$$t_0 = \frac{I_r I_t \omega_0}{T_{rs}I_t + (T_{rs} - T_{ds})I_r} + \frac{(I_r + I_t)\omega_0}{T_{ds}} - \frac{T_{rs}I_t(I_r + I_t)\omega_0}{T_{ds}[T_{rs}I_t + (T_{rs} - T_{ds})I_r]}$$

$$= \frac{T_{ds}I_rI_t + \cancel{T_{rs}I_t(I_r + I_t)} + (T_{rs} - T_{ds})(I_r + I_t)I_r - \cancel{T_{rs}I_t(I_r + I_t)}}{T_{ds}[T_{rs}I_t + (T_{rs} - T_{ds})I_r]}\omega_0$$

$$= \frac{\cancel{T_{ds}I_rI_t} + T_{rs}I_r^2 + T_{rs}I_rI_t - T_{ds}I_r^2 - \cancel{T_{ds}I_rI_t}}{T_{ds}[T_{rs}I_t + (T_{rs} - T_{ds})I_r]}\omega_0$$

$$= \frac{I_r\omega_0}{T_{ds}} \tag{2–49}$$

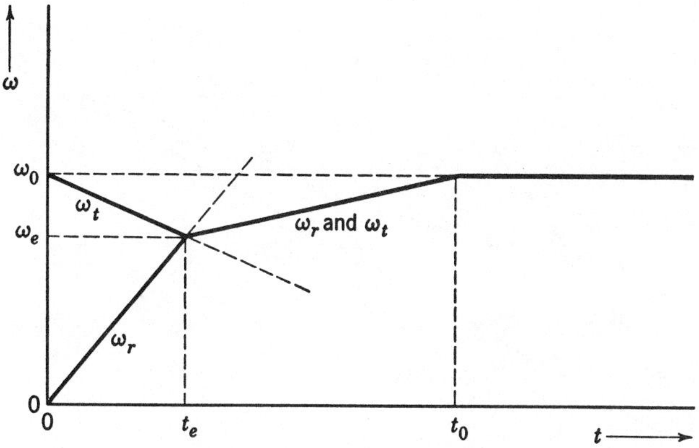

Fig. 2-13. How the speeds vary when the drive slips.

The absence of I_t and T_{rs} from this result is surprising and at first glance possibly erroneous. On the other hand, if the result is correct, its simplicity suggests that it might have been reached by a more direct method.

For a check of this result return to the beginning of case 2. The basic equations were 2–36 and 2–40:

$$T_{rs} = I_r \frac{d\omega_r}{dt}$$

$$T_{ds} - T_{rs} = I_t \frac{d\omega_t}{dt}$$

Evidently T_{rs} can be eliminated by adding these equations, thus:

$$T_{ds} = I_r \frac{d\omega_r}{dt} + I_t \frac{d\omega_t}{dt} \qquad (2\text{–}50)^*$$

* This relationship could have been written more directly from another fundamental principle as explained in section 3–6.

Now integrate over the whole interval from $t = 0$ to $t = t_0$ using the corresponding limits on ω_r and ω_t.

$$\int_0^{t_0} T_{ds}\, dt = \int_0^{\omega_0} I_r\, d\omega_r + \int_{\omega_0}^{\omega_0} I_t\, d\omega_t \tag{2-51}$$

$$T_{ds} t_0 = I_r \omega_0 \tag{2-52}$$

which checks with the former result 2–49.

In spite of this formal confirmation of 2–49 it still is not clear to us physically how the result for case 2 can be independent of T_{rs} and I_t. To

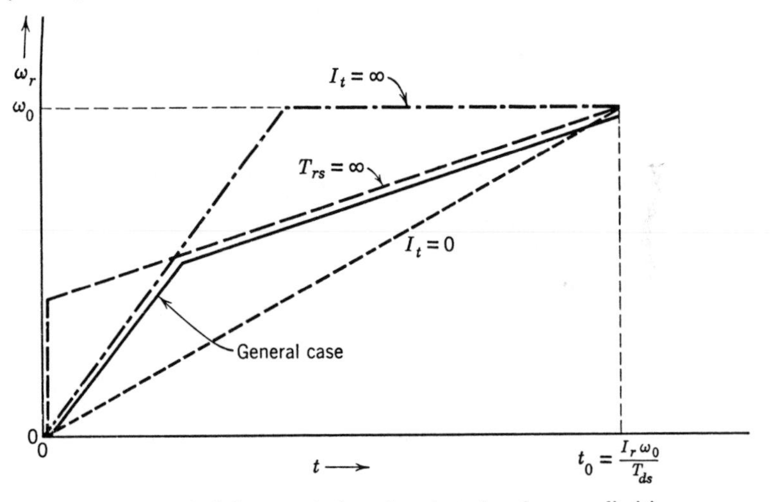

Fig. 2-14. Speed of the record plotted against time for some limiting cases.

gain a fuller understanding of this matter, we consider by reasoning as independent of the foregoing analysis as possible what would have to happen in some special limiting cases in each of which we let either T_{rs} or I_t approach a limiting value while the other parameters remain fixed. The three limiting cases we try are:

$$T_{rs} = \infty \tag{2-53}$$

$$I_t = 0 \tag{2-54}$$

$$I_t = \infty \tag{2-55}$$

In the first two cases it turns out as indicated in figure 2–14 that t_0 for case 2 is indeed independent of T_{rs} and I_t. For the third case, however, $I_t = \infty$, the time to reach playing speed is less than the value given by 2–49. Nevertheless, we are satisfied with 2–49 because, if I_t is not infinite but only very large, we see that playing speed actually will not be reached exactly until the time given by 2–49.

EXERCISE. Deduce the reasoning of these three limiting case checks and carry them out in detail.

Now that we have completed and checked the analysis, we come to the stage where the results are assembled to see what can be learned from them and what generalizations may be possible.

CASE 1. Only the record slips.
Occurs when $T_{ds} > T_{rs}$ (equation 2–39).

$$t_0 = \frac{I_r \omega_0}{T_{rs}}$$

CASE 2. Both record and drive slip.
Occurs when $T_{ds} < T_{rs}$ (equation 2–49).

$$t_0 = \frac{I_r \omega_0}{T_{ds}}$$

We now understand the variation of speed of the record, which was the first goal of our analysis. Then the question arises; how can the design be improved to shorten the time required to bring the record up to speed?

Since I_r and ω_0 are dependent on the type of records to be played, the only factors influencing t_0 which can be altered by design of the changer are T_{rs} and T_{ds}.

If we have case 1, we want T_{rs} as large as possible in order to make t_0 small. But, if T_{rs} exceeds T_{ds}, we have case 2 and then it becomes desirable to increase T_{ds}, which presently returns us to case 1.

The conclusion is that the best that can be done is to have T_{ds} larger than T_{rs} so that only the record ever slips. Moreover, if the record is to rest on the table under its own weight, as is the usual practice, there is little that can be done effectively to increase T_{rs}, so that t_0 is not susceptible to much, if any, reduction by refinement in design of a record changer in which the stored records are stationary. The advantage in this respect of a changer in which the stored records rotate with the table is apparent.

How is the conclusion affected by the simplifications made in the analysis?

In considering the effects of neglecting the various friction torques and the torque to operate the tone-arm mechanism, we find that, if these are included as constant independent of speed, the effect on the result 2–39 is that T_{rs} must be replaced by T_{rs} minus the frictional retarding torques which act on the record. The effect on 2–49 is that T_{ds} must be replaced by T_{ds} minus the sum of the frictional retarding torques which act on the record and on the turntable (exclusive of T_{rs}) and the torque to operate the mechanism.

EXERCISE. Confirm this finding and determine the change, if any, in the boundary between cases 1 and 2.

A major simplification which was made in the foregoing analysis was to consider only the case of the first record to be dropped onto the table. Let us see now how study of the simplified case can help us to generalize to the more complicated case where there are already a number of records on the table. In extending the analysis to the general case we write:

Suppose there are $n - 1$ records already on the table when the nth record is dropped on. Find the time for this record to accelerate to playing speed.

If we apply the law for rotation, $\Sigma T = I\alpha$ to the table and to each record separately, equations 2–34 and 2–35 are expanded to the following set:

$$\left.\begin{aligned} T_d - T_1 &= I_t \frac{d\omega_t}{dt} \\ T_1 - T_2 &= I_1 \frac{d\omega_1}{dt} \\ &\vdots \\ T_{n-1} - T_n &= I_{n-1} \frac{d\omega_{n-1}}{dt} \\ T_n &= I_n \frac{d\omega_n}{dt} \end{aligned}\right\} \qquad (2\text{–}56)$$

The symbols are as already defined except that the subscript r which identified the single record is replaced by subscripts 1 through n for the respective records counting up from the bottom of those on the table.

By analogy with the technique used in getting 2–51, add equations 2–56 together and then integrate over the time interval 0 to t_0. The result is:

$$\int_0^{t_0} T_d \, dt = I_n \omega_0 \qquad (2\text{–}57)^*$$

* This equation is derived differently in section 3–7.

Thus, if the drive slips, $T_d = T_{ds}$, and we get

$$t_0 = \frac{I_n \omega_0}{T_{ds}} \tag{2-58}$$

If the drive does not slip and if the records are assumed to be all alike, only the top record will slip. This is because the torque developed when one record slides past another will be proportional to the force acting normally across the surface between them, that is, to the weight of the records above the joint being considered. If there were slipping of a record other than the top one, the torque on it developed at its bottom surface is positive (in the direction of rotation) and larger than that at its top which is negative (backward); hence slipping could never get started. Consequently the table and $n - 1$ records will be unaccelerated and T_d will have just the value to balance the torque to slip the top record, that is, $T_d = T_{rs}$ (see figure 2-11). Then from 2-57 there results

$$t_0 = \frac{I_n \omega_0}{T_{rs}} \tag{2-59}$$

and the results 2-58 and 2-59 for the general case are the same as 2-49 and 2-39 for the simplified case done first.

Thus we have solved part of the whole problem. We have found under various conditions the time required for a record to reach playing speed starting from rest. We shall not determine here the other components which make up the total time of the record-changer operation.

Turning now to an overall consideration of the foregoing illustration, we may be led to ask the following question: Considering the elegance and brevity of the general solution, why was the problem not solved that way in the first place?

It might indeed have been by one sufficiently versed in the science of mechanics. We chose rather to imagine ourselves possessed of but rudimentary knowledge and experience in this field. Evidently this is really the case whenever you find yourself confronted with a problem which seems to be very complex or difficult. Then you must strip away some of the complexities and simplify the problem to a point where it comes more nearly within range of your understanding. If the simplification is done well, later generalization to include the effects of the complicating factors may be as easy as in the present illustration. How much to simplify a problem at first in order to arrive eventually at a useful solution depends entirely on the difficulty of the problem in relation to your skill. Too much or the wrong simplification may change the basic character of the problem so as to preclude useful generalization. Toward making appropriate simplifications you must be guided by insight, and this kind of insight comes from experience in solving difficult problems. Capacity to make wise simplifications in problem situations and capacity to generalize correctly and effectively from the results of the simplified problem will contribute immensely to your effectiveness in engineering.

2–7 Bimetal Thermostat

As the final illustration of professional method in this chapter, we consider a situation more complex than the earlier examples in that it necessitates taking account of more conditions at once than the others did.

We suppose that we are contemplating the design of a line of thermostats using bimetal as the temperature-sensitive element. Bimetal is composed of flat strips of two metals having different coefficients of thermal expansion, brass and steel for example, with one flat surface of each intimately bonded to the other. This, and a casual acquaintance with its characteristics, we suppose to be all we know about bimetal.

Before attempting to design for the specific applications which we have in mind, we deem it wise to gain understanding by a thorough analysis of a simplified case. We decide to investigate the behavior of an initially straight bimetal strip clamped at one end and unrestrained at the other end even though in practice there usually would be a load due to the mechanism to be actuated. The strip is of rectangular section with the two dissimilar metals equal in thickness.

Determination of the deflection of the free end of the strip as a function of temperature is the problem we define for ourselves, and we wish the solution to be entirely in literal terms, so that we will be able to study the influence of all the factors which enter and so that we will find it as easy as possible later to extend the analysis to more general cases.

Entering the planning stage, we think about how the bimetal works. When the strip is heated uniformly, one side tries to expand more than the other, but it cannot expand as much as it would if free, because of the restraint imposed by the less expansive side. The result must be that the strip bends toward the less expansive side, a compressive force developing in the more expansive side and a tensile force in the less expansive; but since there are no external restraints these internal forces are in equilibrium with each other. The internal forces are the resultants of distributions of stress acting across the section. Since the strip bends, the strain must vary across the section as it does in a bent beam, but in this case the strain will be the resultant of the combined effects of temperature and stress and not just of stress as in the familiar case of the beam. There will be then a state of stress and strain varying somehow across the section of the strip, but there is no reason for this state to differ from section to section along the strip. Accordingly we are led to focus attention on the states of stress and strain at a single cross section at some general place along the length of the strip.

Now that we have some rough ideas of what goes on in the strip, we plan on the following definite steps as the basis for starting a solution.

1. Relate strain to curvature of the strip as is done in the analysis of the bending of beams, assuming as is done there, that cross sections of the strip originally plane remain plane after the strip bends.
2. Using the principle of superposition, express the strain as the sum of two parts: one due to the temperature change, and the other the result of the stress.
3. Use the idea that the internal forces which develop in the component metals of the strip are in equilibrium with each other, evidently a condition to which the stress distribution must conform.

It is to be noted that this is not a comprehensive plan for carrying out an entire solution, for the situation is too complex for us to see all the

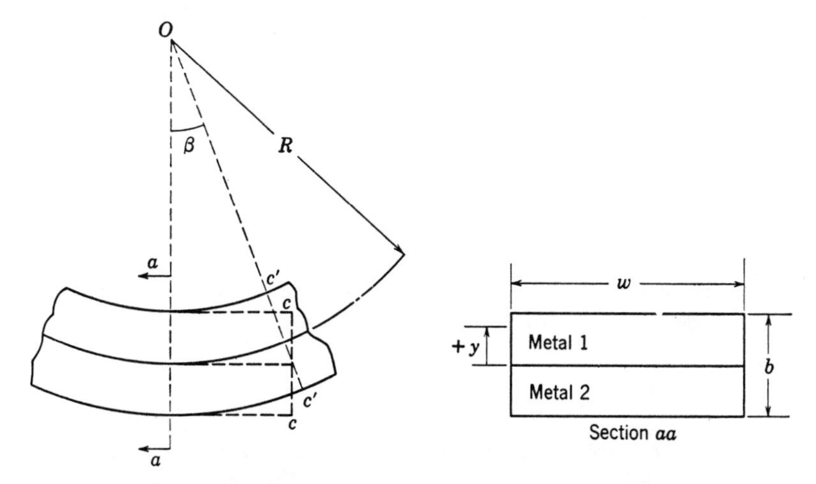

Fig. 2-15. Diagram used in establishing geometric relations for the bimetal strip.

way through. We expect to have to plan further steps when these first ones have been completed and we can see better what is needed next, or perhaps we may find a completely new plan to be necessary.

In execution, then, of the first part of our tentative plan, we make a sketch (figure 2-15) showing at the left a side view of a short length of the strip bent to an exaggerated degree and at the right a cross section *aa* of the strip. Metal 2 is more expansive than metal 1. A cross section near *aa* is shown at *cc* in its position when the strip is cold and at *c'c'* when it is hot, supposing *aa* does not move. Assuming, as already implied by the figure, that plane sections remain plane, we consider a center O about which the region of the strip near *aa* is curved. Each lengthwise fiber of the strip between *aa* and $c'c'$ subtends the same angle at O.

To express this fact mathematically, we let

R = radius of curvature of the interface between the two metals (in.).
e_0 = strain at the interface (change of length per unit length).
y = coordinate locating a fiber with respect to the interface, positive in metal 1 (in.).
e = strain of fiber at y.
λ = original distance between *aa* and *cc* (in.).
β = angle subtended at O between planes *aa* and $c'c'$ (radians).

Then

$$\beta = \frac{(1 + e_0)\lambda}{R} = \frac{(1 + e)\lambda}{R - y} \tag{2-60}$$

Clearing the second equation of fractions, and dividing by λ,

$$R - y + Re_0 - ye_0 = R + Re$$

$$e = e_0 - \frac{1 + e_0}{R} y \tag{2-61}$$

and thus we have completed the first part of our plan by relating strain to curvature, and incidently we have also related strain to position in the cross section.

Now we express the strain as the sum of parts due to temperature change and to stress using the principle of superposition. Thus:

$$e = \alpha T + \frac{s}{E} \tag{2-62}$$

where α = coefficient of thermal expansion, per unit change of length per degree (°F^{-1}).
T = rise above the temperature at which the strip is straight (°F).
s = stress (lb in.$^{-2}$) positive for tension, negative for compression.
E = Young's modulus (lb in.$^{-2}$).

With appropriate constants α and E, 2-62 holds for either metal. In writing 2-62 we assume that α is independent of stress and that E does not depend on temperature.

To carry out the last step of the plan, we will need stress as a function of position in the cross section. To get this, we solve 2-62 for stress, and then for e we substitute the expression 2-61. Thus for $y > 0$:

$$s = E_1(e_0 - \alpha_1 T) - E_1\left(\frac{1 + e_0}{R}\right) y \tag{2–63}$$

and for $y < 0$

$$s = E_2(e_0 - \alpha_2 T) - E_2\left(\frac{1 + e_0}{R}\right) y \tag{2–64}$$

where the subscripts 1 and 2 identify the properties α and E with the respective metals.

Coming to the third step in our plan, the condition that no resultant force acts across the section, we see that this means that the net force or the integral of stress over the section must be zero; hence

$$\int_{-b/2}^{b/2} sw\, dy = 0 \tag{2–65}$$

where b is the thickness of the strip and w its width. We separate this integral into the sum of two parts: from $-b/2$ to 0, and from 0 to $b/2$; then for s substitute the expression 2–63 or 2–64 as appropriate. Upon integration, we get

$$E_1(e_0 - \alpha_1 T)\frac{b}{2} - \frac{E_1(1 + e_0)}{R}\frac{b^2}{8} + E_2(e_0 - \alpha_2 T)\frac{b}{2} + \frac{E_2(1 + e_0)}{R}\frac{b^2}{8} = 0 \tag{2–66}$$

and, collecting terms,

$$(E_1 + E_2)e_0 - \frac{b}{4}(E_1 - E_2)\left(\frac{1 + e_0}{R}\right) = (E_1\alpha_1 + E_2\alpha_2)T \tag{2–67}$$

Having carried our tentative plan to completion, we proceed to check the result 2–67 and then look over what we have done to see if we can learn what to do next.

Each term of 2–67 we observe to be a stress, and therefore the equation is dimensionally correct. Next we note that, if the two metals are the same ($E_1 = E_2$ and $\alpha_1 = \alpha_2$), we get $e_0 = \alpha T$, which evidently is correct. Although this is by no means a complete check, we are satisfied for the time being.

From the standpoint of learning, we see that the result 2–67 relates e_0, strain of the midsurface of the strip, and R, the radius of curvature, to temperature rise T, the other quantities being properties of the metals. But to describe the behavior of the strip we need to know e_0 and R each separately in terms of T. Studying what we have done, we see clearly that it is not possible from what we have to derive another independent relation involving e_0 or R. Therefore, we are forced to seek an additional physical condition for incorporation in the analysis, and so we return to the planning stage.

Bearing in mind the similarity of what we are doing to the familiar analysis of the bending of beams, it is apparent, as you already may have realized, that the needed relation involves the bending moment transmitted across the section. Indeed, part 3 of our original plan should have been more comprehensive. We should have invoked the general conditions for static equilibrium including equilibrium of moments as well as of forces. Why did we not think of this in the beginning and incorporate it in the original plan? We might have if we had been a little more astute, but in new and complicated situations it is unusual to do everything perfectly at the first trial.

Since the strip is subject to no externally applied forces or moments, it must be that there is no bending moment transmitted across the section. Thus mathematically the additional relation is

$$\int_{-b/2}^{b/2} swy\,dy = 0 \tag{2-68}$$

Separating the integral into two parts as we did with 2-65, substituting for s from 2-63 and 2-64, and integrating, we get

$$E_1(e_0 - \alpha_1 T)\frac{b^2}{8} - E_1\frac{(1 + e_0)}{R}\frac{b^3}{24} - E_2(e_0 - \alpha_2 T)\frac{b^2}{8} - E_2\frac{(1 + e_0)}{R}\frac{b^3}{24} = 0 \tag{2-69}$$

and this is the additional independent relation connecting e_0, R, and T which is needed. Collecting terms to facilitate simultaneous solution with 2-67 and for convenience setting the latter down again:

$$(E_1 - E_2)e_0 - \frac{b}{3}(E_1 + E_2)\left(\frac{1 + e_0}{R}\right) = (E_1\alpha_1 - E_2\alpha_2)T \tag{2-70}$$

$$(E_1 + E_2)e_0 - \frac{b}{4}(E_1 - E_2)\left(\frac{1 + e_0}{R}\right) = (E_1\alpha_1 + E_2\alpha_2)T$$

These equations might be solved simultaneously for e_0 and R, but it appears easier to solve them first for e_0 and $(1 + e_0)/R$. Thus by multiplying 2-70 by $-3(E_1 - E_2)$ and 2-67 by $4(E_1 + E_2)$ and then adding the resulting equations, we get

$$e_0 = \frac{-3(E_1 - E_2)(E_1\alpha_1 - E_2\alpha_2) + 4(E_1 + E_2)(E_1\alpha_1 + E_2\alpha_2)}{-3(E_1 - E_2)^2 + 4(E_1 + E_2)^2}T \tag{2-71}$$

By multiplying 2-70 by $-(E_1 + E_2)$ and 2-67 by $(E_1 - E_2)$ and then adding:

$$\frac{1 + e_0}{R} = \frac{-(E_1 + E_2)(E_1\alpha_1 - E_2\alpha_2) + (E_1 - E_2)(E_1\alpha_1 + E_2\alpha_2)}{\frac{b}{3}(E_1 + E_2)^2 - \frac{b}{4}(E_1 - E_2)^2} T \tag{2-72}$$

Noticing that, if the factors in these equations were to be multiplied out a number of cancellations and combinations might be possible, we try multiplying out in the hope of simplifying the expressions. The results of this effort justify our hope and are:

$$e_0 = \frac{\dfrac{E_1}{E_2}\alpha_1 + \dfrac{E_2}{E_1}\alpha_2 + 7(\alpha_1 + \alpha_2)}{\dfrac{E_1}{E_2} + \dfrac{E_2}{E_1} + 14} T \tag{2-73}$$

$$\frac{1 + e_0}{R} = \frac{24(\alpha_2 - \alpha_1)}{b\left(\dfrac{E_1}{E_2} + \dfrac{E_2}{E_1} + 14\right)} T \tag{2-74}$$

We choose these forms out of a number of possibilities because they are particularly convenient for calculation. In addition they make the dimensional correctness of the results especially evident.

By an obvious substitution of 2-73 into 2-74, we could arrange matters so that R could be calculated directly without first finding e_0. We do not bother to do this, however, because we suspect that e_0 will be very small compared to unity and we have in mind neglecting it entirely in 2-74. To see whether this simplification is legitimate, we scan a table of coefficients of thermal expansion of metals and find the coefficients to be of the order of 10^{-5} per °F. Thus even if the temperature rise were some hundreds of degrees, e_0 would be but a fraction of 1 per cent, and we will therefore ignore it in using 2-74. Incidentally, for calculation of stresses from 2-63 and 2-64 we note that $(1 + e_0)/R$ is the factor we really need and not $1/R$.

Now before using the results to see what we can learn about our problem we examine them critically for reasonableness.

In the first place having the central extension e_0 and the curvature $1/R$ proportional to temperature rise T is as would be expected since no external forces or moments are imposed on the strip, and temperature is the only disturbing influence. Also, we observe that, if T is negative, that is, a temperature drop below that for a straight strip, the extension and curvature will be reversed in direction, and this is as it should be.

Having the coefficients of expansion α_1 and α_2 in the numerators of 2–73 and 2–74 is reasonable, as we would expect the deformation of the strip to be affected directly by them. Moreover, if $\alpha_2 > \alpha_1$ the curvature is positive, which is in accordance with the way the original mathematical expressions were set up; but if $\alpha_2 < \alpha_1$ the curvature is negative as it should be because then the strip would bend the other way.

The thicker the strip, the less we would expect it to bend, other things remaining constant, and so b in the denominator of 2–74 seems right.

Now suppose the two metals were identical so that in effect there is a homogeneous piece instead of a strip of bimetal. Then there should be no curvature, and the central extension should be simply αT, and this is what we find by putting $E_1 = E_2$ and $\alpha_1 = \alpha_2 = \alpha$ in 2–73 and 2–74. Also, if $E_1 = E_2$, $e_0 = \frac{1}{2}(\alpha_1 + \alpha_2)T$, that is, average α times T, and this seems reasonable.

EXERCISE. Show that 2–73 and 2–74 also are correct for the case where either of the two metals is inifinitely stiff elastically.

Satisfied with these checks, we go on to the learning stage to see what the analysis can tell us about our general problem. We start by considering a numerical example, often a useful device for getting an idea of magnitudes in an unfamiliar field.

We assume a bimetal $\frac{1}{32}$ inch thick made of brass and steel, and take the following approximate values for the properties of these metals:

Steel: $\alpha_1 = 0.6 \times 10^{-5}$ per °F $\qquad E_1 = 28 \times 10^6$ lb in^{-2}

Brass: $\alpha_2 = 1.2 \times 10^{-5}$ $\qquad E_2 = 14 \times 10^6$

For a temperature rise of 100°F, we calculate from 2–73 and 2–74 respectively:

$$e_0 = 0.87 \times 10^{-3} \tag{2–75}$$

$$\frac{1 + e_0}{R} = 0.028 \text{ in}^{-1} \tag{2–76}$$

Thus such a strip will bend in the arc of a circle of radius $(0.028)^{-1} = 36$ inches for a temperature rise of 100°F. Evidently to get substantial movement with a small temperature change a strip of considerable length is needed; and this explains why thermostats for control of house heating often are made with a long strip of bimetal coiled up in a spiral.

Since internal stresses induced in the bimetal are important to the action, we decide to calculate them for our example using 2–63 and 2–64. We get for $0 < \frac{y}{b} < \frac{1}{2}$, the steel side,

$$s = 7600 - 24{,}400\,\frac{y}{b} \tag{2-77}$$

and for $-\frac{1}{2} < \frac{y}{b} < 0$, the brass side,

$$s = -4600 - 12{,}200\,\frac{y}{b} \tag{2-78}$$

These are used to draw to scale the stress diagram figure 2-16.

Exercise. Check figure 2-16 for consistency with equations 2-65 and 2-68.

An important fact which we learned in making the stress diagram is that the magnitudes of the stresses are independent of the thickness of the

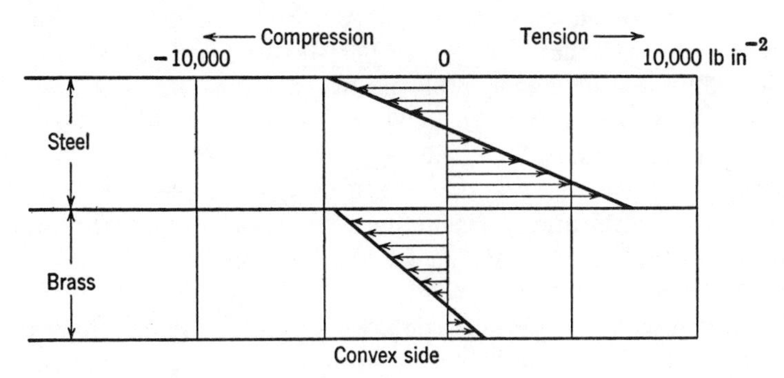

Fig. 2-16. Stress diagram for the bimetal strip of steel and brass at $T = 100°$F.

strip. This should be evident from inspection of 2-63, 2-64, and 2-74; nevertheless we did not observe it until we used the equations for calculation. Alerted now to the possibilities of important findings without further calculation, we study 2-63, 2-64, 2-73, and 2-74 further, and make the following valuable observation:

The stress at every point varies directly as T, the points of zero stress remaining fixed. Thus, for our example, it is easy to use the diagram to estimate at what temperature any specific maximum stress will be reached.

Next, we consider what we have learned that will be useful in the general problem of bimetal thermostat design. In a practical thermostat, it is unlikely that the strip would be unrestrained as assumed here. Rather there would be a load at the end imposed perhaps by the pressure of an electric contact or by some mechanism that is to be actuated thermally. A load of this sort would result in there being a moment to be transmitted across each section of the strip, and so the integral 2–68 would be equated not to zero, but to this moment, which in general would be a function of position along the length of the strip. Consequently, we see from the solution already done that the curvature in the loaded case would vary along the length, but that in spite of this complication the analysis for any prescribed loading would be perfectly straightforward.

Toward solution of the general problem, we find another useful fact from 2–74. To choose metal pairs for greatest curvature, difference in expansion coefficients is what counts and not ratio, as one might guess, and modulus of elasticity has little influence on curvature because the greatest ratio of moduli that can be obtained for any actual metal pair is only 3 or 4 and this is but a fraction of the constant number 14 to which the ratio must be added.

From the solution, there doubtless is more to be learned about the behavior of bimetal, but our concern here really is with developing proficiency in solving engineering problems in general and not with designing thermostats, so we review the highlights of what we have done with respect to the professional method.

In the first place, as in section 2–6, a complex problem was attacked first through simplification (cf. function 1 in section 1–4); an unrestrained bimetal was studied even though we knew full well that in a practical thermostat there would be restraints.

Then, even with this simplification, we found the problem so complex that we could plan the solution only in part and had to go through the stages of executing and checking the partial plan before we learned how to plan the remainder of the solution. In the planning, too, we made use of insight as well as analysis (cf. function 2 in section 1–4), the insight coming from experience with the somewhat related analysis of beams.

Again, in deciding to ignore e_0 in 2–74, we performed the function of simplification, aided by insight, this time an inkling that e_0 would be very small, and this perhaps was instrumental in our arriving at useful conclusions, for, if one blindly follows the logical algebraic process of eliminating e_0, the resulting expression becomes quite awkward and hard to interpret.

Finally, we learned by studying the simplified case how we might modify and extend the solution to cover the more generally useful case of a bimetallic strip restrained by external loads.

2–8 Applying Professional Method: Questions to Ask Yourself

In this chapter, we have presented some typical problems in the way in which they might confront you early in your career and how using good professional method you might analyze them. Such problems are very different from those often given in engineering textbooks where commonly a situation is clearly defined with certain things specified to be found in terms of given data, the very nature of which may be strongly suggestive of the way to proceed. In the real practice of engineering, your problems will usually have to be untangled from confused situations, and to solve them you will have to rely heavily on your own powers. This may seem to be a discouraging prospect, but you need not be discouraged for, as you learn to deal with problems professionally in the manner which has been illustrated, you will find that your effectiveness is surprisingly great and that it grows with practice.

The professional method which has been summarized in section 1–4 is in a way no more than well-organized common sense, but even so it sometimes is not easy for the beginner to grasp. Its five stages are not a rigid sequence always to be followed straight through in order. Rather they are in the nature of widely spaced guideposts pointing the way through a tangled forest; and, as in this analogy, when further progress is stopped by an obstruction you must be ready to retrace your steps carefully to an earlier stage to make a new attempt.

Application of the professional method may be facilitated if at each stage you ask yourself a question whose nature depends on the stage and on the problem, and then when a satisfactory answer is obtained proceed to the next stage, or if no satisfactory answer can be found return to an earlier stage. The kind of question to ask at each stage is suggested by the list which follows.

Defining the problem

What is the real problem or issue?

What is the question to be answered?

What are the pertinent facts?

Have I collected enough of the pertinent facts so that I can start planning?

If there are several interrelated problems, which should I attack first?

Planning the treatment

How can I solve the problem I have set myself?

What fundamental principle must be satisfied?

What general truths do I know that will help toward a solution?

What is my plan for using what is known to get what I want?

Is my plan sufficiently complete for me to start its execution?

Executing the plan

What is the result of my plan?

How do I get a useful conclusion from the principle I have applied?

Where am I in respect to my plan?

Checking

Is the work correct in every detail?

Are my assumptions reasonable?

Have I taken into account all the important factors?

In terms of things I know to be true, does the result make good sense?

Learning and generalizing

What have I found out?

What does the result tell me about the answer to the original question?

What does the result mean, and what is its interpretation in common terms?

How may the result be affected by my assumptions?

What have I found out that may be useful in future problems or in the remainder of this problem?

Is the result good enough to act on, or must I refine the solution or perhaps make an entirely different one?

You can never be sure you have answered one of these questions well enough to be ready to go on to the next stage. For example, it may often not be certain that, through ignorance, an important consideration has not been omitted altogether. On the other hand, there is always a danger of being too thorough and becoming mired in a morass of unessential detail. Every now and then in the course of an analysis, it may pay to stop and try to answer questions of the nature:

What am I doing now? Why?

How will it help me toward my real objective?

Judgment plays an important part in determining the kind of an analysis to make under given circumstances. You must take care not to embark on a study so detailed that a result is not likely to be obtained in time to be useful. Similarly, it is pointless to spend time on calculations that are more accurate than warranted by the data or by the use to which the answer is to be put. Judgment or intuition, call it what you will, is often decisive in determining when to terminate a pencil-and-paper analysis and start an experimental one. These intangible powers, and similar ones like insight and a capacity for invention, come with experience, particularly with experience in dealing with problems requiring you to extend your knowledge and skills to the limit.

Working in good professional style reduces the uncertainties which must be resolved by exercise of judgment, and this is particularly so if you start always from fundamental knowledge of which you are perfectly sure. What is meant by such fundamental knowledge is the subject of chapter 3.

CHAPTER THREE

The Understanding of Principles Fundamental to Engineering

3–1 Working from Fundamentals

Engineers who have carefully considered education for their profession generally agree that one of the most valuable abilities for the young engineer to acquire is an understanding of fundamental principles so thorough that he can use them with assurance in dealing with situations new to him. What is meant by fundamental principles and understanding them is the subject of this chapter. The task is not easy because the word fundamental has different shades of meaning, depending upon the user. A fundamental principle is a principle which serves as the foundation of a system of knowledge, and from which other principles may be derived. Thus, if we are considering mechanics, suppose we think of two principles: Newton's law $f = ma$ applied to a body, and the law for free fall of a body under gravity $s = \frac{1}{2}gt^2$. Since the second is derived from the first for a special case in which f is restricted to the weight of the body, the first is more fundamental than the second, is deeper in the foundation on which the structure of mechanics is built. Clearly the term fundamental principle is relative; we can often say that one principle is more fundamental than another and we mean the opposite of special or restricted in application. Of course, there are pairs of principles in which either may be derived from the other; these are equally fundamental. In such a case, the engineer may arbitrarily choose one as a foundation stone for his thinking and regard the other as a derived relation. Also, since some pairs of principles are quite unrelated, for example, Newton's law $f = ma$ and the law of action and reaction, an attempt to say that one is more fundamental than the other is meaningless.

The term fundamental principle is relative in another way depending upon the system of knowledge, and hence upon the observer. You would hardly say that the science of mechanics is built upon the relation $s = \frac{1}{2}gt^2$ for a falling body, and hence you would not regard this as a fundamental principle in general mechanics. However, in a more restricted field of knowledge, that of falling bodies only, the relation $s = \frac{1}{2}gt^2$ might well be taken as the basis of the subject and hence a fundamental principle. Such a difference in point of view between the engineer in a specialized field and another who is a general consultant is often the basis for a difference in the use of the term.

Our philosophy is to beware of the restricted formula; it is so easy to forget the restrictions that have been placed upon it when it is used only infrequently. By their very nature, well-understood restricted formulas apply to old and well-known situations; on the other hand, to think through an unfamiliar problem carefully in terms of more fundamental principles is to achieve a power for dealing with new problems. To make this matter clearer consider some instances from the problems in the preceding chapters.

In treating the liquid-level accelerometer, section 1–2, the engineer based his analysis on the application of Newton's law $f = ma$ to the liquid in the horizontal tube. There are, however, other principles from which a useful result could have been reached, perhaps more quickly. For example, the engineer might have used the principle that when a liquid undergoes uniform rectilinear acceleration a in a horizontal direction, its free surface is inclined to the horizontal at an angle θ given by

$$\theta = \tan^{-1} a/g \tag{3-1}$$

In the case of the accelerometer, the free surface comprises the tops of the two vertical columns of liquid. The difference in height H of these surfaces, figure 1–3, and the horizontal distance L between the tubes are related to the angle θ by

$$\tan \theta = \frac{H}{L} \tag{3-2}$$

Combination of 3–1 and 3–2 gives the result of the former analysis, equation 1–7. Thus the principle about behavior of liquids undergoing uniform horizontal acceleration leads directly to a more elegant solution to this problem than does Newton's law: it is more elegant in that the accelerometer did not have to be separated into horizontal and vertical tubes; neither horizontal nor vertical tubes need to have the same or even uniform cross sections; and fewer symbols need to be defined and used. However, and this is the important point, Newton's law $f = ma$ is a basis for analyzing an immensely larger class of problems, *all* those involving acceleration of mass; you will do well to regard it as fundamental knowledge to be thoroughly mastered. The principle about the free surface of an accelerating liquid, on the other hand, for most engineers is worthy of no more than passing notice; it should be regarded merely as an interesting and occasionally useful consequence of Newton's law. If you chance in later years to specialize in the study of liquid fuel systems for vehicles which accelerate rapidly, you may find this principle derived from Newton's law very useful. But if you need it, there will be plenty of time to derive it and to find what limitations it may have. When as a specialist you have mastered the derived principle, you may rightly regard it as fundamental to your particular work.

As another instance, consider the problem on the electrical accelerometer, section 2–2. Our solution is based on the use of Kirchhoff's voltage law, but, as pointed out already, an engineer specializing in the field of electric computing devices might have built his solution from a knowledge of circuits for differentiating electrically. Such special knowledge is something extra to be acquired during professional specialization; it is not useful for analyzing every circuit. For our purpose Kirchhoff's voltage and current laws are fundamental; they are among the things to be understood thoroughly.

As a matter of fact, Kirchhoff's laws can be derived from still more fundamental relations, the Maxwell electromagnetic field equations. These are most appropriate as basic principles in dealing with electric- and magnetic-field problems, especially when the fields vary at high frequency, but to use them in analyzing ordinary circuit problems would be as impractical as starting every problem involving the rotation of a rigid body about a fixed axis by applying Newton's laws to the motion of every particle in the rotating body.

In the instances cited, the analysis is based on a principle having a very wide range of usefulness. Study of the other problems analyzed in this book will show the same thing: the principles used are general, they are fundamental principles in contrast to ones applying only under narrowly restrictive conditions. On the other hand, the principles are not, in every case, the most general that might be employed; the principles used have been selected with a view to their convenience and utility as well as to their generality. Working thus from fundamentals is characteristic of the competent professional man. This means that he works from knowledge which is well established and which he thoroughly understands, and not indiscriminately from formulas without regard for their limitations. He does this for two reasons: first, the professional man must be sure he is right and, to leave room for the necessary understanding, he cannot afford to burden himself with many formulas having restricted usefulness; and second, in a new situation there may in fact be no formula, but there is likely to be a fundamental principle which governs. Also, the very speed and elegance of the derived formula may lead the engineer to overlook simplifying assumptions upon which the formula is based, and which may not be legitimate in the particular situation he is trying to analyze.

In thinking over what you know, you may have difficulty in distinguishing what you should regard as fundamental, for your knowledge having been acquired over a period of years and from a variety of sources may be jumbled. This state of affairs is a perfectly natural consequence of the educational process, even of good education. But if you aspire to professional success you must take steps to organize what you know, and you should not expect to finish the organization for a long time. Indeed, the process of assimilating and ordering knowledge should continue throughout your professional life. In studying the problems worked out in this book, you should consider carefully the principles and concepts which are used and be sure that you understand them in relation to things already known. But, even more important, you should analyze by yourself situations new to you in terms of fundamental principles. As you apply these principles, you should make a deliberate effort to increase your understanding of them. To make clearer what we mean, the remainder of this chapter is devoted to discussions of a few of the fundamental principles and concepts which are particularly useful in engineering. The topics selected for these discussions represent an important but quite small segment of the fundamental knowledge you need to have at your command. It is assumed that you already have a good working understanding of the common concepts such as force, voltage, heat, and so forth, and indeed that you already have met most of the principles which are discussed. The purpose is to bring into sharper focus important things already known and to help you organize some of this knowledge so that you will better appreciate its power as well as its limitations.

3-2 Conservation of Matter

We begin our discussion of principles with a consideration of conservation of matter for two reasons: first, because it brings us face to face at once with the fact that physical laws and concepts in general may be limited in accuracy, and second, because this principle is so much a part of common every-day experience that its power and usefulness in analyzing engineering problems is easily overlooked.

The principle of conservation of matter is an expression of the idea that matter, that is, mass, can be neither created nor destroyed but only changed in form. But this idea we know now to be incorrect; for example, in reactions involving atomic nuclei mass is converted into energy. Also, in accord with relativity theory, the mass of a particle is known to increase with velocity, the amount of increase being large at speeds near that of light. Such speeds, however, are not yet of concern to engineers except in the case of motions of electrons and other atomic particles in such devices as electron tubes and particle accelerators. If nuclear reactions and high velocity particles are excluded, it can safely be said with a very high order of accuracy that matter *is* conserved.

The question may arise: why be content with this somewhat restricted principle when more accurate, more general principles are available? We would answer such a question on the grounds of convenience and cost to the problem solver. The more accurate principles are more complicated, and a thorough understanding in terms of relativity theory is too high a price for most engineers to pay. The situation is somewhat similar to using Kirchhoff's laws for circuit analysis rather than Maxwell's field equations, or $T = I\alpha$ for the rotation of a rigid body rather than Newton's law $f = ma$ applied to every particle in the body.

Conservation of matter was used implicitly in analyzing the liquid-level accelerometer in section 1–2 where following equation 1–7 it was stated that, if the two vertical tubes have the same area, the liquid rise in one tube must equal the fall in the other. To examine the situation critically, suppose we focus our attention on some part of the U tube which remains filled with liquid, for instance, the horizontal part. If the walls of the tube are rigid and the liquid is incompressible, then the mass of liquid in the horizontal part of the tube must remain constant. Consequently, if there is a fall of liquid in one vertical tube, it must mean that a mass of liquid moves into one end of the horizontal tube and an equal mass moves out at the other end. Thus it follows that if the cross-sectional areas of the vertical tubes are equal and uniform the rise of liquid in one must equal the fall in the other.

Notice that if the liquid were compressible the reasoning would have to be somewhat different, for then the mass of liquid in the horizontal tube would not necessarily be constant. But, by the conservation of matter, we could say that in any interval of time the inflow of mass through one end of the horizontal tube minus the outflow through the other must equal the increase of mass within the horizontal tube. By arguments precisely like this, equations of continuity which are basic for analyzing problems of fluid flow may be derived from conservation of matter.

Conservation of matter has another important and very different use; it is in part the basis for the quantitative aspect of chemical equations. For instance, the equation describing the burning of carbon monoxide

$$2CO + O_2 = 2CO_2$$

expresses not only the proportions in which the substances react but also the fact that the mass of carbon monoxide plus the mass of oxygen which combine equals the mass of carbon dioxide which is formed. Used in this way, conservation of matter is the basis for the material balance which is a part of the usual analysis of problems involving combustion and other chemical processes.

From this discussion, it should be clear that useful results may be obtained from the conservation of matter by applying it to a space or region. Then the principle says that, in any interval of time, the quantity of matter (mass) which accumulates in the space must equal what moves into the space less what moves out in the given time interval. If the process is a steady one involving no change of mass within the space, the mass entering per unit time must equal that leaving per unit time.

3-3 Conservation of Energy

The principle of conservation of energy is an expression of the idea that energy can be neither created nor destroyed but only changed in form. For the sake of complete generality and accuracy, this principle, like the conservation of matter, should be modified in the light of the equivalence of energy and mass. However, for present-day engineering work, apart from nuclear reactions, the principle that energy by itself is conserved is extremely useful, and in fact there is probably no principle more all-pervading in engineering.

A typical use of conservation of energy was made in analyzing the non-fogging mirror, section 2-3. There, to find how the temperature of the mirror would vary with time, the engineer set up the differential equation 2-21. This equation expresses the fact that the power flowing into the glass from the heater must equal the sum of the power dissipated from the glass to its surroundings and the time rate of increase of energy stored within the glass. Equation 2-21 is a relation among amounts of power; it may be rearranged to express a relation between differential amounts of energy. Thus

$$(q - hAT)\,dt = C\rho AL\,dt \qquad (3\text{-}3)$$

which expresses the application of conservation of energy a little differently. The left-hand member is a differential quantity of heat which flows into the glass, and the right-hand member is the corresponding differential increase in energy stored within the glass.

As applied in 3-3 the conservation of energy is identical with the first law of thermodynamics, a familiar statement of which is that the heat added to a system equals the work done by the system plus the increase of energy stored within the system. The heat added to the system, which comprises the glass of the mirror, is that flowing in an incremental time from the heater to the glass less that dissipated from the glass to the surroundings in the same time. Work done by the system would include work done by the glass in expanding against the atmosphere and the restraints imposed by the supports, but this was tacitly assumed to be negligibly small. In trying to express the increase of energy stored within the glass, our young engineer, however, actually accounted for the work done by the glass in expanding against the atmosphere, for he used a value of specific heat of glass taken from an ordinary table of physical constants, and these specific heats are determined in such a way that the energy change accounted for includes the work done as the substance expands against atmospheric pressure; that is, they are specific heats at constant (atmospheric) pressure. Consequently, 3–3 accounts correctly for the work done by the glass in expanding against atmospheric pressure, and accordingly the term on the right of 3–3 is not the increase in internal energy of the glass but its increase in enthalpy. In the case of ordinary solids, however, the difference is trivial.

In using conservation of energy, attention must be focused first on defining the system to which the principle is to be applied. One procedure is to choose a system which includes the same material substances throughout the period considered and which need not be at rest. Another procedure is to take as a system a carefully defined region in space and then to account for changes of energy stored within this space and for transfer of energy through the surfaces bounding the space by such means as the work done by a force which acts through the bounding surface, the flow of heat by conduction through the bounding surface, the flow of electric power into or out of the region, and the transport of stored energy by virtue of matter entering or leaving the space.

To apply conservation of energy, an exchange of energy need not actually occur; the principle can be applied usefully in a situation which is really static but which for purposes of analysis is imagined to undergo a small displacement. Conservation of energy is used to relate the transfers of energy that would occur for an incremental displacement of some sort and then the displacement is made to approach zero agreeing with reality. This is the method of virtual displacement or virtual work which we used to find the force on the iron core of a solenoid and which we discussed at some length in section 1–3.

Conservation of energy is a powerful tool for analyzing problems and for checking solutions arrived at by other means. Its implications are very simple, and if you keep your wits about you it is easy to apply the principle correctly. However, workers in different fields have tended to adopt different points of view in their conventional uses of conservation of energy, and these differences can lead to trouble. For instance, in dealing with a freely falling body, you can say no work is being done and that only a change of stored energy from potential to kinetic form is involved, or you can say that the force of gravity is doing work on the body and that kinetic energy is being stored. These statements are each true and would in fact lead to identical mathematical equations, but trouble would come if you blindly adopted both points of view at once and had the falling body both losing potential energy and receiving work through the force of gravity at the same time. You must define the system carefully and then use it consistently in applying the principle. In studying a steam turbine, for instance, you must decide if your system is going to be the space within the turbine itself, the whole power plant, or a pound of steam to be followed through the plant; and then, when you have decided which of these to adopt, you must be consistent in its use.

3–4 Newton's Laws of Motion

These are usually stated as three laws with the first as a special case of the second, so we are really concerned with only the last two, one of which is often stated mathematically as

$$\mathbf{f} = m\mathbf{a} \tag{3–4}$$

or

$$\Sigma\mathbf{f} = m\mathbf{a} \tag{3–5}$$

and the other usually in words as

Action and reaction are equal and opposite (3–6)

These are extremely important laws to the engineer. Let us examine them as fundamental principles to find out just what they mean.

In the first place we should recognize that a mathematical equation is a very compact expression; one of the reasons that it is useful is that it is so compact and can be manipulated with other compact expressions. Yet understanding an equation is generally a matter of unraveling the compactness so that the full meaning is clear; usually sketches help greatly in such clarification. Equations 3–4 and 3–5 are illustrated in figure 3–1, where $\mathbf{f}$, $\mathbf{f}_1$, $\mathbf{f}_2$, $\mathbf{f}_3$, etc., designate forces, m designates mass, and $\mathbf{a}$ acceleration.

Consider these quantities separately. In equation 3–4, $\mathbf{f}$ is the single force applied to mass m; in 3–5 it is shown as a summation, which in this case must be taken with due regard to the vector nature of force. In any case the left side of 3–4 or 3–5 is the resultant vector force applied to mass m. To emphasize that the forces and the acceleration in 3–4 and 3–5 are vectors we use bold-face type for their symbols.

The acceleration $\mathbf{a}$, which results from the resultant force $\mathbf{f}$, is a vector whose direction is the same as that of **f** and whose magnitude is proportional to the magnitude of **f.** The acceleration must always be measured with respect to some reference—will any reference satisfy 3–4? No; it has to be a non-accelerated reference, presumably one that has no acceleration with respect to the center of mass of the universe. Fortunately for most engineering work, the earth is a satisfactory reference. Indeed, the

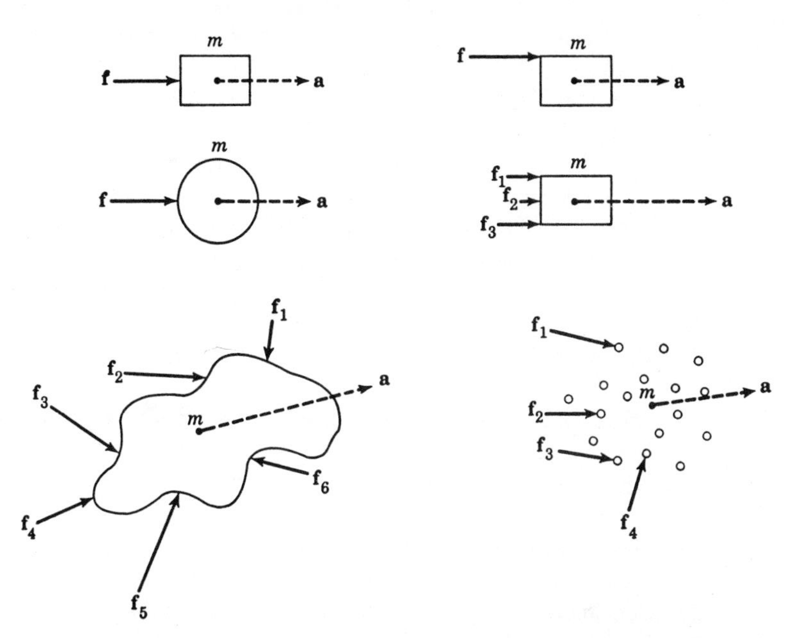

Fig. 3-1. Illustrating the acceleration law of Newton. In each case the acceleration of the center of mass of the body or group of particles is given by equation 3-5.

philosophical difficulties inherent in the use of the center of mass of the universe as a reference led to a recognition that 3–4 is not an entirely satisfactory law and in its place a relativistic theory was developed; but only in rare cases of interest to the engineer must this be used instead of Newton's laws.

What about the mass m? Is it necessarily small for equations 3–4 and 3–5 to hold; must it be spherical, or rectangular, or can it have any shape? Must m be rigid? In the most fundamental conception, the law holds for a particle, that is, a body or portion of a body so small that rotation can be neglected. In general if the motion of any body is to be found we can conceive of finding it by dividing the body into elements each one of which is small enough that at any instant the acceleration is sensibly the same throughout the element. Then 3–4 or 3–5 is applied to each such element and the result summed over the body as a whole. This is one of the fundamental techniques of mechanics.

Using this technique we can show that 3–4 or 3–5 holds for a body of any size or shape or even for a swarm of particles with or without ties between them. Thus the acceleration law holds for both rigid and non-rigid solid bodies and for gaseous and liquid bodies; in section 1–2 it was applied correctly to a liquid body, the liquid in the horizontal tube of the accelerometer. In every case the acceleration **a** is that of the center of mass of the body or of the swarm of particles. The mass m is the total mass of the body or the sum of the masses of the particles whose motion is being considered. The force **f** in 3–4 or the vector summation $\Sigma\mathbf{f}$ in 3–5 is the resultant of all of the forces acting on the particles comprising m; in evaluating this resultant, however, only the forces applied by external means need be considered because the internal forces of action and reaction exerted between the particles of m cancel each other in pairs. The law holds regardless of how or where the forces are applied. It is true whether the resultant force passes through the center of mass or not; if not, the body will rotate as well as translate, but in any case the acceleration of its center of mass is given by 3–4 or 3–5. In the top two cases of figure 3–1, if **f** and m are the same, **a** is the same in spite of the fact that one mass rotates. In the last case of figure 3–1 the positions of the particles may be changing with respect to one another, that is, the shape of the aggregation may be changing but the center of mass is accelerating in accordance with Newton's law. In summary, the acceleration law, equation 3–4 or 3–5, holds for any body, rigid or non-rigid, and this is why it is such a powerful fundamental principle.

Incidentally, from childhood we have known, that to move a body we have to push or pull it, that is, to apply a force; the law 3–4 tells us that force (more carefully defined than the child's concept) does not produce displacement directly, nor for that matter velocity, but acceleration. The law also implies that the force has to be applied to the body in order to cause it to accelerate; one of the silly errors that sometimes occurs in writing the acceleration law for a complicated case is that of wrongly including a force that doesn't act on the mass. To avoid such mistakes is one reason for always drawing a free-body diagram.

Equation 3–4 holds in various systems of units among which are those shown in the accompanying table:

f	m	a
Dynes	Grams	Centimeters (seconds)$^{-2}$
Newtons	Kilograms	Meters (seconds)$^{-2}$
Pounds	Slugs or pounds (seconds)^{2} (feet)$^{-1}$	Feet (seconds)$^{-2}$
Poundals	Pounds	Feet (seconds)$^{-2}$

The possibility for confusion inherent between the last two systems with pounds used either for force or mass is a monumental example of poor standardizing of names.

Before leaving Newton's acceleration law we must mention a special form derived from it which unfortunately sometimes leads to misunderstanding. If in equation 3–5 the term $m\mathbf{a}$ is transposed to the left-hand side there results

$$\Sigma \mathbf{f} - m\mathbf{a} = 0 \tag{3–7}$$

Now $(-m\mathbf{a})$ may be treated as if it were one of the forces acting on the body, and then 3–7 becomes identical with the principle of statics which states that for a body in equilibrium the vector sum of the forces is zero. The term $(-m\mathbf{a})$ in 3–7 is then given a name such as the inertial reaction, the inertia force, or the reversed effective force. In the authors' opinion the supposed advantage of 3–7 in making certain problems in dynamics equivalent to ones in statics is outweighed by the possibility of error in establishing the proper sign for the term $m\mathbf{a}$.

Let us now turn to the action-reaction law, 3–6, and try to really understand it. In the first place the words "action" and "reaction" mean nothing more than force. Whereas the acceleration law is usually applied to a single body, or a group of bodies treated as a single one through use of a common center of mass, the action-reaction law always involves exactly two bodies. "Action" in the law is the force exerted by the first body on the second; "reaction" is the force exerted by the second on the first. This is shown in figure 3–2*a* where the bodies are labelled 1 and 2 and $\mathbf{f}_{12}$ is the force on body 1 caused by body 2 and $\mathbf{f}_{21}$ is the force on body 2 caused by body 1. The positive directions of the forces are shown by the arrows in the figure. The action-reaction law can thus be written mathematically as

$$\mathbf{f}_{21} = -\mathbf{f}_{12} \tag{3–8}$$

which like 3–4 and 3–5 is a vector equation. Evidently the terms action and reaction are quite interchangeable; either one of the pair of forces may be called the action and the other the reaction. On the other hand it *is* important to preserve the identities of the forces in an analysis since the two are oppositely directed and act on different bodies. If the two bodies in figure 3–2*a* are in contact and fastened by glue, a positive value for $\mathbf{f}_{21}$ resulting from an analysis would mean that the glued joint is in compression; a negative value for $\mathbf{f}_{21}$, on the contrary, would signify tension in the glue. To avoid confusion the body on which each force acts, as well as the body exerting the force, must be carefully designated; a convenient means for accomplishing this is double subscript notation and careful definition.

The action-reaction law is true regardless of the mechanism by which the force is exerted; the mechanism may be impact or the attraction between electric charges of opposite signs, as suggested in figure 3–2*b* and *c*, or any other means whatever. The forces of action and reaction are always collinear; if not, as shown in figure 3–2*d*, the two bodies taken together would be acted on by a self-generated torque and so constitute a beautiful perpetual motion machine!

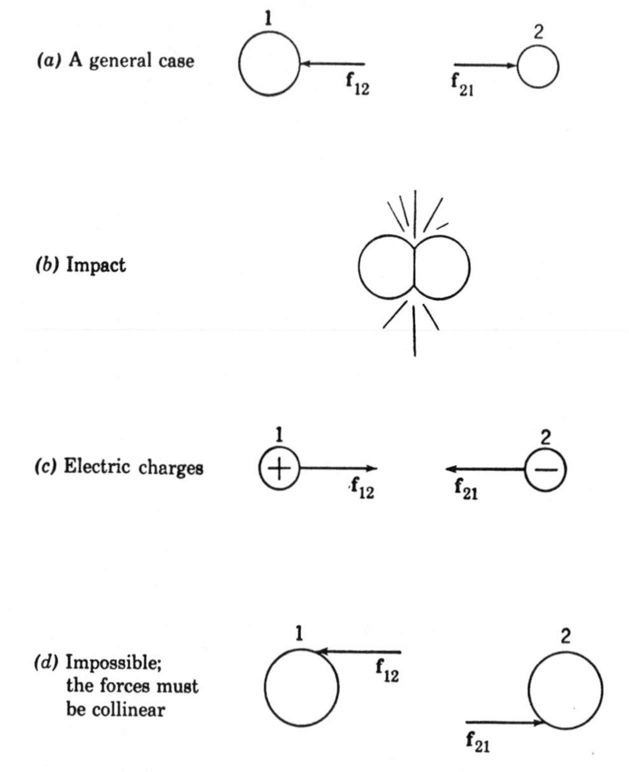

Fig. 3-2. Illustrating the action-reaction law of Newton.

A common misconception about the action-reaction law may be cleared up by considering figure 3–3, which shows a book at rest on the center of a table and free-body diagrams showing the forces on the book and on the table. The forces acting on the book are its weight $\mathbf{w}_b$ and the force $\mathbf{f}_{bt}$ exerted on it by the table top. The forces acting on the table are its weight $\mathbf{w}_t$, $\mathbf{f}_{tb}$ caused by the book, and the force $\mathbf{f}_{tf}$ exerted by the floor on the table, one-fourth on each leg. The forces $\mathbf{w}_b$ and $\mathbf{f}_{bt}$ are equal in magnitude but not by the action-reaction law. They are equal by the acceleration law, for the acceleration of the book is zero and hence the resultant force on the book is zero, from which it follows that $\mathbf{w}_b$ and $\mathbf{f}_{bt}$ are equal. What the action-reaction law says about this situation is that $\mathbf{f}_{tb} = -\mathbf{f}_{bt}$ and that there is a force on the earth equal and opposite to $\mathbf{w}_b$. Also, it is not correct to say that $\mathbf{f}_{tf}$ is equal to $\mathbf{w}_t + \mathbf{w}_b$ by the law of action and reaction; again the acceleration law must be invoked. The action-reaction law applies when two bodies are involved; the acceleration law when there is only one.

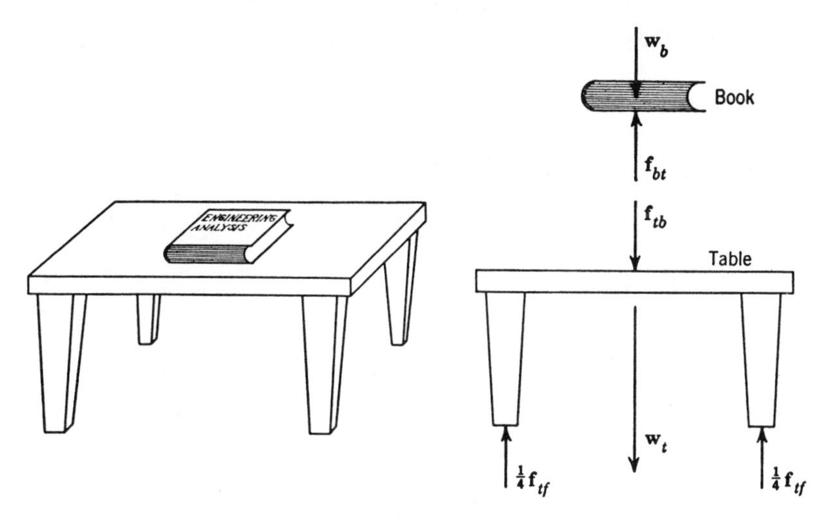

Fig. 3-3. Forces acting on a book at rest on a table.

The principal value of the action-reaction law is in the analysis of mechanisms and structures where for simplicity separation of the whole into parts or free bodies is desirable. Then the forces which act through the separating boundaries are dealt with systematically by this simple law.

3–5 The Rotation Law $T = I\alpha$

In section 2–6 dealing with the phonograph record changer we made use of the law $T = I\alpha$, relating torque T to angular acceleration α. This law is a rotational analog of $f = ma$. Let us examine the rotational law to see where it comes from and what its limitations are.

The rotational law can be derived by first dividing a rotating body such as the record or the turntable into small particles, next applying the acceleration law $f = ma$ to each one, and then manipulating and interpreting appropriately. Let us examine these steps in more detail. Suppose we imagine a body like the record turning about an axis. The axis does not move with respect to the earth, and thus for practical purposes is near enough to being fixed in space. The record is shown in figure 3–4.

Suppose it is turning in the usual direction for records, that is, clockwise, as we look down on it, with angular velocity ω radians per second, which may be changing as the record comes up to speed. As pointed out in section 2-6, the record is acted on by various torques such as the torque of friction with the turntable, the drag torque of surrounding air, and friction with the central pivot.

Now we imagine the record to be divided into small particles by means of radial lines and concentric circles, and we focus our attention on a single particle, the ith one as shown in figure 3-4. The torques that act on the

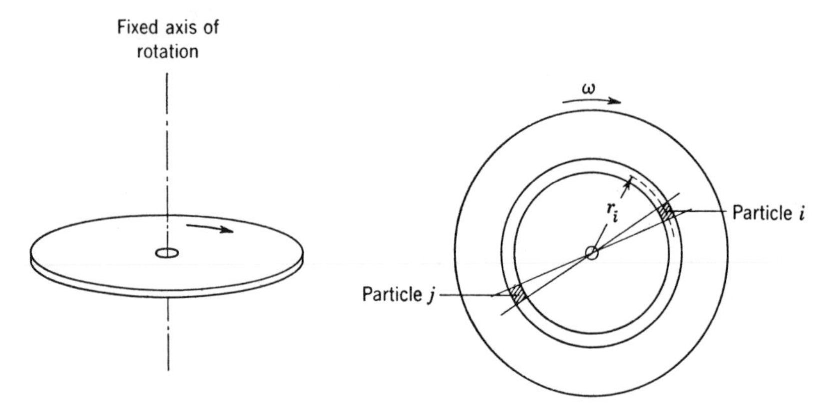

Fig. 3-4. A phonograph record illustrating the rotation of a rigid body.

record originate as forces acting on the individual particles. We have mentioned the torques that come from the external forces. There are internal forces, too, of one particle on another, forces that hold the particles together as a rigid body. Whatever the forces may be that act on a particle like the ith one, their resultant must be related by $f = ma$ to the linear acceleration of the particle, which we shall express in terms of ω and its time rate of change.

If the ith particle moves with constant angular velocity ω in a circle of radius r_i, the acceleration is toward the center and equal to $\omega^2 r_i$. However, if ω is changing but r_i is fixed, the acceleration of the particle has two components, one toward the center, which is the same as before, $\omega^2 r_i$, and the other tangential and of magnitude $r_i \dfrac{d\omega}{dt}$.

Exercise. Prove that the components of acceleration are as given above and also find them when both r and ω vary with time (the body is no longer rigid).

The components of acceleration are shown in figure 3-5, and they are added vectorially to show the resultant acceleration. However, it is more convenient to work with the components of the resultant force as shown in the figure. By the acceleration law, equation 3–5, the vector resultant

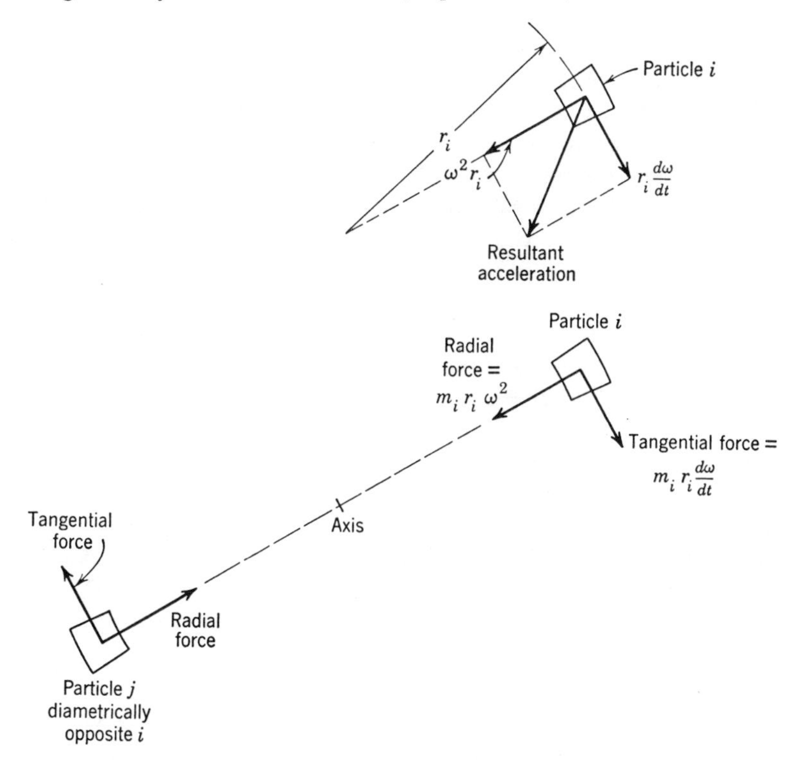

Fig. 3-5. Components of acceleration and of resultant force for individual particles of the phonograph record.

force on the particle equals the mass of the particle multiplied by its resultant vector acceleration, or

$$\mathbf{f}_i = m_i \mathbf{a}_i \tag{3–9}$$

Likewise for the radial component

$$f_{iR} = m_i a_{iR} = m_i r_i \omega^2 \tag{3–10}$$

and for the tangential component

$$f_{iT} = m_i a_{iT} = m_i r_i \frac{d\omega}{dt} \tag{3–11}$$

where the subscripts R and T are added as needed to show radial and tangential components of the vector quantities.

We have an expression for the resultant force on a general particle, the

ith; let us sum over all the particles to see what the resultant force must be. There is an easy way to do this: think of another particle, the jth, bounded by the same circles and radial lines but diametrically opposite the ith particle and with the same mass. The forces on j are equal and opposite to those on i; hence the sum of forces applied to the two is zero. Moreover, all the particles of which the record is comprised come in pairs like i and j, since the record is symmetrical and not eccentrically placed on the axis. Therefore the resultant force on the record turning about its fixed axis at either constant or variable angular velocity is zero. On reflection we decide this is consistent with equation 3–5, which told us that any resultant force on any body had to be accompanied by a linear acceleration of its center of mass. The center of mass of the record in this case is not accelerated; hence there can be no net force. If we tried to apply a force to the record the mounting of the pivot will either move or if it is fixed, as we postulated, will develop an equal opposing force so that the resultant is zero.

Actually, summing up forces did not yield an equation pertaining to rotation. Let us look at torques.

The torque, or turning effort, of any force about an axis is the tangential component of that force multiplied by the distance from its point of application to the axis. Thus radial components of force like 3–10 yield no torque, but tangential components like 3–11 do. In fact the torque of the resultant force on the ith particle is

$$T_i = r_i f_{iT} = m_i r_i^2 \frac{d\omega}{dt} \tag{3-12}$$

Now this must be a torque in the direction of ω. Moreover every particle, the jth and all the others comprising the record, requires a torque in the same direction; thus the sum of the torques applied to all of the constituent particles is the sum of terms like 3–12 which we may write as

$$\sum T_i = \sum m_i r_i^2 \frac{d\omega}{dt} \tag{3-13}$$

On both sides of the equation the summation is to be taken over all particles, and a typical term, that for the ith particle, is shown. Since the body we are considering is rigid each particle has the same angular acceleration $d\omega/dt$ and we may write

$$\sum T_i = \frac{d\omega}{dt} \sum m_i r_i^2 \tag{3-14}$$

Let us examine first the summation of torques that appears on the left side of the equation. As written it includes torques that are attributable to internal as well as external forces. However, it is relatively easy to show that the summation of torques attributable to internal forces is zero and hence the sum of torques in 3–14 need include only external torques. This is fortunate since internal forces might be hard to determine. Internal forces come in pairs. Any particle *i* may exert a force on an adjacent particle *h*, as shown in figure 3–6, and then by the action-reaction law *h* exerts an equal and opposite force on *i*. Clearly when *internal* forces are

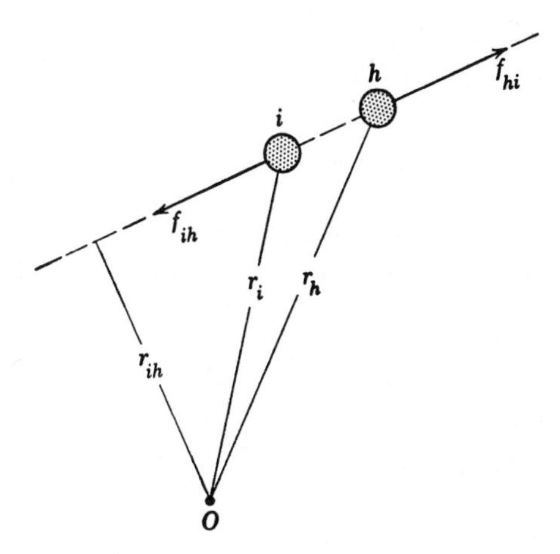

Fig. 3-6. A pair of internal forces has neither resultant force nor resultant torque about an axis through *O*.

summed up over all the particles, as we did earlier, the sum must consist of pairs each of which is zero and hence the whole resultant for internal forces is always zero. Also the pair of forces f_{ih} and f_{hi} are collinear and hence have the same perpendicular distance r_{ih} from axis *O* in figure 3–6 to the line of action of either force. Thus the torque from such a pair of internal forces is also zero. Hence the sum of torques that appears in 3–14 need include only external torques; that is, torques that arise from forces which act on the body from outside.

In the summation on the right side of 3–14 each term consists of the mass of a particle multiplied by the square of its distance from the axis. This sum is of great significance in mechanics problems that involve rotation; it is the moment of inertia, frequently designated by the symbol *I*. Thus

$$I = \Sigma m_i r_i^2 \tag{3–15}$$

gives the moment of inertia of a body about a specified axis, the sum being taken over all the particles of which the body is composed. Combining 3–15 and 3–14 we have

$$\sum T = I\frac{d\omega}{dt} \tag{3–16}$$

or

$$\sum T = I\alpha \tag{3–17}$$

where α is angular acceleration and the sum of torques now includes only external ones.

The summation in 3–15 may be replaced by an integral if we think of the particles as being elements of mass dm; thus

$$I = \int r^2\, dm \tag{3–18}$$

We may find the moment of inertia for a symmetrical body of cylindrical form like the record by forming the sum indicated in 3–15, and incidentally show how it is equivalent to an integral like 3–18. Referring to figure 3–4 let us first sum the terms of 3–15 around the circular region between concentric circles separated by distance Δr. All the particles of this ring have the same radial distance r from the axis, hence the sum becomes $r^2\Sigma m_i$ taken around the circle. The sum of the masses between the circles is the mass density multiplied by volume. Denote the density by ρ; the volume is $2\pi r(\Delta r)b$ to a good approximation if Δr is small enough; b is the thickness. Thus 3–15 becomes

$$I = \Sigma\rho(2\pi r\,\Delta r\,b)r^2 \tag{3–19}$$

taken over all such rings into which the body is divided. If $\Delta r \rightarrow 0$ the sum over all the particles becomes the integral

$$I = \int_0^{r_0} \rho(2\pi rb\,dr)r^2 \tag{3–20}$$

r_o being the outermost radius. The similarity to 3–18 is apparent since $\rho(2\pi rb\,dr)$ is an element of mass dm. If we put 3–20 into suitable form and perform the integration we have

$$I = 2\pi\rho b\int_0^{r_0} r^3\,dr = (\pi/2)\rho b r_o^4$$

This may be expressed also as

$$I = (\pi r_o^2 b\rho)(r_o^2/2) = Mr_o^2/2 \tag{3–21}$$

that is, as the mass M of the body multiplied by $r_o^2/2$.

Sometimes it is convenient to say that the moment of inertia of a body is given by its mass multiplied by the square of its radius of gyration R. The latter is thus a kind of average radius. If all the mass were at this distance R from the axis then the body would have the same moment of inertia as it actually has with spread out mass. Since

$$I = MR^2 = \Sigma m_i r_i^2$$

$$R = \sqrt{(1/M)\Sigma m_i r_i^2} \tag{3–22}$$

and we see that the radius of gyration of a body is the root mean square of radius weighted according to the mass of the constituent particles. For a cylindrical body like the record we can see from 3–21 that

$$R = \sqrt{0.5}\, r_o = 0.707 r_o \tag{3–23}$$

Instead of moment of inertia some engineers use the expression WR^2, which is the weight W of the body, rather than its mass, multiplied by the square of the radius of gyration R. Evidently in these terms

$$I = \frac{WR^2}{g} \tag{3–24}$$

where g is the acceleration of gravity.

The purpose of this section was to investigate the rotational law $T = I\alpha$ to discover its origins and limitations. We found it could be derived by applying both the acceleration law and the action-reaction law of Newton, and no other principles of physics were needed, although we did manipulate mathematically. We derived the law for the case of a body which must be rigid and whose angular acceleration α is about a fixed axis. We found that the sum of the externally applied torques T is equal to $I\alpha$ where I, the moment of inertia, is a quantity that characterizes the body as mass does, and depends both on mass and its geometric distribution. In our derivation we assumed the body to be symmetrical about the axis of rotation, although this is not a necessary restriction; the same law applies regardless of symmetry provided the axis is fixed. If the center of mass is not on the axis of rotation the resultant of the external forces, however, is not zero as in our derivation but is equal to the centripetal force.

Other derivations of $T = I\alpha$ may be made in which the axis may move in certain ways if, for example, it passes through the center of mass of the body or through the point of contact of a body rolling on a surface. It is well to have the derivation clearly in mind, however, before using this, or any other, equation.

3–6 Rotation of Non-Rigid Systems

Sometimes we may want to analyze the rotation of a non-rigid system about a fixed axis, for instance, a rotating body of liquid or a mechanism comprising a number of rigid members not rigidly connected together.

In such cases the law $T = I\alpha$ for a rigid body does not apply. The needed law may be derived, however, by a similar method and the result is

$$\sum T = \frac{dH}{dt} \tag{3-25}$$

where again on the left side we have the sum of the external torques applied to the system of particles (or non-rigid body) about some fixed axis. On the right is the time rate of change of H, the angular momentum of the

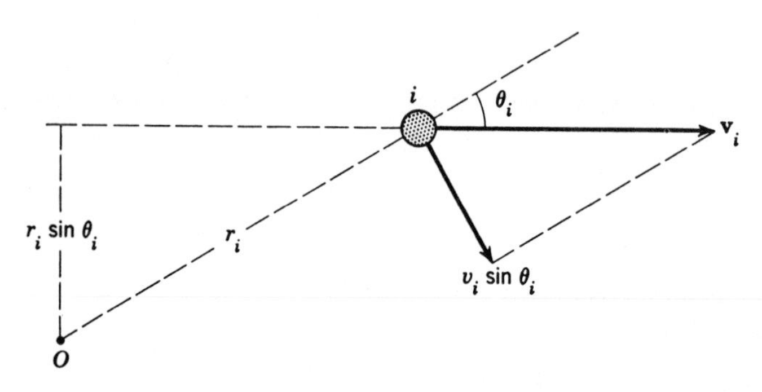

Fig. 3-7. Pertaining to the angular momentum of a particle i about a fixed axis through O.

system, which is the sum of the angular momenta of each of the constituent particles. For one of these, say the ith, the angular momentum H_i is the moment of its linear momentum, that is, the perpendicular distance $r_i \sin \theta_i$ to the line of motion multiplied by the linear momentum $m_i v_i$. Thus, referring to figure 3-7,

$$H_i = m_i v_i r_i \sin \theta_i = m_i r_i (v_i \sin \theta_i) = m_i r_i^2 \omega_i \tag{3-26}$$

where we recognize $v_i \sin \theta_i / r_i$ as the instantaneous angular velocity ω_i of the particle i about the axis. The total angular momentum is found by summing 3-26 over all the particles comprising the system; thus

$$H = \Sigma H_i = \Sigma m_i v_i r_i \sin \theta_i = \Sigma m_i r_i^2 \omega_i \tag{3-27}$$

For the case of a rigid body it is interesting to see whether 3-25 reduces as it should to 3-16. If the body is rigid r_i is constant and in figure 3-7 each v_i necessarily has to be at right angles to r_i ($\theta_i = 90°$). Then, too, ω_i has to be the same for all particles, and from 3-27, 3-26, and 3-15,

$$H = \Sigma H_i = (\Sigma m_i r_i^2)\omega = I\omega \tag{3-28}$$

and substituting this in 3–25, recognizing that I is constant for a rigid body, we have

$$\sum T = I\frac{d\omega}{dt}$$

which is 3–16, and thus we have found agreement.

Equation 3–28, derived above for a rigid body characterized by the constant I, also holds with I treated as a variable quantity for a particular

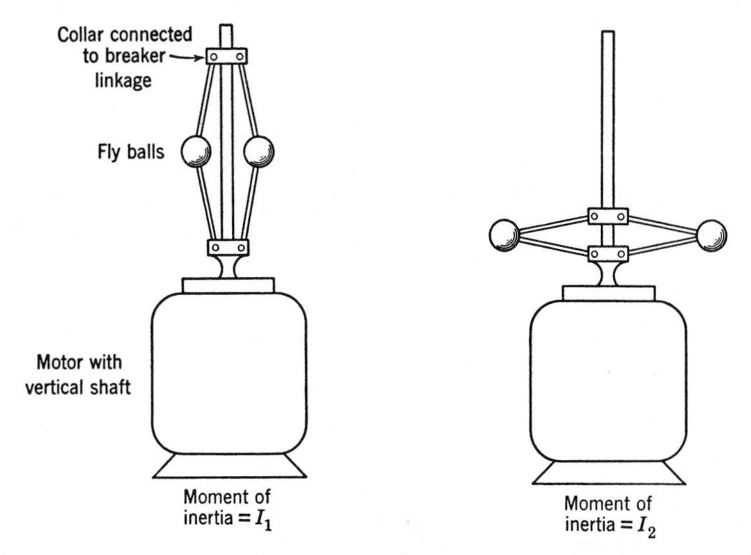

Fig. 3-8. Centrifugal mechanism for closing a circuit breaker.

kind of non-rigid system, one in which all particles have a common instantaneous angular velocity. An example of such a body is illustrated by figure 3–8, which shows a centrifugal mechanism sometimes employed for closing circuit breakers. A small motor drives a fly-ball mechanism like that of an engine governor. Initially the balls are close to the axis and are at rest. To close the breaker the motor is energized and the balls rotate. Under the influence of centrifugal forces they also move outward when sufficient speed is attained and the upper collar is pulled down, exerting the force which closes the breaker and stores energy in a spring which then is available for opening the breaker again. In this mechanism the moment of inertia I of the rotating system is not constant but varies with time as the balls move radially, and it is clear that 3–16 does not apply because the rotating system is not rigid. The relation between torque and angular velocity is found from 3–25 and 3–28 with I now variable, and the result is

$$\sum T = I\frac{d\omega}{dt} + \omega\frac{dI}{dt} \tag{3–29}$$

in which ΣT is the resultant of the torque supplied by the motor and any frictional torques which may be present. Use is made of equation 3–29 in section 3–7.

As another example of the utility of 3–25 let us return again to the phonograph record changer. Consider the situation when the record is slipping with respect to the turntable and the turntable is slipping with

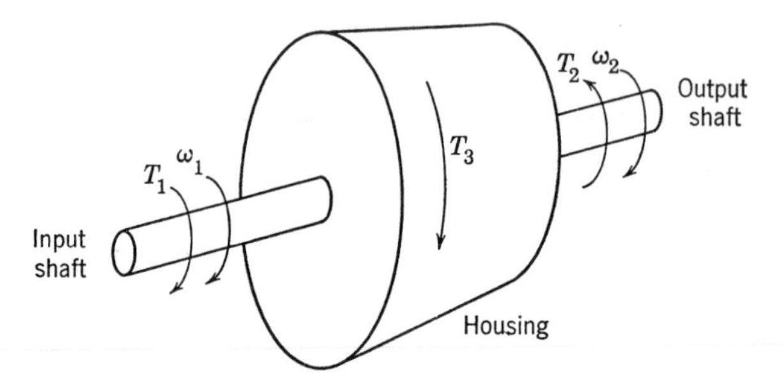

Fig. 3-9. A torque converter.

respect to the motor. The record and turntable then comprise a non-rigid system. Its angular momentum may be expressed as the sum of the separate angular momenta for record and turntable each of which is a rigid body. Thus using symbols defined in section 2–6 and with the aid of 3–28 we have

$$H = I_r\omega_r + I_t\omega_t \tag{3–30}$$

The torque applied externally to the system is T_{ds} exerted by the drive; hence, using 3–25,

$$T_{ds} = \frac{dH}{dt} = I_r\frac{d\omega_r}{dt} + I_t\frac{d\omega_t}{dt} \tag{3–31}$$

since the moments of inertia, I_r and I_t, are each constant. This result is identical with 2–50, which was found by applying $T = I\alpha$ to record and turntable separately.

The law expressed by equation 3–25 is also helpful in determining torque reactions in rotating machinery of various kinds. For instance, suppose we have a torque converter such as is used in many automobiles to couple the engine to the drive shaft. The engine shaft turning at angular velocity ω_1 exerts torque T_1 on the input shaft to the converter (see figure 3–9), and the load exerts a larger opposing torque T_2 on the output shaft which turns at a lower speed ω_2. That is, the output torque is larger than the input torque, and we wish to find the relation between the torque T_3 necessary to keep the housing stationary and the torques T_1 and T_2 which are presumed to be known.

The torque converter works by the action of oil circulating through a pump impeller fastened to the input shaft and a turbine runner fastened to the output shaft. In its passage from the turbine back to the pump the oil is deflected by stationary vanes fixed to the housing. We might solve our problem by dealing with these internal actions, but there is a much simpler way.

If the torque converter is running steadily, that is, ω_1 and ω_2 are constant, then the oil flow will also be steady and we can say that the angular momentum of the system comprising the whole converter about any fixed axis is constant. Consequently dH/dt is zero, and from 3–25 it follows that the resultant of torques applied externally to the system is zero; that is,

$$T_1 + T_3 - T_2 = 0 \tag{3–32}$$

or

$$T_3 = T_2 - T_1 \tag{3–33}$$

This system is not at rest, and the sum of the externally applied torques is zero only because the total angular momentum of the system is constant, and this, of course, is also the basis for the law of static equilibrium. It is important to note, however, that, if the angular momentum of the system is changing, as would be the case if the engine were accelerating, dH/dt would not be zero and the externally applied torques would not be in equilibrium with each other as in 3–32.

3–7 Space and Time Integrals of the Acceleration Laws

Of great usefulness in solving engineering problems are the relations between mechanical work and kinetic energy. These are, for translation of any body,

$$\int_{s_1}^{s_2} f\, ds = \tfrac{1}{2}m(v_2^2 - v_1^2) \tag{3–34}$$

and for rotation of a rigid body about a fixed axis

$$\int_{\theta_1}^{\theta_2} T\, d\theta = \tfrac{1}{2}I(\omega_2^2 - \omega_1^2) \tag{3–35}$$

The first expresses the fact that the work done by a force as it moves a body from s_1 to s_2 along a path s, the component of the force along s being f, is equal to the gain of the kinetic energy of translation $\frac{1}{2}mv^2$ between the points s_1 and s_2. Similarly 3–35 expresses the relationship between work done by a torque as it turns a rigid body between angles θ_1 and θ_2 about a fixed axis and the gain of kinetic energy of rotation.

Understanding these relationships is mostly a matter of realizing that they are merely integrals of the acceleration laws with respect to distance or angle. Thus suppose we have a body of mass m moved along a straight

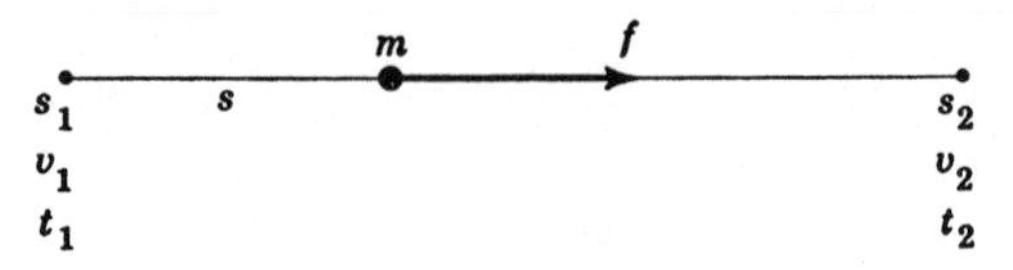

Fig. 3-10. Integration of $f = ma$ with respect to s and with respect to t.

line s from s_1 to s_2, as shown in figure 3-10, by a force whose component along s is f. By the acceleration law

$$f = ma = m\frac{dv}{dt} \tag{3-36}$$

If we integrate both sides with respect to s between limits s_1 and s_2

$$\int_{s_1}^{s_2} f\,ds = m\int_{s_1}^{s_2} \frac{dv}{dt}\,ds \tag{3-37}$$

But velocity $v = ds/dt$ or $ds = v\,dt$, whence

$$\int_{s_1}^{s_2} f\,ds = m\int_{t_1}^{t_2} \frac{dv}{dt}\,v\,dt = m\int_{v_1}^{v_2} v\,dv = \tfrac{1}{2}m(v_2^2 - v_1^2) \tag{3-38}$$

In the above t_1 and v_1 are time and velocity corresponding to s_1; t_2 and v_2 to s_2.

EXERCISE. In a manner similar to the above, derive equation 3-35 from $T = I\alpha$.

Another useful law is that relating impulse to momentum; for translation of any body it is

$$\int_{t_1}^{t_2} f\,dt = m(v_2 - v_1) \tag{3-39}$$

and for rotation of a rigid body about a fixed axis

$$\int_{t_1}^{t_2} T\,dt = I(\omega_2 - \omega_1) \tag{3-40}$$

The first is merely a time integral of $f = ma$. Thus

$$\int_{t_1}^{t_2} f\,dt = m\int_{t_1}^{t_2} \frac{dv}{dt}\,dt = m\int_{v_1}^{v_2} dv = m(v_2 - v_1) \tag{3-41}$$

and relation 3-40 is obtained similarly as a time integral of $T = I\alpha$. These relations express the fact that the impulse of a force, defined as the left side of 3-39, is equal to the increase of the linear momentum mv, and that the impulse of a torque is equal to the increase of angular momentum $I\omega$ of a rigid body. If f is zero the integral on the left of 3–39 is zero and we have a statement of the principle of conservation of linear momentum.

For a non-rigid system, 3–25 leads to

$$\int_{t_1}^{t_2} T\,dt = H_2 - H_1 \tag{3–42}$$

where T is the resultant of the externally applied torques, and the angular momentum H of the particles of the system replaces the quantity $I\omega$ in 3–40.

To show a use of the time integral 3–42 let us look again at the phonograph record changer problem. Near the end of section 2–6 we treated the case where the turntable was carrying $n - 1$ records and an nth record was dropped on, and we wished to find how long it would take the nth record to reach playing speed. The method we employed in section 2–6 was to apply $T = I\alpha$ separately to the turntable and each record, thus obtaining 2–56, a set of $n + 1$ differential equations, one for each rigid body involved. We then manipulated the $n + 1$ equations so as to eliminate numerous unknowns and finally found the result desired. The argument using 3–42 is much simpler and goes as follows: Before the last record is dropped, the table and the other records are all turning at the playing speed ω_0 and the total angular momentum of the system comprising them and the nth record which is still at rest is H_1. The nth record is dropped, and in general for a short time all the records and the drive may slip, one with respect to another. Finally the whole system is back up to speed ω_0 and the final angular momentum is

$$H_2 = H_1 + I_n\omega_0$$

where I_n is the moment of inertia of the nth record. During the time interval $t = 0$ to $t = t_0$, while this change of momentum was occurring, the only torque acting on the system from outside was T_d, that exerted by the drive. The impulse of this torque then equals the increase in the angular momentum, or in accordance with 3–42

$$\int_0^{t_0} T_d\,dt = H_1 + I_n\omega_0 - H_1 = I_n\omega_0$$

which is identical with the former result 2–57.

To further illustrate the time integral 3–42 consider again the centrifugal mechanism for closing a circuit breaker described in section 3–6 and shown in figure 3–8. One way of using the device is to keep the balls at their inner radius until maximum speed is attained, after which the power supply to the motor is cut off, and then the balls are allowed to fly out. Neglecting friction, the torque acting becomes zero and either 3–25 or

3–29 tells us that as the balls move outward the angular momentum H must remain constant; that is,

$$H = I\omega = \text{Constant} \tag{3–43}$$

If I_1 is the moment of inertia and ω_1 the angular velocity with the balls all the way in, and I_2 and ω_2 the values when the balls are all the way out, we have

$$I_2\omega_2 = I_1\omega_1 \tag{3–44}$$

which gives ω_2 in terms of known quantities I_1, I_2, and ω_1. The work W available for closing the breaker may be found from conservation of energy as the decrease in rotational kinetic energy of the system as the balls move outward. Thus

$$W = \tfrac{1}{2}I_1\omega_1^2 - \tfrac{1}{2}I_2\omega_2^2 \tag{3–45}$$

and using 3–44 to eliminate ω_2

$$W = \frac{1}{2}I_1\omega_1^2\left(1 - \frac{I_1}{I_2}\right) \tag{3–46}$$

showing that a positive quantity of work W is available since I_2 is clearly greater than I_1.

It is noteworthy that the rotational kinetic energy of the system changes while the angular momentum remains constant; kinetic energy and momentum are different properties of a system, an important fact sometimes overlooked.

Where the space and time integrals are applicable they often simplify analysis, or in other cases provide checks. When an equation is set up by using either $f = ma$ or $T = I\alpha$ it must usually be treated as a differential equation since acceleration is either the first derivative of velocity, or the second of displacement, and solving a differential equation means integrating it. On the other hand, work-kinetic energy and impulse-momentum relationships are preformed integrals; if we can use them they lead to equations that do not contain derivatives. We have in effect integrated our basic principle before applying it.

3–8 Electric Circuit Principles

In the simpler electric circuit problems we have systems consisting of generators (or batteries), resistors, capacitors, inductors, and mutual inductors connected so that they form one or more closed circuits. If there are several circuits they may be coupled by conductive, inductive, or capacitive links. Examples are shown in figure 3–11 in which E denotes a battery, e a source of variable voltage, i a current generator, R resistance in ohms, L inductance in henries, M mutual inductance in henries, C capacitance in farads, and S a switch. The problem ordinarily is to find the currents with the source voltages given, or to find the voltages when the source currents are given.

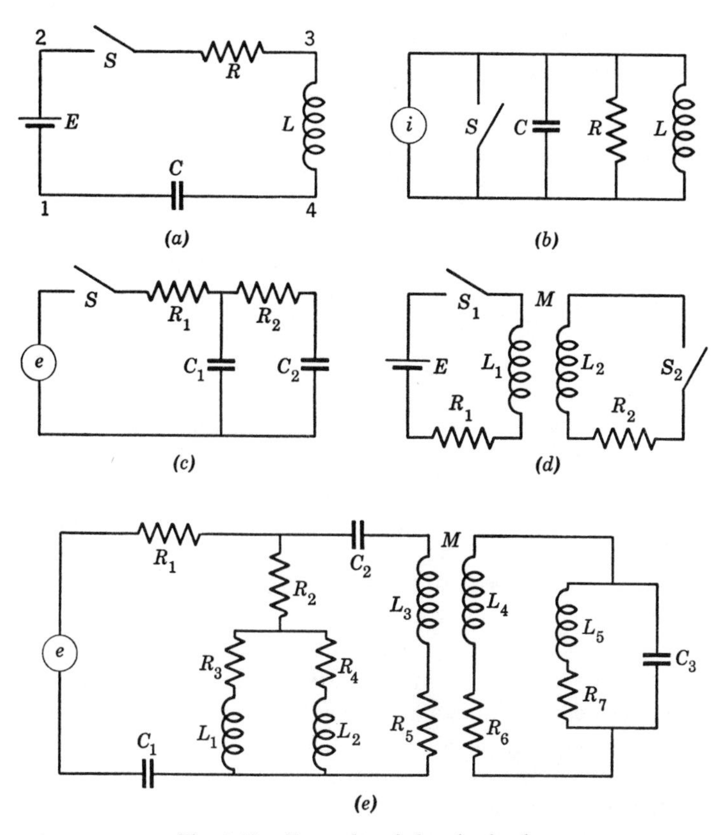

Fig. 3-11. Examples of electric circuits.

In solving an electrical problem there is already considerable progress when a circuit diagram has been drawn and values have been assigned to the circuit parameters. That is, coils, resistors, and leads have distributed capacitance; coils and leads have resistance which varies with temperature, and sometimes with frequency, and so on; and decisions have to be made about what to include and what to neglect before a circuit can be drawn. This is especially true when constant parameters are used. And the fact that we can obtain excellent agreement between experiment and calculation when high-quality components are employed and when great care is taken leads us to forget that often in practice the real circuit elements are not as represented in the idealized circuit diagram, and, indeed, the values of the parameters may not be very accurate.

After we have determined the circuit to represent a given situation to a suitable accuracy we usually deal with it by means of Kirchhoff's laws:

The algebraic sum of the voltage drops around any closed circuit in a specified direction is zero or

$$\Sigma v = 0 \tag{3–47}$$

and, *the algebraic sum of the currents flowing away from any circuit junction is zero* or

$$\Sigma i = 0 \tag{3–48}$$

In the case of 3–47 we may deal equally well with voltage rises instead of drops, and in 3–48 with currents flowing toward instead of away from a junction.

For example, in circuit *a* of figure 3–11 with the switch closed the voltage law 3–47 means: the fall of voltage from point 1 to point 2 plus the fall of voltage from 2 to 3 plus the fall of voltage from 3 to 4 plus the fall of voltage from 4 to 1 equals zero at all instants of time. It is a part of our circuit convention that no drop occurs along the symbols for connections; nor is any charge lost or piled up as it flows along a connection. Some of the drops must be negative numbers, some positive, since they add up to zero, but this is a matter that is taken care of by algebra. However, we must be careful to express all the voltage differences as *drops*, or all as rises, and beginners may find this harder than it looks.

For this same case the current law states that, if any point such as 1, 2, 3, or 4 is regarded as a junction, the net current out is zero, or as much current flows in to that junction from one side as flows out at the other. From this we conclude that at any one instant of time the current at every point of the circuit is the same.

Suppose in circuit *a* of figure 3–11 we know the battery voltage E, together with its polarity, and want to find the current that flows when the switch is closed. We shall use the implications of both laws. The current law tells us that the unknown current can be designated by a single symbol for every point of the circuit. We use i and define it as current in amperes. However this is not a complete definition; i, being an algebraic quantity, may be either positive or negative, which corresponds to the fact that current can flow in either of two directions, and we have to specify which direction is positive. This is done by drawing an arrow on the circuit diagram, say as shown in circuit *a* of figure 3–12. Later we shall see the effect of drawing the arrow the other way, as in *b*. The choice of the arrow direction is purely arbitrary—we do not mean that current actually flows in the direction indicated; we mean that, if it does, it is represented by a positive number. Indeed, the current may oscillate, that is, flow first one way and then the other periodically, and then it will be expressed sometimes by a positive number, sometimes by a negative one. Defining the direction of positive current is just as arbitrary as defining the direction of positive displacement in a mechanical problem.

Referring again to figure 3–12*a* we also mark the known polarity of the battery. What is indicated means that the positive terminal is connected through the switch to the resistor and the negative terminal to the condenser. The positive terminal of a battery is the one out of which the

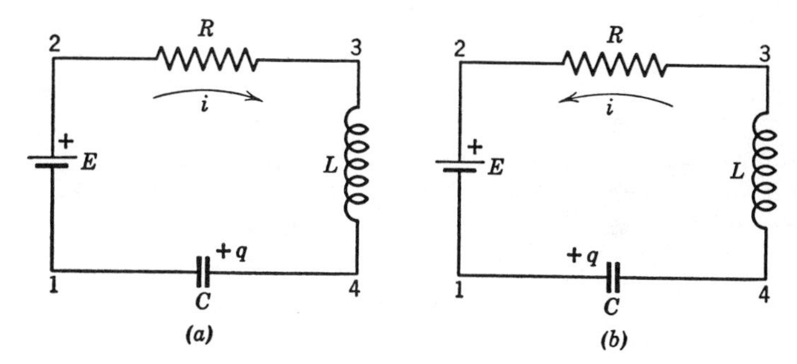

Fig. 3-12. Circuit *a* of figure 3-11 with different sign conventions.

battery tends to send current; the terminal marked + on an ammeter, on the other hand, is the one into which current must flow to make the pointer move up-scale.

Now we are ready to use the voltage law, carefully writing the voltage drops in terms of the unknown current. With the battery polarity as indicated the voltage drop from 1 to 2 is $-E$, the negative sign because the voltage actually *rises* from negative to positive terminal as we go through the battery. The drop from 2 to 3 through the resistor is $+Ri$. Observe that if the current is positive the voltage of 2 is above that of 3; there is a drop from 2 to 3. If on the other hand i is negative, current flows from 3 to 2 and the voltage of 3 is above that of 2. The voltage drop from 2 to 3 is negative but is still correctly given by $+Ri$ since i is now negative. Thus $+Ri$ is the proper expression for the drop from 2 to 3 no matter what the direction of current flow provided $+i$ means current flow from 2 to 3 through the resistor as it does in this case.

The drop from 3 to 4 through the inductance is $+L\,di/dt$ since di/dt designates an increasing current which requires that voltage be applied to the coil with a drop in the direction from 3 to 4. On the other hand if the current is decreasing di/dt is negative, the drop becomes a rise in fact, and we see that the inductor acts like a battery. A little study shows that whether i is $+$ or $-$, is increasing or decreasing, $L\,di/dt$ properly gives the drop from 3 to 4, if $+i$ means current flowing through the coil from 3 to 4.

The drop from 4 to 1 is that through the condenser, equal to $1/C$ times the charge q on one plate (the other has an equal and opposite charge). It is really a drop if the right-hand plate is positive. So the drop from 4 to 1 is q/C, q being the charge on the plate into which the current flows. How is q related to i? In this case with $+i$ flowing into the right-hand plate of the condenser

$$i = \frac{dq}{dt} \tag{3-49}$$

or

$$q = \int_0^t i\,dt + q_0 \tag{3-50}$$

That is, q at any instant of time is the time integral of i from 0 to t added to whatever charge q_0 was on the condenser at $t = 0$.

Now that all the voltage drops have been expressed in terms of i, E, and the circuit parameters, we may express the voltage law as

$$-E + Ri + L\frac{di}{dt} + \frac{q}{C} = 0 \tag{3-51}$$

or, using 3-49 and its derivative,

$$-E + R\frac{dq}{dt} + L\frac{d^2q}{dt^2} + \frac{q}{C} = 0 \tag{3-52}$$

which is generally written in terms of descending derivatives with the term independent of q on the right side:

$$L\frac{d^2q}{dt^2} + R\frac{dq}{dt} + \frac{q}{C} = E \tag{3-53}$$

Sometimes it is more convenient to differentiate 3-51 and then use 3-49 with the result

$$L\frac{d^2i}{dt^2} + R\frac{di}{dt} + \frac{i}{C} = 0 \tag{3-54}$$

In either case to find q or i we must solve a differential equation; this is discussed in chapter 5.

If we were dealing with rises instead of drops but still going from 1 to 2, 2 to 3, 3 to 4, and 4 to 1, we would write the same equation as 3-51 except that the algebraic sign of each term would be changed, equivalent to the same equation.

Suppose we had chosen positive i in the other direction as shown in figure 3–12*b*. The drop from 1 to 2 is $-E$ as before. From 2 to 3 the drop is $-Ri$ since we are going against the positive direction of current. From 3 to 4 the drop is $-L\,di/dt$ because if i is increasing in its positive direction the potential of 4 must be higher than that of 3. If we again take $+q$ as the charge on the plate into which $+i$ flows, this time the plate on the left, the drop from 4 to 1 is $-q/C$. Then, writing the voltage law,

$$-E - Ri - L\frac{di}{dt} - \frac{q}{C} = 0 \tag{3–55}$$

Charge and current are still related by 3–49, and so finally we have, on rearranging,

$$-L\frac{d^2q}{dt^2} - R\frac{dq}{dt} - \frac{q}{C} = E \tag{3–56}$$

which is the same as 3–53 except for the reversal of sign of each term containing q or its derivatives, and this corresponds as it should to the reversal of the arrow direction which defines $+i$ and $+q$. If current at some instant actually flows clockwise in the circuit, solution of 3–53 will lead to a positive number for i at that instant, whereas solution of 3–56 will result in a negative number.

Let us see how Kirchhoff's laws apply to circuit *b* in figure 3–11. Here the generator supplies a current i, and to do so develops whatever voltage may be necessary. In general i may be a function of time, but for simplicity let us consider the case when it is constant, equal to I. With the switch closed the voltage of the generator is zero and all the current flows through the switch; with the switch open the generator voltage appears across each one of the circuit elements. We can obtain this by inspection, or, more formally by Kirchhoff's voltage law. Let us designate this voltage by e, and of course we must specify the generator polarity that corresponds to positive values of e. This is shown in figure 3–13 together with the branch currents i_C, i_R, and i_L, and their arbitrarily assigned positive directions. According to Kirchhoff's current law 3–48 the net current flowing away from junction J is zero, or

$$-I + i_C + i_R + i_L = 0 \tag{3–57}$$

The quantity I, the output of the current generator, is known; i_C is found by taking the derivative of the charge on the top plate of the condenser, which, with due regard for sign, is

$$i_C = \frac{d}{dt}(Ce) = C\frac{de}{dt} \tag{3–58}$$

Also as before

$$i_R = \frac{e}{R} \tag{3-59}$$

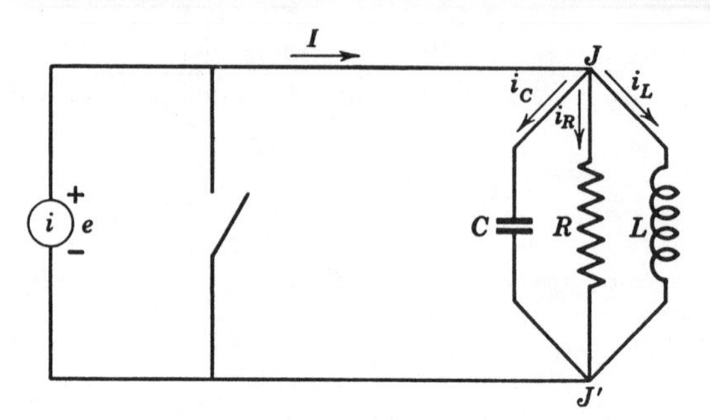

Fig. 3-13. Circuit b of figure 3-11 showing sign conventions.

We may find i_L from the expression

$$L\frac{di_L}{dt} = e \tag{3-60}$$

which on integration gives

$$i_L - i_{L0} = \frac{1}{L}\int_0^t e\,dt \tag{3-61}$$

with i_{L0} representing the current i_L at the instant $t = 0$ when the switch is opened. Thus

$$i_L = \frac{1}{L}\int_0^t e\,dt + i_{L0} \tag{3-62}$$

and when we substitute 3–58, 3–61, and 3–62 into 3–57

$$-I + C\frac{de}{dt} + \frac{1}{R}e + \frac{1}{L}\int_0^t e\,dt + i_{L0} = 0 \tag{3-63}$$

We may eliminate the integral by differentiation; on doing this we have

$$C\frac{d^2e}{dt^2} + \frac{1}{R}\frac{de}{dt} + \frac{1}{L}e = 0 \tag{3-64}$$

which is a differential equation to be solved for e.

With circuit b of figure 3–11 we used Kirchhoff's current law somewhat as the voltage law was used with circuit a, and the same care had to be taken with signs. A comparison of 3–54 and 3–64 shows a strong similarity. Indeed, the roles of voltage and current are interchanged as are the roles of C and L, while R is replaced by $1/R$. The series connection is replaced by a parallel connection, and a closing switch by an opening switch. There is a kind of analogy between these circuits, and each one is called the *dual* of the other.

The application of electric circuit principles will be illustrated by yet another case, c in figure 3–11, which is redrawn in figure 3–14 to show current symbols and sign conventions. We may suppose that the generator voltage $e(t)$ is known as a function of time, and we want to find the current

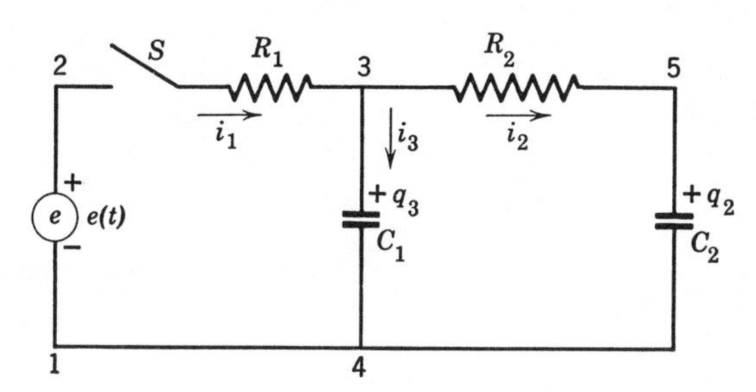

Fig. 3-14. Circuit c of figure 3-11 showing sign conventions.

i_1 that it supplies. To do this we have to deal with the three currents i_1, i_2, and i_3, all of which are unknown. We may set up one equation by applying the voltage law around the mesh including e, R_1, and C_1; and another equation similarly for the mesh including R_2, C_2, and C_1. We might also write the voltage law for the mesh comprising e, R_1, R_2, and C_2, but it turns out that the relation so obtained is not independent of the equations already described. A third, and independent, relation may be had by applying the current law at either of junctions 3 or 4, the relation being the same at both.

According to the voltage law the algebraic sum of the voltage drops from 1 to 2, 2 to 3, 3 to 4, and 4 to 1 is zero. Likewise the algebraic sum of the drops from 4 to 3, 3 to 5, and 5 to 4 is zero. And by the current law the sum of the currents leaving junction 3 is zero. In figure 3–14 we have shown condenser charges q_2 and q_3 as positive on the plates into which the respective currents i_2 and i_3 flow when they have their arrow or positive directions. Then our three relations become

$$-e(t) + R_1 i_1 + \frac{1}{C_1} q_3 = 0 \tag{3–65}$$

$$-\frac{1}{C_1} q_3 + R_2 i_2 + \frac{1}{C_2} q_2 = 0 \tag{3–66}$$

$$-i_1 + i_2 + i_3 = 0 \tag{3–67}$$

EXERCISE. Explain in detail why the signs are as written in equations 3–65, 3–66, and 3–67.

We have also the relations between currents and condenser charges:

$$i_2 = \frac{dq_2}{dt} \tag{3–68}$$

$$i_3 = \frac{dq_3}{dt} \tag{3–69}$$

Thus we have five equations to describe this circuit with five unknowns, i_1, i_2, i_3, q_2, q_3; it should be possible to eliminate unknowns and find a single equation in one unknown. We may well expect it to be a differential equation because of the presence of derivatives.

We might first differentiate 3–65 and 3–66 and use 3–68 and 3–69 to eliminate q_2 and q_3, obtaining

$$-\frac{d}{dt}e(t) + R_1\frac{di_1}{dt} + \frac{1}{C_1}i_3 = 0 \tag{3–70}$$

$$-\frac{1}{C_1}i_3 + R_2\frac{di_2}{dt} + \frac{1}{C_2}i_2 = 0 \tag{3–71}$$

Then use 3–67 to eliminate i_3 from the last two:

$$-\frac{d}{dt}e(t) + R_1\frac{di_1}{dt} + \frac{1}{C_1}i_1 - \frac{1}{C_1}i_2 = 0 \tag{3–72}$$

$$-\frac{1}{C_1}i_1 + \frac{1}{C_1}i_2 + R_2\frac{di_2}{dt} + \frac{1}{C_2}i_2 = 0 \tag{3–73}$$

We now have two equations in the two unknowns i_1 and i_2. Solve 3–72 for i_2 and substitute it in 3–73 differentiating where necessary:

$$i_2 = i_1 + R_1C_1\frac{di_1}{dt} - C_1\frac{d}{dt}e(t) \tag{3–74}$$

$$R_2\frac{d}{dt}\left(i_1 + R_1C_1\frac{di_1}{dt} - C_1\frac{d}{dt}e(t)\right) + \left(\frac{1}{C_1} + \frac{1}{C_2}\right)\left(i_1 + R_1C_1\frac{di_1}{dt} - C_1\frac{d}{dt}e(t)\right) - \frac{1}{C_1}i_1 = 0 \tag{3–75}$$

After differentiating, rearranging terms, and multiplying by C_2:

$$R_1R_2C_1C_2\frac{d^2i_1}{dt^2} + (R_1C_1 + R_2C_2 + R_1C_2)\frac{di_1}{dt} + i_1$$
$$= R_2C_1C_2\frac{d^2}{dt^2}e(t) + (C_1 + C_2)\frac{d}{dt}e(t) \quad (3\text{–}76)$$

Since $e(t)$ is known this equation can be solved for i_1. We need, however, also to have two initial conditions (see chapter 5); that is, the values of i_1 and di_1/dt at $t = 0$. Suppose we start with known values of the charges q_2 and q_3 at $t = 0$ and suppose furthermore that these are zero. We can use the equations that have been written to help obtain the initial conditions.

First we must consider how the charges q_2 and q_3 may change during the brief time that the switch is closing. By integrating 3–68 we can write

$$\Delta q_2 = \int_{t'}^{t''} i_2\,dt \quad (3\text{–}77)$$

where Δq_2 is the change in the charge q_2 occurring in a time interval $t'' - t'$. Suppose that the switch is closed within this interval and that the times t'' and t' are made to approach each other. Then we can see from 3–77 that, if the change in charge Δq_2 is other than zero, i_2 momentarily must be infinite. In our case if i_2 were infinite there would have to be infinite voltage drop across R_2, and since there is nothing to produce this we conclude that q_2 has the same value just after the switch is closed as it did before, namely, zero. A similar argument shows that q_3 also is continuous through the instant of switching.

Since 3–65 holds for any instant, it holds as t approaches zero from the positive side and, thus, with q_3 zero, as we have shown,

$$(i_1)_{t=0} = \frac{e(0)}{R_1} \quad (3\text{–}78)$$

We can easily verify this from physical considerations as follows: If we look at figure 3–14 and consider it at the instant after S closes, the condensers have zero charge, hence zero voltage drop, and can be regarded as short-circuited connections for a brief initial instant, as shown in figure 3–15. Then since the generator is connected directly across R_1 the current i_1 is as given by 3–78, incidentally correct also as to sign.

Similarly, we can find the other initial condition. From 3–66 with q_2 and q_3 at their initial values of zero at the first instant it follows that $i_2 = 0$ initially, and this agrees with our picture of the circuit in figure 3–15 since R_2 is short circuited. If for the instant of zero time we substitute $i_2 = 0$ and 3–78 into 3–74 we obtain

$$\left(\frac{di_1}{dt}\right)_{t=0} = \frac{1}{R_1}\left(\frac{d}{dt}e(t)\right)_{t=0} - \frac{e(0)}{R_1{}^2C_1}$$

and are in a position to proceed with the solution.

In this section we have tried to show the common electric circuit principles and how they are applied in dealing with problems. The commonest

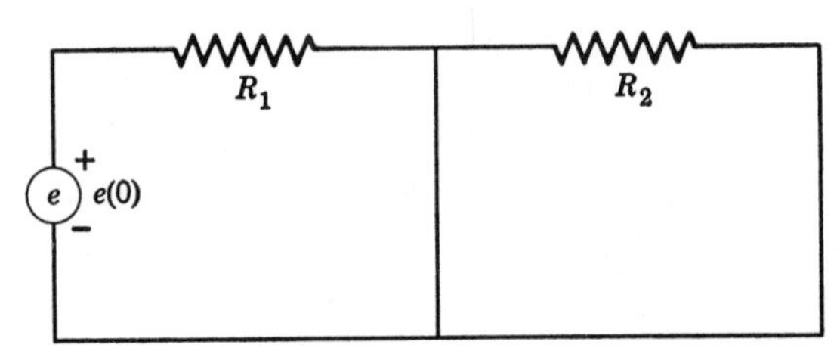

Fig. 3-15. Circuit equivalent to figure 3-14 when $t = 0$ after closing switch, condensers initially uncharged.

pitfalls are those of determining the proper algebraic sign for each term in the equations, and accordingly we have emphasized the thinking that must underlie good usage. It is our experience that little difficulty with signs will occur if care is taken to be clear about what is meant by the positive direction for each quantity as it is introduced.

3–9 Conservation of Magnetic Flux Linkages

In the same way that in mechanics the law of conservation of momentum (section 3–7) is a useful concept, there is an analogous useful principle for electric circuits, that of the conservation of magnetic flux linkages. Furthermore, it is found in the same way, by taking a time integral, in this case of the Kirchhoff voltage law.

To gain an understanding of this let us begin by dealing with a particular case, circuit *d* of figure 3–11, which we may think of as a transformer whose primary winding is energized by a battery and whose secondary winding can be short circuited. Such a situation arises when the resistances of transformer windings are being measured during tests by the manufacturer.

Suppose the transformer we are dealing with is large, of 10,000 kva rating or higher, and thus is one in which the winding resistances are very low and the self inductances high. The primary has been closed for some time, and the current in it has reached the steady-state value, E/R_1, the secondary circuit being open. Then it happens that the secondary is short circuited, and afterward the primary switch is opened. We want to find the current induced in the closed secondary circuit by the rapid decay of primary current.

Let us suppose that the switch is opened at $t = 0$ and the primary current has fallen to zero by time T. Throughout the time interval from 0 to T the secondary circuit is closed, and referring to figure 3-16 we may apply the Kirchhoff voltage law as follows: the voltage drop from 1 to 2 through the transformer secondary plus the voltage drop from 2 to 1 through the resistance R_2 is equal to zero. In the circuit as drawn the transformer is regarded as having windings of zero resistance, such small winding resistance as there is being shown as R_2 in the secondary circuit. Similarly, R_1 represents the primary winding resistance in addition to any external resistance that may be present. Positive directions for the currents are defined by the arrows in figure 3-16.

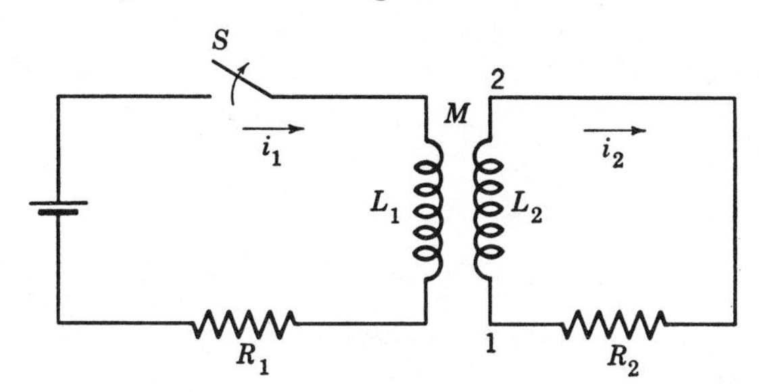

Fig. 3-16. Transformer circuit.

The voltage drop through the secondary winding from 1 to 2 results from the time rate of change of the total magnetic flux linking the winding, and the flux is caused by current in either the primary or in the secondary itself. As far as the secondary is concerned the drop is $L_2\, di_2/dt$, as we have seen in section 3-8. The voltage drop from 1 to 2 that comes from the changing primary current is written as $M\, di_1/dt$ with a positive sign.

Whereas L_2 or L_1 or any self-inductance is always a positive number, mutual inductance M may be either a positive or negative number. The sign of M depends upon the direction of winding of the coils in the transformer; or, perhaps it is more accurate to say, depends upon the sense of the windings in relation to the directions defined as positive for the currents. After the positive directions for the currents have been established by drawing arrows on the physical circuits (possibly only mentally, but at least indicating them on the circuit diagram) then the sign of M is determined from the construction of the transformer. If positive currents in both windings produce fluxes in the same direction through the windings, M is a positive number. If positive currents produce opposing fluxes M is a negative number. The method of determining the sign of M is shown in figure 3-17. In the left-hand part of the figure the transformer windings are such as to make M positive, and in the right hand part the winding direction of the secondary coil has been reversed, making M negative. The sign of M would also be changed by reversing the direction defined to be positive for either current. Reversing the positive directions for both currents, however, leaves the sign of M unchanged.

EXERCISE. Is M a positive or a negative number in figure 3–19?

Going on with the transformer problem, we may write the voltage drop from 1 to 2 through the transformer as

$$v_{12} = L_2 \frac{di_2}{dt} + M \frac{di_1}{dt} \tag{3–79}$$

Adding the drop through the resistance we have, by Kirchhoff's voltage law,

$$L_2 \frac{di_2}{dt} + M \frac{di_1}{dt} + R_2 i_2 = 0 \tag{3–80}$$

These equations assume that the flux-current relations are linear; indeed, there is no generally useful and simple definition of inductance for the non-linear case.

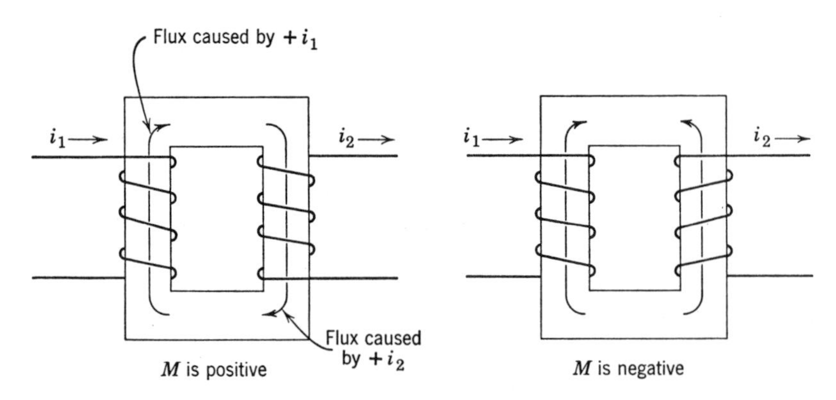

Fig. 3-17. Sign of mutual inductance.

If we know i_1 as a function of time during its decay period, we can use 3–80 to find i_2 as a function of time. To simplify matters we will assume that the term $R_2 i_2$ is negligible and omit it from the equation. This is reasonable for large power transformers, and in both conception and order of accuracy is like ignoring friction in mechanisms designed to run freely. Indeed it is exactly similar to neglecting friction in section 3–7 where we applied the law of conservation of angular momentum to the centrifugal mechanism for closing circuit breakers. Dropping the resistance term from 3–80 gives

$$L_2 \frac{di_2}{dt} + M \frac{di_1}{dt} = 0 \tag{3–81}$$

Let us try to integrate this with respect to time over the period 0 to T during which the primary current decays from I_{10} to 0 and the secondary current rises from 0 to a value I_{2T} that we seek to determine:

$$L_2 \int_0^T \frac{di_2}{dt}\, dt + M \int_0^T \frac{di_1}{dt}\, dt = 0 \tag{3–82}$$

Changing variables we have

$$L_2 \int_0^{I_{2T}} di_2 + M \int_{I_{10}}^{0} di_1 = 0 \tag{3–83}$$

from which

$$L_2 I_{2T} - M I_{10} = 0 \tag{3–84}$$

and

$$I_{2T} = \frac{M I_{10}}{L_2} \tag{3–85}$$

This is the answer to the particular problem we set out to solve, namely, finding the current induced in the short-circuited secondary when the primary current is interrupted, subject to the assumptions of zero secondary resistance and linearity between current and flux.

The careful interpretation of our result is worth while; it leads to an important generalization. In trying to understand what we have, we look at 3–83 and observe that during the interval of time 0 to T the increase of the quantity $L_2 i_2$ plus the increase of the quantity $M i_1$ is zero. In other words the increase of $L_2 i_2$ is equal to the decrease of $M i_1$. Evidently the quantity $L_2 i_2 + M i_1$ is a quantity that does not change although both i_1 and i_2 are changing. Indeed, we could rewrite 3–81 as

$$\frac{d}{dt}(L_2 i_2 + M i_1) = 0 \tag{3–86}$$

which again shows that the quantity in parentheses is constant in time, no matter how i_1 and i_2 change. If we force i_1 to decrease, as we did by opening the switch in the primary circuit, i_2 increases to keep the quantity in parentheses constant.

We may well ask what is the quantity in the parentheses of 3–86? It is the magnetic flux linkage of the secondary circuit, a concept that we encountered in connection with equation 1–12 and figure 1–7 in section 1–3 when we were determining the force exerted by a solenoid on its iron core. When all the N turns of a coil link the same magnetic flux ϕ, the flux linkage is $N\phi$. When the flux changes, the voltage induced in the coil expressed as a drop in the direction of positive current, is given by

$$v = N\frac{d\phi}{dt} = \frac{d}{dt}(N\phi) \tag{3-87}$$

provided that the number of turns is not being changed, as by sliding contacts.

We also have the expression for the voltage drop in the secondary of the transformer as

$$\begin{aligned} v &= L_2\frac{di_2}{dt} + M\frac{di_1}{dt} \\ &= \frac{d}{dt}(L_2 i_2 + M i_1) \end{aligned} \tag{3-88}$$

Equations 3–87 and 3–88 are different expressions for the same voltage, that which comes from a changing magnetic flux which links the secondary. A comparison of 3–87 and 3–88 shows that

$$N\phi = L_2 i_2 + M i_1 \tag{3-89}$$

Strictly speaking, the comparison of 3–87 and 3–88 shows that the two sides of 3–89 can differ by a constant but the constant is zero if the flux depends only upon the currents and is zero when the currents are zero, as for example in the case of figure 1–7.

Thus the flux linkages of a coil or a circuit can be found by multiplying flux by turns or from the inductances and currents by 3–89. We may compare this expression with equation 1–12 by imagining the primary removed so that $M = 0$ and then $N\phi = L_2 i_2$.

We may also think of flux linkages when each turn does not link the same flux, and hence the expression $N\phi$ is not applicable. If we have a number of turns N the first of which links flux ϕ_1, the second ϕ_2, and so on up to ϕ_N, the voltage induced would be

$$\begin{aligned} v &= \frac{d\phi_1}{dt} + \frac{d\phi_2}{dt} + \cdots + \frac{d\phi_N}{dt} \\ &= \frac{d}{dt}(\phi_1 + \phi_2 + \cdots + \phi_N) \end{aligned} \tag{3-90}$$

$$= \frac{d\psi}{dt} \tag{3-91}$$

where we define ψ by the relation

$$\psi = \phi_1 + \phi_2 + \cdots + \phi_N \tag{3-92}$$

and, comparing 3–91 with 3–87, we see that ψ plays the same role as $N\phi$ in the earlier case; again it is the magnetic flux linkage of the coil or circuit. If there are several coils in a closed circuit the total linkage is the algebraic sum of the individual linkages of the coils. The sign of flux linkage is determined after the direction for positive current has been chosen. Flux linkage is positive if it threads the circuit in the same direction as flux caused by positive current.

In our transformer problem the flux linkage equation for the secondary circuit is

$$\psi_2 = L_2 i_2 + M i_1 \tag{3–93}$$

and 3–86 shows that ψ_2 remains constant as long as the circuit is closed and if the resistance of the circuit is zero. In fact we may say that, initially,

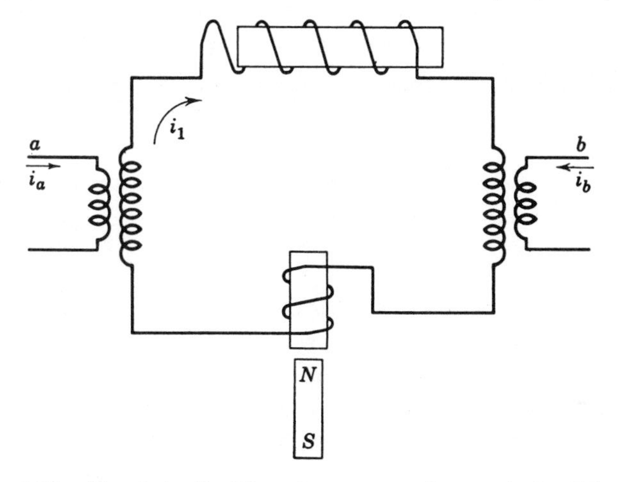

Fig. 3-18. Closed circuit with various sources of magnetic flux linkage.

when i_2 was zero and we closed the switch, linkages caused by the primary current i_1 were established or "trapped" in the secondary circuit. When we reduced i_1 the linkages of the closed secondary circuit could not change and so i_2 rose to whatever value was necessary to keep them constant.

Compare this with what happens in the flyball system used to close circuit breakers treated in sections 3–6 and 3–7. There, in accordance with equation 3–43, angular momentum equal to $I\omega$ must be constant. As I increases, ω must decrease and by just the right amount to keep the momentum unchanged.

Let us see if we can generalize further the concept of conservation of flux linkages. Consider the closed circuit in figure 3–18 which contains four coils and is of zero resistance. The current is i_1 and has the arbitrarily defined positive direction shown by the arrow. The circuit is magnetically coupled to neighboring circuits a and b of which only the inductive parts are shown; a and b may contain resistors, capacitors, and sources as well. There is a movable permanent magnet N-S in the vicinity of one of the coils, and a movable iron core in another of the coils.

Magnetic flux linkages in the closed circuit are attributable to

1. The current i_1 flowing through the four coils.
2. The currents i_a and i_b in the neighboring coils.
3. The permanent magnet N-S.

We may write the total linkage of the circuit as

$$\psi_1 = L_1 i_1 + M_{1a} i_a + M_{1b} i_b + \psi_{1p} \tag{3-94}$$

where L_1 is the total self-inductance of the closed circuit, taking account of the four coils and possible intercouplings between them; M_{1a} and M_{1b} are the mutual inductances and may have either positive or negative numerical values; and ψ_{1p} is the linkage from the permanent magnet.

EXERCISE. Assuming the flux distribution around the permanent magnet to be known explain how to determine ψ_{1p} using 3-92. As figure 3-18 is drawn, what is the sign of ψ_{1p}?

Now consider the means by which any part of the linkage with the closed circuit can change. The first component in equation 3-94 can change if either i_1 or L_1 changes, and L_1 can change if either the core of the upper coil or the permanent magnet moves, since the magnet not only supplies flux but serves as a core for the coil with which it is coupled. The next two terms in 3-94, the components of linkage originating in mutual inductance, can change if i_a, i_b, or both, change, or if there is relative motion between any of the mutually coupled coils. Finally, ψ_{1p} can change if the permanent magnet moves.

The Kirchhoff voltage law, applied to the closed circuit, requires that the voltage drop around the circuit be zero, which by 3-91 becomes

$$\frac{d\psi_1}{dt} = 0 \tag{3-95}$$

because the resistance and hence the resistance drop is zero, and therefore

$$\psi_1 = \text{Constant} \tag{3-96}$$

That is, in general, for any closed, purely inductive circuit such as shown in figure 3-18, the linkages are constant at the initial value that exists at the instant at which the circuit is closed, or

$$L_1 i_1 + M_{1a} i_a + M_{1b} i_b + \psi_{1p} = \text{Constant} \tag{3-97}$$

and all the quantities in 3-97 (inductances, currents, and linkage from permanent magnets) may be functions of time. If we think of changes in all these quantities except i_1 as being forced by external means, then i_1 assumes at every instant whatever value may be necessary to satisfy 3–97. We may generalize by making the following statement:

In any closed circuit containing only inductances, the resistance being zero, the total of the magnetic flux linkages remains constant at the value it had at the time the circuit was closed.

This principle, the conservation of flux linkages, or, as it is sometimes called, the *constant flux linkage theorem*, can be used to calculate currents if the resistance is small enough to be neglected. This is often the case for large a-c motors and generators when short circuited. In such cases the constant flux linkage theorem provides not only a means for calculating the short-circuit currents but also a physical principle that makes it easy to understand what is happening when self- and mutual inductances of windings vary with time as the machine rotates.

So far we have restricted our discussion to the case where the closed circuit is purely inductive, that is, has no resistance, series capacitance, or generators. Let us see what happens if this restriction is removed. Suppose we imagine that resistance R, capacitance C, and a source e are inserted in series with the four coils in the closed circuit of figure 3–18. Instead of 3–95 we now have, by the Kirchhoff voltage law,

$$\frac{d\psi_1}{dt} + Ri_1 + \frac{q_1}{C} - e = 0 \tag{3–98}$$

where q_1 is the charge on the capacitance. Then, over any interval of time T during which we have the general changes that have been described,

$$\psi_{1T} - \psi_{10} + \int_0^T Ri_1\,dt + \int_0^T \frac{q_1}{C}\,dt - \int_0^T e\,dt = 0 \tag{3–99}$$

where each term in 3–98 has been integrated, and ψ_{1T} is the magnetic flux linkage at the end and ψ_{10} at the beginning of the time interval T. Thus the change of linkage is

$$\psi_{1T} - \psi_{10} = \int_0^T e\,dt - \int_0^T Ri\,dt - \int_0^T \frac{q_1}{C}\,dt \tag{3–100}$$

which checks with our former result 3–96 if the source voltage and the resistance are removed by making them approach zero, and the capacitance is eliminated by making it infinite.

Equation 3–100 tells us that linkages are conserved even with R, C, and e present if T is sufficiently small, for then the integrals on the right may be small enough to neglect. Thus equation 3–96 holds either for a circuit which is purely inductive, or, approximately, for any circuit if a short enough time is being considered. Equation 3–100 provides a criterion for determining whether the assumption of constant flux linkages is sufficiently accurate in any given case. We may calculate the current assuming constant flux linkages, and also the charge q_1 if there is a condenser, and then evaluate the integrals in 3–100 and so obtain a first approximation to how much the linkages actually change and the error that such a change introduces.

We are led by the foregoing to another generalization, namely:

The total magnetic flux linkages of a closed circuit cannot change instantaneously.

This statement of the principle of conservation of flux linkage is sometimes used to determine conditions after a sudden switching operation or a

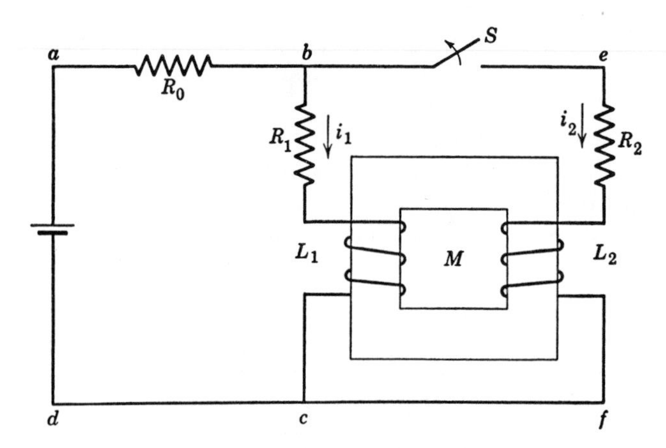

Fig. 3-19. Circuit in which switch is suddenly opened.

sudden change of inductance when these occur in a time so small that the change may be assumed to be instantaneous. Suppose, for example, we have the circuit of figure 3–19 with switch S closed for a long period of time so that steady-state conditions exist. Then suppose that the switch is suddenly opened. We wish to find the currents immediately after the switching process to use as initial conditions for calculation of current during the following interval.

We may say that the linkages of the circuit *abcda* remain constant during the switching because this is a closed circuit throughout the switching period. The linkage in any circuit which includes *ef* is not necessarily constant because this part of the circuit does not remain closed. If the currents before switching are i_{10} and i_{20}, and afterward are i_{1T} and i_{2T}, then the linkage of circuit *abcda* before switching is

$$\psi_{10} = L_1 i_{10} + M i_{20} \tag{3–101}$$

and after switching is

$$\psi_{1T} = L_1 i_{1T} + M i_{2T} = L_1 i_{1T} \tag{3–102}$$

since $i_{2T} = 0$ if the switch does its work. Then, by the conservation of flux linkages,

$$\psi_{1T} = \psi_{10} \tag{3–103}$$

or

$$L_1 i_{1T} = L_1 i_{10} + M i_{20} \tag{3–104}$$

and

$$i_{1T} = i_{10} + \frac{M}{L_1} i_{20} \tag{3–105}$$

which gives the current i_1 immediately after switching in terms of the initial currents. The result 3–105 is not restricted by any assumption about the values of the resistances; they may have any values, but the switch is assumed to interrupt i_2 instantly.

EXERCISE. In order to find i_{1T}, is it necessary that the switch be closed long enough so that steady-state conditions exist?

It is true that we need not use conservation of flux linkages to calculate i_{1T}; instead we might write the voltage equation around the closed path *abcda* and integrate it over the very short time interval of the switching operation, just as we did in deducing the principle itself, yet the principle gives a direct key to understanding the results. Knowing that i_1 must jump up suddenly to replace the linkage lost by the sudden reduction of i_2 to zero gives us a physical picture which is most helpful.

3–10 Heat Flow

Heat flow is likely at some time or other to be of importance to nearly every engineer. Heat may be generated where it is not wanted and must be dissipated without causing temperatures high enough to damage materials, particularly ones like electrical insulation, brake linings, and lubricating oils. On the other hand, the problem may be to maintain a particular temperature with the least expenditure of energy, as in a furnace or in a cathode heater for an electronic tube. To cope successfully with such situations requires a basic understanding of the laws of heat transfer.

In practice heat flow calculations are not as accurate as those in some other kinds of problems and often errors of 10 per cent or more may be expected and tolerated. There are two main reasons for this: it is not easy to confine the flow of heat to the particular paths for which calculations are made; and some of the laws are so complicated that practical expediency demands simplification of their form even at the sacrifice of accuracy. Moreover, the laws and coefficients are not known as precisely as in some other fields.

Analysis of problems in heat flow must take into account the fact that heat can be transferred by three very different physical processes: conduction, convection, and radiation, and frequently all three may occur simultaneously. For instance, in the problem on the non-fogging mirror, section 2–3, heat starts from the electric heating element on the back of the

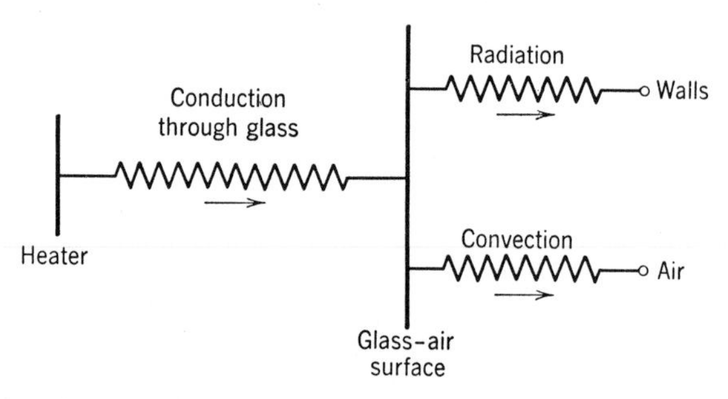

Fig. 3-20. Thermal circuit diagram for the non-fogging mirror.

mirror and flows through the glass by thermal conduction to the front surface of the glass. From there the heat flows to the surroundings by two other, and entirely different, processes: part of it is transferred to the walls of the room by radiation, and the remainder goes into the air by convection. For understanding such situations a thermal circuit diagram is useful, as illustrated in figure 3–20 for the non-fogging mirror problem. Here the symbols drawn like electrical resistances represent thermal resistances. The flow of heat through each thermal resistance is a function of the temperature difference across its terminals; the relationship may be a simple one of direct proportionality, or it may be more complicated. Thus temperature drop is analogous to voltage drop and flow of heat to flow of electric charge or current. The analogy with electric circuits is, however, only partial; thermal circuits do not appear closed like electric circuits but start from a source of thermal energy and end in a sink. A closer analogy is to liken the thermal system to a hydraulic system in which water flows by gravity from a reservoir to a sink at lower level through a system of pipes having appropriate hydraulic resistances. Here pressure drop is analogous to temperature drop, and water flow in mass per unit time to the flow of heat power in units such as watts or British thermal units per hour.

A thermal circuit diagram is useful in systematizing the analysis, just as a free-body diagram is in mechanics problems. For example, figure 3–20 implies that heat conducted through the glass divides at the glass-air surface into two parts which flow to different sinks, the walls and the air, by two different processes. The diagram also makes clear that the whole temperature drop from heater to walls is the sum of the drop through the glass and the drop from the glass-air surface to the walls. Figure 3–20 is drawn for a steady-state case since it makes no provision for accumulation of thermal energy in the glass. The diagram could be modified to take such storage into account, at least approximately, by separating the thermal resistance of the glass into a number of sections and inserting elements having thermal capacity. These thermal capacitances would be like storage tanks in the hydraulic analogy, or like capacitors in the electric circuit.

To make calculations after deciding upon the thermal circuit appropriate to describe a given situation, it is necessary to deal with the thermal resistances themselves and this means understanding the laws governing the transfer of heat by conduction, convection, and radiation.*

Conduction

Conduction is a process by which thermal energy is transmitted through matter by virtue of vibrations of the molecular structure assisted greatly in the case of metals by agitation of the free electrons. Because of the part played by the free electrons, thermal conduction in metals is closely related to electrical conduction; indeed from metal to metal the thermal and electrical conductivities are in a nearly constant ratio.

Thermal conduction obeys Fourier's law, which is much like Ohm's law for electrical conduction. A simple mathematical statement of Fourier's law can be made for the geometric configuration shown at the top of figure 3–21. There heat energy flows down a brick (more formally, a rectangular parallelepiped) from one end to the other, each of the ends being maintained at a constant temperature and the sides being thermally insulated so that no heat is lost from them. For this case Fourier's law becomes:

$$q = kA\,\frac{T_1 - T_2}{L} \tag{3–106}$$

* We have found the following references to be helpful: W. H. McAdams, *Heat Transmission*, Second Edition, McGraw-Hill, New York, 1942; W. J. King, "The Basic Laws and Data of Heat Transmission," *Mechanical Engineering*, 1932 pp. 190–194, 275–280, 347–353, 410–415, 492–497, 560–565; Max Jakob, *Heat Transfer*, Wiley, New York, 1949; Marks, *Mechanical Engineers' Handbook*, McGraw-Hill, New York, 1951 pp. 366–392.

in which q is the flow of heat per unit time from the end at higher temperature T_1 to the one at lower temperature T_2. The law states that q is proportional to temperature difference and cross-sectional area A, and inversely proportional to length L. The dependence on area and length

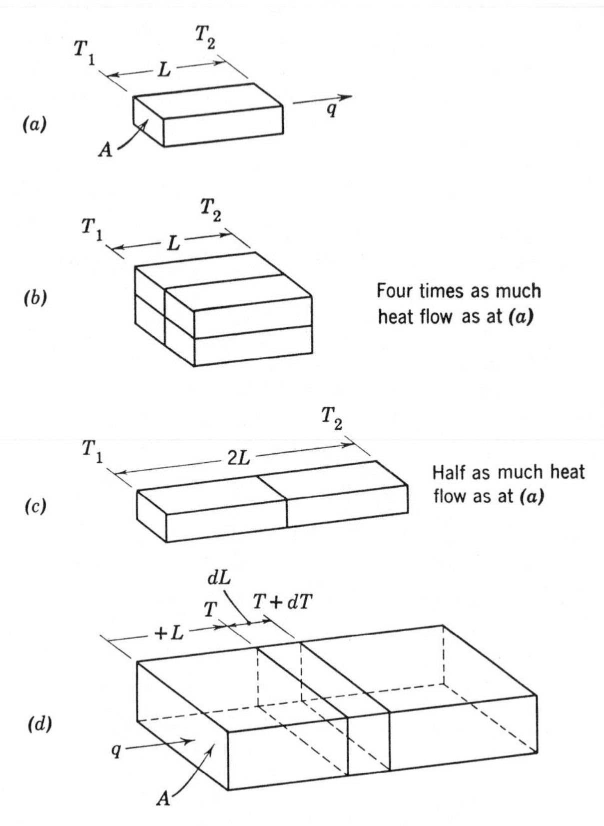

Fig. 3-21. Illustrating Fourier's law of heat conduction.

seems reasonable if we think of four bricks combined in parallel as at b in figure 3–21 or two in series as at c. The factor of proportionality k is the thermal conductivity of the material. Various systems of units are used in equation 3–106. Four common ones are shown in the accompanying table.

q	T	A	L	k
Btu hr^{-1}	°F	ft^2	ft	Btu hr^{-1} °F^{-1} ft^{-1}
Btu hr^{-1}	°F	ft^2	in.	Btu hr^{-1} °F^{-1} in. ft^{-2}
Cal sec^{-1}	°C	cm^2	cm	Cal sec^{-1} °C^{-1} cm^{-1}
Watts	°C	in.^{2}	in.	Watt °C^{-1} in.$^{-1}$

We have already seen a use of Fourier's law in the non-fogging mirror problem, section 2–3, where the law is stated symbolically in equation 2–19, which happens to be solved for temperature drop instead of for heat flow.

Another mathematical statement of Fourier's law may be obtained by considering a path of differential length dL as shown at d in figure 3–21, which is an enlarged picture of the brick. Then, substituting T for T_1, $T + dT$ for T_2, and dL for L in equation 3–106, we obtain

$$q = -kA\frac{dT}{dL} \tag{3–107}$$

Here dT/dL is the temperature gradient. The minus sign means that, if the gradient is positive, that is, if temperature T increases with L, the heat flow will be opposite to the positive direction of L. In other words if q is in the direction shown in the figure, dT/dL must be negative, that is, temperature falls with increase in L. Equation 3–107 is more general than 3–106 for it is not restricted to paths where q and A are constant along a given length as in the cases shown in figure 3–21; nor need k be independent of temperature.

Evidently, when k is constant, Fourier's law of conduction gives a linear relation between flow of heat and temperature difference; the thermal resistance in a diagram like figure 3–20 is constant. In many problems, perhaps most, k is nearly enough constant so that we do have linearity, but in a few materials account must be taken of the variations of k with temperature, especially when a great variation of temperature is encountered, and thus k at a point of high temperature is significantly different from k at a point of low temperature.

Let us next consider the loss of heat from the surface of a body by convection and radiation, dealing with radiation first. After we have considered these two processes separately we shall see that in some situations it is convenient and reasonable to treat the two together as total surface loss by means of the approximation known as Newton's law of cooling, according to which the heat energy lost per unit of time from unit area of the surface is directly proportional to the rise of temperature of the surface above its surroundings.

Radiation

Radiation is the process by which heat is transmitted through space by electromagnetic waves excited by molecular vibrations in the radiating matter. The intensity of the radiation depends on the fourth power of the temperature of the radiating body and on the emissivity of the radiating surface. Heat waves behave in general like light; indeed, light comprises that small range of frequencies in the total radiation spectrum which happen to affect human eyes. As the temperature of a body is raised an increasing amount of its thermal radiation is clearly visible, and is what makes us say a body is red hot or white hot.

Calculation of radiation heat transfer was illustrated in the non-fogging mirror problem, section 2–3, for a simple, but none the less important, case. Equation 2–18 for that case is repeated here in slightly different form:

$$q = \sigma A \epsilon (T_1^4 - T_2^4) \tag{3–108}$$

This expression gives the radiation heat transfer between a convex body and a concave enclosing body assuming they act as gray bodies and that the surface area of the enclosing body is very much larger than that of the body enclosed. The heat power transferred is q and is positive in the direction from the body at absolute temperature T_1 to the body at absolute temperature T_2. A is the surface area of the small enclosed body, ϵ is the emissivity of this surface, and σ is the Stefan-Boltzmann constant.

The assumption of gray-body radiation is justified in many engineering situations. It means what the word gray implies in the case of visible radiation: that there is no selective absorption or reflection of certain frequencies of radiation such as is responsible for visible color. A body which absorbs all incident radiation is said to be a black body, and such a body is also an ideal emitter of radiation; at a given temperature it radiates at the highest possible rate. A gray body emits a fraction of the radiation it would if it were black, the fraction being the emissivity ϵ. The frequency distribution of the radiation from a gray body is the same as for a black body. The fraction of incident radiation absorbed by a gray body is also equal to its emissivity.

Equation 3–108 assumes that no radiation is absorbed by the gas in the space between the bodies. This assumption is justified for air and many gases but not when there are large amounts of carbon dioxide or water vapor present. The presence of luminous gases or flames also makes equation 3–108 inapplicable.

In general, the calculation of radiation heat transfer between bodies involves the surface areas and emissivities of all bodies present and not just those of one body as in equation 3–108. However the radiation in general is always proportional to the difference in the fourth powers of the absolute temperatures.

Convection

Convection is a process in which a liquid or a gas in motion conveys thermal energy to or from the surface of a solid body. In the cooling of a solid by this process heat is conducted through a thin layer or film of fluid. This warms the moving fluid and thus the heat is carried away as an increase in the internal energy or sensible heat of the fluid. If the solid body is cooler than the fluid the heat transfer is in the reverse direction. The motion of the fluid may be self-induced by the buoyancy of hotter fluid, and the convection is then said to be *free* or *natural;* or the motion may be caused by a fan or pump, and this is called *forced* convection.

In contrast to conduction and radiation which obey well-known and relatively simple laws, our knowledge of convection is less exact and contained in empirical formulas.

The basic relationship in convection heat transfer is often written as

$$q = hA(T_s - T_f) \tag{3–109}$$

where q is the heat flow, positive when it is in the direction from solid surface to fluid, h is an empirical coefficient to be discussed below, A is the area of the surface exposed to the fluid, and T_s and T_f are the temperatures of solid surface and fluid respectively.

The factor h in 3–109 is called the heat transfer coefficient, or sometimes the film coefficient in recognition of the fact that most of the temperature drop $T_s - T_f$ occurs in a thin film of fluid adjacent to the heat transfer surface. Also, because temperature differences from point to point in the fluid outside the thin surface film are small compared to that across the film, it does not matter much where in the bulk of the fluid the temperature T_f is measured.

Empirical formulas for the convection heat transfer coefficient h pertaining to various important cases are available. We have met an example in the mirror problem where 2–17 gives h for free convection from vertical plane surfaces in air. As may easily be imagined from our description of the convection process, h depends on the thermal conductivity and the specific heat of the fluid. It also depends, less obviously, on the size and geometry of the heat transfer surface. Since motion of the fluid is involved, density and viscosity of the fluid also influence h. In free convection h depends upon the coefficient of volume expansion and upon the temperature difference $T_s - T_f$; in forced convection h does not depend on these quantities but does depend on the velocity with which the fluid is forced past the surface. Empirical formulas more fundamental than 2–17 take explicit account of the various factors affecting h. Also, the methods of dimensional analysis referred to in section 7–9 have proved to be very useful in correlating and expressing the data.

Convection heat transfer may involve a change of state of the fluid, as for instance in a steam boiler or in a machine for manufacturing ice. Such situations are more complicated than those discussed above but are also handled empirically in the same general fashion.

Newton's Law of Cooling

The total loss of heat from a surface may be expressed over a moderate temperature range as a linear law, at some sacrifice of accuracy, but with a great gain in simplicity. Let us see how this works for a particular case, that of the mirror surface of section 2–3, for which loss by radiation is given by 3–108 and loss by convection by 3–109. Employing numerical values from section 2–3 and using equation 2–17 for h we have for the two components of loss, in Btu hr^{-1} ft^{-2},

$$\frac{q_r}{A} = 119.4\left[\left(\frac{T_s + 460}{525}\right)^4 - 1\right] \tag{3–110}$$

$$\frac{q_c}{A} = 0.27(T_s - 65)^{1.25} \tag{3–111}$$

in which T_s is the temperature of the mirror surface in °F and the temperatures of the surrounding air and walls are both taken as 65°F. These components and their sum are plotted as functions of T_s in figure 3–22, and this shows how nearly linear the relationship is over the whole range shown, of 65 ± 60°F. In general the smaller the range of temperature the better the approximation. Thus we often find it convenient to use a linear relation for surface loss

$$q = hA(T_s - T_0) \tag{3–112}$$

which is like equation 3–109 except that in 3–112 h is regarded as a constant and T_0 is the temperature of the heat sink. This is Newton's law of cooling. When forced convection predominates in surface loss the law holds with relatively great accuracy for then h is independent of temperature difference, and 3–109 is a linear law, instead of the five-fourths power law that is shown in figure 3–22.

EXERCISE. Explain why the fourth power radiation law when plotted in figure 3–22 is so nearly a straight line.

Processes in Combination

As we have seen, conduction, radiation, and convection all may occur in a single situation.

In solids the only process which needs to be considered is conduction. Most glass is relatively opaque to the total thermal radiation in spite of its visual transparency. Porous insulating materials may have radiation and convection within the interstices, but the tabulated values of conductivity account for these internal processes as well as for the true conduction within the solid itself. However, when one conductor is in ordinary contact with another and heat flows across the interface it is likely that the inevitable small air film, and perhaps also an oxide film if we deal with a

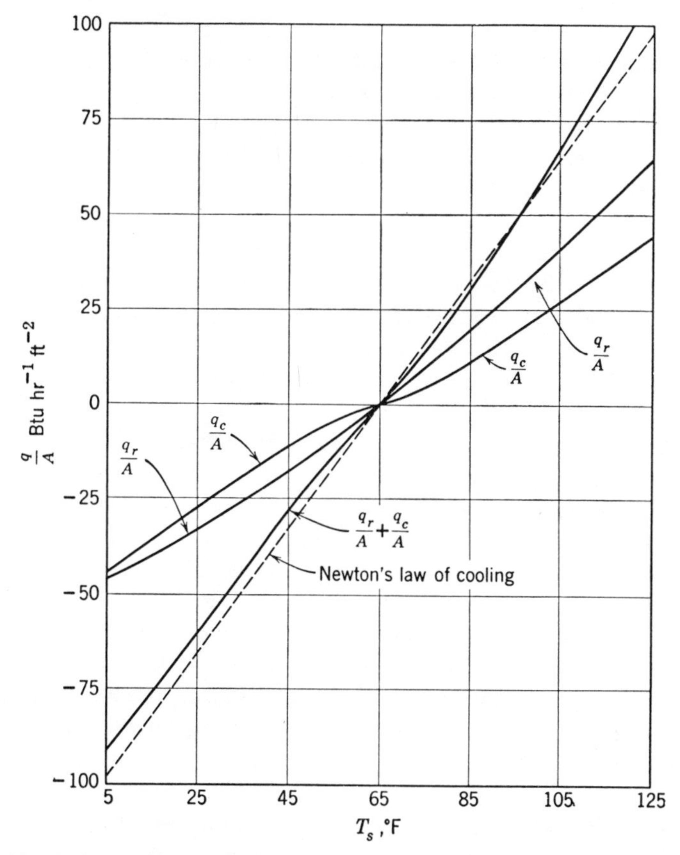

Fig. 3-22. Newton's law of cooling; convection and radiation loss from glass surface approximated by the dashed straight line.

butt joint between metals, will introduce an appreciable thermal resistance which must be taken into account.

Through spaces occupied by gases, radiation and convection both take part in the heat transfer, and either may be the more important depending on the conditions. In liquids convection is the only process which normally needs to be considered. Except when the fluid is present only in a very thin layer, conduction need not be considered separately in gases and liquids; it takes place, of course, as part of the convection and is accounted for in the formulas for that process.

3–11 Importance of Economic and Human Factors

In this chapter we have tried to do two things: show the kind of understanding that an engineer must have of the principles he uses in solving problems, and in so doing actually review principles of especially wide application in dynamics, electric circuits, and heat flow. There are, of course, other principles that must be understood, and even in the fields discussed we have given only a few examples.

This chapter has emphasized physical principles, but the great importance of economic and human factors in engineering problems must not be overlooked. The engineer working in industry must solve his problems in such ways as to yield profits and maximize them for his company: he must design or help to produce a product which the potential customer will buy, and he must strive toward reduction of manufacturing costs by saving material and labor or by simplifying design. Even in public works and on military projects the ultimate purpose of engineering is still the same: to produce something of value at minimum cost, the value and cost in such cases being to the whole community rather than to a private enterprise. No matter how well a solution to a problem may satisfy all the physical conditions which are imposed, the solution is worthless if the result is not expected to be a net gain to the business or the community commensurate with the costs likely to be incurred. Judging whether or not a solution has economic value may be difficult for a young engineer who has not yet achieved a place in the organization from which he can clearly see the enterprise as a whole; yet he must learn to make such judgements.

Engineering analysis must also take human behavior into account. Are the controls on the machine placed so the operator can manipulate them easily, and can he see the meters clearly? Will Mr. America remember to turn on his non-fogging mirror? And if he forgets to turn it off, is little Johnny America going to burn his fingers? Is the girl who tests relay heaters going to get her hand caught in the rig of figure 2–9 if her foot slips off the pedal? Will the music-loving public accept a phonograph in which playing begins before the record is quite up to speed? Questions such as these must be considered in connection with a surprisingly large proportion of all engineering problems.

Human and economic factors also enter engineering analysis in another way. When the engineer has finished his technical analysis of a problem he still has to convince others of the value of his decisions and recommendations. It is disappointing but true that results rarely sell themselves to the engineer's associates, his superiors, and the public; and it is a further disappointment to find that there is no one who has the background to do the selling as well as the engineer himself. In presenting his conclusions to other people he must take into careful account their interests, points of view, educational background, and in short everything about them that may bear on his efforts to communicate his findings and win acceptance for his recommendations.

Finally, in dealing with a problem the engineer must consider the cost of his own time. He must work to no greater accuracy than the data and end results warrant, and he must expend no more and no less of his own time and effort than are appropriate to the ends he expects to achieve.

CHAPTER FOUR

Translation into Mathematics

4–1 The Place of Mathematics in Problem Solving

After planning to apply some fundamental principle to the solution of a problem comes the necessity for executing the plan so as to draw a correct conclusion. Correctness of conclusions requires correct application of principle followed by flawless reasoning. If the application of principle to the situation at hand can be reduced to mathematical terms, the burden of making the deductions so as to reach logical conclusions may be reduced enormously; indeed, without mathematics, making correct deductions often would be impossible. Mathematical symbolism provides compactness and accuracy of expression, and the rules of mathematics aid tremendously in the reasoning.

Thus mathematics has inestimable power for increasing your capacity for making logical deductions from a set of facts, but it cannot supply the facts nor can mathematics distinguish between right facts and wrong ones. For the mathematics to be useful the original formulation must be a true expression of the right facts. Moreover, after the mathematical manipulations are done correctly, the mathematical conclusion must be interpreted properly in the language appropriate to the occasion.

This chapter is concerned with translation into mathematics, that is, with setting up equations; carrying out some kinds of mathematical operations and interpreting mathematical results are discussed in chapters 5 and 7 respectively.

4–2 Exact Statements in Words

Unless one is sure, beyond a shadow of doubt, that a familiar routine method is applicable, he should never attempt to go directly from the general concept of a plan of solution to a mathematical formulation. A vitally important intermediate step is to make a clear verbal statement in correct English of just exactly what the principle or concept on which the plan is based implies about the situation at hand. Examples of such statements occur throughout this textbook. Wherever there is a mathematical analysis, you will find just preceding its first equation or perhaps a few sentences earlier a statement in one or more sentences of just how some fundamental law or generalization applies in the particular problem under study. The first such statement occurs in the paragraph of section 1-2 preceding equation 1-1, the very first equation in the book. It states how one of Newton's laws of motion applies in the problem of the liquid-level accelerometer. It is important to note that the statement does more than say merely that $\Sigma f = ma$ applies; it says how this law applies, carefully identifying the forces, the mass, and the acceleration all in the specific application.

The reason for preceding mathematical formulations with clear and exact word statements is that careful writing is a powerful aid to clear thinking. The act of trying to write out in words clearly and unambiguously what you think you know to be true serves not only to clarify your thoughts but often to correct them. Trying to do this in mathematical symbolism during the first consideration of a problem involves several distinct but interdependent steps simultaneously: there are, for example, the implication of the principle in general, its application to the situation at hand, a consideration of what simplifying assumptions may be desirable, the decision as to what quantities to represent by symbols, the selection and definition of symbols, and the choice of coordinates with their positive directions. Making the best choices among the several sets of alternatives and keeping everything straight is likely to be too much to do correctly all at once except in situations which either are very simple or very familiar. Moreover, once a mathematical analysis is set in motion, you are apt to become so engrossed with the manipulations that effective thinking about the realities of the problem may cease. For this reason, it is often desirable to "talk your way" into a problem as far as possible before resorting to mathematics. By so doing a fuller understanding of the situation is likely to be gained than would be so with a more immediate formulation in mathematical terms.

Although making exact verbal statements has been discussed as a preliminary to a mathematical formulation, the import of the discussion is far more general than this. Many problems, often the toughest ones an engineer meets, are incapable of mathematical formulation. In such cases, you will be forced, if the problem is to be solved at all, to reason your way through from one carefully written verbal statement to another; and this may be extremely difficult to do when the necessity arises if you have not already formed the habit of using careful verbal expression to help and to consolidate your thinking.

In making these exact statements in correct English, you should remember that you are not doing it for a reader but for yourself; you are telling yourself precisely what the truth must be in the situation. A sure way to a fall is to assume that you know so much that such self-counseling is not needed.

4-3 Simplifying Assumptions

In analyzing any engineering problem, assumptions are made, consciously or unconsciously, to facilitate the analysis without restricting its generality too much. Studying the illustrations in this book, you will find assumptions being made sometimes at one stage of the analysis and sometimes at another, and this is characteristic of actual professional practice. What happens is that, as an analysis proceeds, difficulties or complications are perceived which make it necessary or desirable to simplify the situation or to limit the treatment, and this need may become apparent at almost any stage.*

Assumptions are sometimes made to simplify a complicated situation so as to make comprehension easier for the problem solver. The idea is to solve first a problem simpler than the real one in the hope that later through learning from the simplified case it will be possible to extend the solution to include the situation of real interest. For instance, in section 2–6, the phonograph-record-changer problem was solved first for the case of a single record dropping on the empty turntable, and then the analysis was extended to include the more general case where any number of records are already on the turntable. Again, in section 2–7, the behavior of an unrestrained bimetallic strip was analyzed as a preliminary to the study of the more practical case where restraints are present. Such simplifications made for the purpose of aiding comprehension are most likely to be introduced in the defining stage of the professional treatment of a problem.

More frequently simplifying assumptions are made to facilitate the mathematical work. For instance, a factor known to vary may be assumed constant simply because the mathematics can be carried through for the constant but not for the varying factor. Similarly, the omission of a factor entirely may make the difference between being able to get a result and not getting one at all. Clearly great care must be exercised in such cases to study the analysis critically so as to estimate the effects of the assumptions.

Need for assumptions to simplify the mathematics may become apparent in the defining or in the planning stages if, as is often the case, you are thinking ahead to what the mathematical solution will be like. On the other hand, such simplifying assumptions are often not introduced until the need actually arises during the mathematical execution of the plan.

* Cf. Function 1, *Simplification,* in section 1–4.

It must be remarked that all too often you may make important assumptions in an analysis without realizing it, and thus be led to think the solution is more general than it really is. This is particularly apt to happen when you plunge hastily into a mathematical formulation without having considered the true situation with sufficient care.

Making assumptions is an important part of the process of translation into mathematics. Its purpose is to bring the situation within range of your mathematical tools. Although such simplifications affect the accuracy of the results, they are nearly always necessary to at least some degree if results are to be had at all; and results with known uncertainties are better than no results.

4-4 Selection of Variables and Parameters

Translation of a physical idea into mathematics involves a decision as to what you wish to express in terms of what. For instance, in an electric network, you may want to find the current through the terminals, and must decide whether you want the current at every instant (that is, as a function of time), or, if it is an alternating-current case, whether the effective (root-mean-square) value of current as a function of effective voltage applied will suffice. Moreover, if the network has more than one path, several applications of Kirchhoff's circuit laws may be needed, and the original equations may have to involve currents in addition to the one sought. In such a case, you have another decision to make: whether to set up the equations in terms of branch currents or loop currents, or perhaps in terms of node voltages. The circuit equations can be set up correctly in accordance with any of the above choices of variables, but the forms of the equations and the difficulty of their solution in a particular case may depend on the choice.

Thus, even after you have decided on a principle and have stated precisely how it applies in the situation, there may still be considerable room for choice. As another instance, this one from mechanics, consider a system consisting of a number of rigid masses all restrained to motion in one straight line and coupled together by springs, and suppose it is desired to find the motion of one of the masses as a function of time after the system has been subjected to some disturbing force. As a basis for solution, it is planned to apply Newton's law of motion separately to each of the masses. In execution of the plan, word statements of the application of this law to each of the masses have been written; for a particular mass identified as number one, the statement is "the only forces acting on mass No. 1 are that due to gravity and that due to the one spring connected to No. 1; the resultant of these forces equals the product of mass No. 1 and its acceleration with respect to a fixed reference." But the decision remains to be made on the way to express the spring force and the acceleration. Should you locate each mass with respect to its original position or with respect to its equilibrium position, or would it be better to locate one mass with respect to some fixed reference and then locate the other masses relative to this one? Or perhaps it would be best to choose the forces in the springs as variables. As in the electrical example above any of these choices can lead to correct results, but the forms of the equations, the ease of solution, and the immediate usefulness of the result may depend importantly on which is decided upon. Experience may aid considerably in making the decision, but it is certainly not essential because part way through the mathematics you may learn from what you have done that another choice would have been better and that it may be worth while to start over again.

Besides the selection of variables, you must decide what parameters will be involved, and often there may be a choice. For instance, suppose one of the springs in the dynamical system discussed above is a cantilever beam. Should the parameters pertaining to this spring be the dimensions of the beam and its material properties; or should all these be lumped together as a spring constant, and if so, how should it be defined? Clearly if there is much mathematics to be done, the use of a single spring constant reduces the chance of misplacing a symbol in going from equation to equation and saves wear and tear on the pencil. If the effect of beam dimensions on the final result is desired, the dimensions can be introduced near the end of the mathematical work.

Sometimes, thinking too hastily, you may choose a parameter which is altogether inappropriate to the situation. A case in point is an electric circuit containing a coil. You might decide to characterize the coil by its resistance and inductance, and indeed, under many circumstances this might be quite proper, but not if the relation between current and magnetic flux in the coil is non-linear, for then, in general, the inductance will not be a constant, and it may be better to work with the flux linkage as a function of current. Surely you must take time to think out carefully whether or not a quantity has meaning before deciding to use it. In this connection, it often helps to try to answer the question, "How would I proceed to measure this quantity as used in this particular situation?"

Thus even after you have made the decisions on the principle to apply and on the way to apply it, you have still to decide what quantities are to be used in the mathematical formulation. Often there are a number of possibilities, and among those which are appropriate to the situation some may lead to simpler or more useful solutions than others. Great care should be taken to define precisely what is meant by each quantity you decide to use; care in definition can uncover wrong thinking.

4–5 Choice of Symbols

After you decide what quantities are to enter a mathematical formulation, you must select symbols to represent them. While this may perhaps seem to be trivial, it has some very important practical aspects. In the first place, there are well-established conventional uses of letters to represent most of the quantities met in engineering;* for example, t for time, m for mass, i for current, p for pressure.

Since most writers tend to follow these conventions, it is desirable for you to do likewise for this will make your technical reading easier and will facilitate checking your own work, as well as make it easier for others to follow.

Unfortunately, the standard letter symbols are not wholly satisfactory because there are not enough letters to go around. The difficulties are particularly apt to occur in the very problems with which this book deals, problems in new areas or in areas where two or more established fields merge. For instance, as has been said already, t is the conventional symbol for time, but some people like to use this same letter for temperature and others for torque. You may then decide that temperature can be θ and perhaps that L will do for torque. But L is already well established for inductance and θ for angle. A single problem concerning an electric regulating mechanism to control the flow of a fluid conceivably may involve all these quantities: time, temperature, torque, inductance, and angle, and perhaps also magnetic permeability and viscosity (both represented conventionally by μ) and electrical resistance and Reynolds number (both R in common usage). Thus it is obvious that you cannot rely entirely on the conventions; in each problem, you should use your own judgment; be conventional where you can and make your own choices where you must.

In a given problem, it is good practice to identify all quantities of one kind by a single letter and to distinguish between them by subscripts or primes. Thus a number of torques might be T_1, T_2, and so forth. Be systematic in assigning subscripts. For instance, the resistance, inductance, and capacitance of a single part of a network all should be identified by the same subscript. A thermodynamic substance in a particular state should have all its properties tagged similarly. Orderly and systematic assignment of letter symbols and subscripts is a great help in carrying out the mathematics and in the checking stage, as illustrated in sections 4–9 and 6–6.

* See, for example, the American Standards Association publications *Letter Symbols for Mechanics* (Z10.3), *for Hydraulics* (Z10.2), *for Thermodynamics Including Heat Flow* (Z10.4), *for Electrical Quantities* (Z10.8).

Subscripts which are the initial letters of quantities are sometimes easier to remember and thus more helpful than numerals, particularly in a lengthy analysis. This is illustrated in the record changer problem, section 2-6. There the subscript r was used for quantities pertaining to the record and the subscript t for the turntable. This made it far easier to keep in mind the physical meaning of the mathematics as it developed than would have been so with a colorless choice of subscripts such as 1 and 2 or a and b.

Double subscripts can be very useful in quick identification of quantities, and are especially valuable in defining mutual electrical quantities that belong to two different circuits or mechanical quantities that pertain to two different masses. A resistance traversed by two loop currents i_1 and i_2 may well be designated by R_{12}. The constant of a spring whose ends are fastened to masses a and b might be called K_{ab} to distinguish it from springs connected between other points. Obviously such usage serves no purpose if there is only one resistance or one spring present.

Use of double subscripts also is a device to show direction. Thus V_{10} may represent potential of point 1 with respect to point 0. With this convention $V_{01} = -V_{10}$.

Subscripts comprising whole words or abbreviations and sub-subscripts are used sometimes but unless clearly helpful they should be avoided. Their use certainly tends to vitiate the whole purpose of symbolism, which is compactness.

Besides using symbols to represent single factors as implied in the discussion so far, it is sometimes expedient to represent a combination of symbols by a single new symbol. Done wisely this can lead to helpful simplification. Done haphazardly, the result may instead be greater complexity in the end, or worse; the whole physical meaning of the mathematics may become so obscured as to render the work valueless.

A combination of factors which constitutes a dimensionless group may have an important physical significance in itself. A familiar example is the Reynolds number which occurs in the analysis of fluid flow. Such a combination of factors may often be represented usefully by a single symbol at least during the mathematical manipulations. Even in the result, it may have greater significance than the group of factors replaced.

Another instance of a group which may be replaced to advantage by a single symbol is the combination of resistance, inductance, capacitance, and frequency defined as alternating-current impedance.

Replacing a series of terms by a single letter merely because it is long is not justifiable except temporarily as during the clearing of an expression of fractions. It is highly probable that several long and complicated groups of terms will combine ultimately to some simpler form. Replacing the combinations by single letters may conceal the chance for such simplification. If real simplification of the end result is not possible, you will want to know it. It may mean that the original choice of variables and parameters was unwise, in which case you should try again; or it may mean simply that the result *is* complicated.

4-6 Constructing Equations

Translation into mathematics implies the forming of an equation of some sort. Doing this includes in general all the functions discussed in the preceding sections: making an exact verbal statement of principle as applied to the situation, perhaps restricted by simplifying assumptions; deciding on the factors in terms of which to express the quantities involved; and choosing symbols to represent the factors selected. But there is something more which has not been discussed so far; determining the algebraic signs of the terms in the equation.

Construction of the equation is bound up with the definitions of the factors and depends very much on the physical nature of the problem. The chief guide in the process is the verbal statement of principle as applied in the particular situation. Usually also of great help in determining the signs is some sort of diagram to represent the relationships among factors which are to enter the equations, for instance, a free-body diagram to show forces, a circuit diagram to show currents, or a diagram to show a thermodynamic system.

The process of constructing equations is best discussed further in terms of examples. Moreover, what is really meant by the preceding sections needs illustration. Accordingly the remainder of the chapter is devoted to illustrative problems, one each from three different fields: dynamics, heat flow, and electric circuits.

The illustrations are fragmentary in that the complete analyses of the problems are not given. The statements of the problems are in terms which could be reached in reality only after considerable thought. Similarly the deliberations leading to plans for solution are abbreviated. Elaboration commences in each case with the execution stage, and the thoughts pertaining to the translation into mathematics are given fully and explained at length. The first two examples are carried only to the point where an equation is formed. In the third case, however, for reasons made clear there, the mathematics is carried through to a result.

4-7 A Mass-and-Spring System with Coulomb Friction and Linear Damping

For the purposes of this illustration, we suppose that the motion of a machine part is being studied and that the analysis has proceeded to the point where the actual mechanism has been idealized by the simplified arrangement shown diagrammatically in figure 4-1. It consists of a rigid mass constrained to vertical motion by guides, the motion being restrained by a linear spring and a dashpot. The dashpot gives a damping force directly proportional to the velocity of the mass. Such a device is said to be linear because the damping force varies as the first power of the velocity (i.e., linearly with velocity). Also, there is friction between the mass and its guides which is constant in magnitude and independent of the velocity except that it changes sign when the direction of motion changes. The mass is set in motion by displacing it up or down and then releasing it from rest. It is desired to find the position of the mass as a function of time after the mass is released.

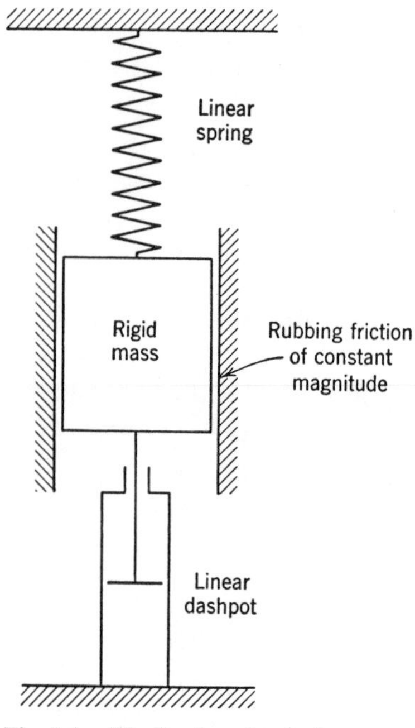

Fig. 4-1. Idealized mechanical system.

From a preliminary consideration, it is clear that there may be a wide variation in the kinds of motion that can occur, depending on the magnitudes of the quantities which characterize the system. Thus, in a limiting case where both friction and damping are reduced to zero, there would be left only a mass suspended by a spring, and the motion would be simple harmonic, that is, sinusoidal with respect to time. On the other hand, if the friction were made great enough, there might be no motion at all, especially if the spring was relatively weak. We are interested in the general case including both these extremes as limiting conditions.

To break into an analysis, we assume that the mass indeed does describe some sort of up-and-down motion. If the mass moves when released from rest, it must be accelerated. Also, if it moves up and down, its velocity sometimes must be increasing and sometimes decreasing, and this implies varying acceleration. Moreover, the acceleration results from the forces which act, and we see that at least some of these depend on the position or velocity of the mass. Thus from this point of view also we anticipate a varying acceleration. Consequently, for a quantitative analysis of the motion, we are led to apply Newton's law to the motion of the rigid mass so as to obtain a relation between the forces which act and the acceleration which results. Applying the law to this situation, we carefully write out the following statement:

> At every instant in the motion, the resultant of the vertical forces acting on the mass equals the product of the mass and its acceleration with respect to a fixed reference. The acceleration is in the direction of the resultant force. There are four forces acting on the mass: those due respectively to the actions of gravity, the spring, the dashpot, and friction with the guides.

To translate this verbal statement into mathematics, we first must decide what variables and parameters are appropriate.

The quantity we want to find is the vertical position or displacement of the mass with respect to some reference point at any instant of time. The acceleration called for in the verbal statement is the second derivative with respect to time of vertical distance of the mass from any fixed point. Thus we choose as a variable the vertical displacement of the mass with respect to a fixed reference. We reserve until later the decision as to what fixed point to choose and whether to take the positive direction up or down.

Since we seek position of the mass as a function of time and since time also enters the acceleration, we take time as another variable and decide to measure it from the instant the mass is first released.

Now we consider in what terms to express the forces which act. The force due to gravity is simply the weight of the mass, and therefore a constant independent of the motion.

The force due to the spring is directly proportional to the change in length of the spring from that when relaxed; to fix this idea we make a sketch, figure 4–2. The slope of this diagram is what is variously known as the constant or gradient of the spring. Clearly it is one of the parameters of our problem. Also, we note that extension of the spring and displacement of the mass become identical if the reference for measuring position is where the mass is when the spring is relaxed. To take advantage of the resulting simplicity in the expression for spring force, we decide to adopt as the fixed reference the position of the mass where the spring is relaxed.

Turning now to the force due to the dashpot, we construct another sketch, figure 4–3. The slope of this diagram is a constant which evidently can be another parameter. Furthermore, the piston velocity relative to the cylinder is the same in magnitude as the absolute velocity of the mass. Thus the dashpot force can be expressed simply in terms of the time rate of change of the displacement of the mass.

As the last of the four forces involved, we consider the friction with the guides. This has been assumed constant in magnitude, but clearly its

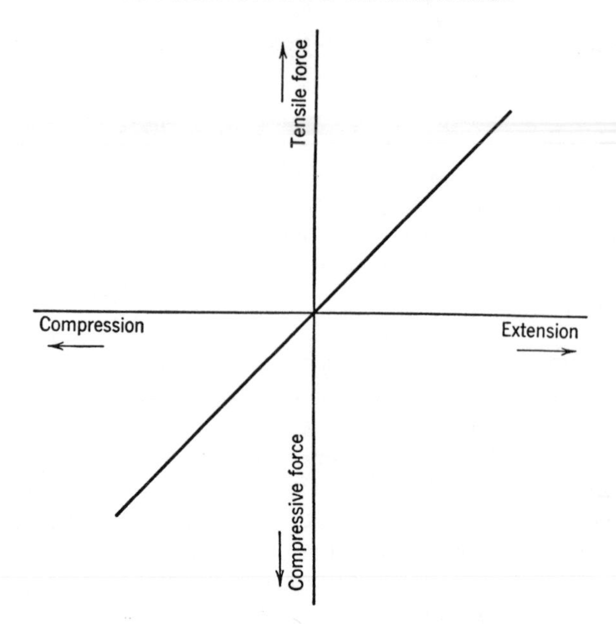

Fig. 4-2. Characteristic of the linear spring.

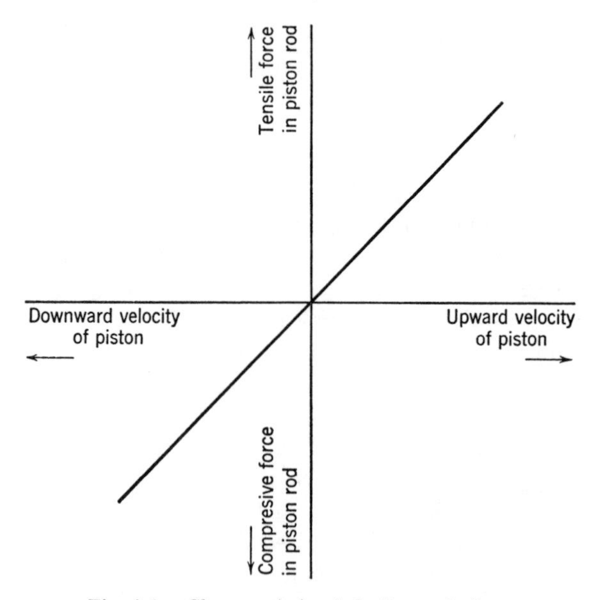

Fig. 4-3. Characteristic of the linear dashpot.

direction is such as to oppose the motion; therefore its direction depends on the direction of the velocity of the mass relative to the fixed guides. The friction force thus is a function of velocity, as shown in figure 4-4. This characteristic is like those already met in the record-changer problem, section 2-6. As explained there, in this case also, the friction at zero

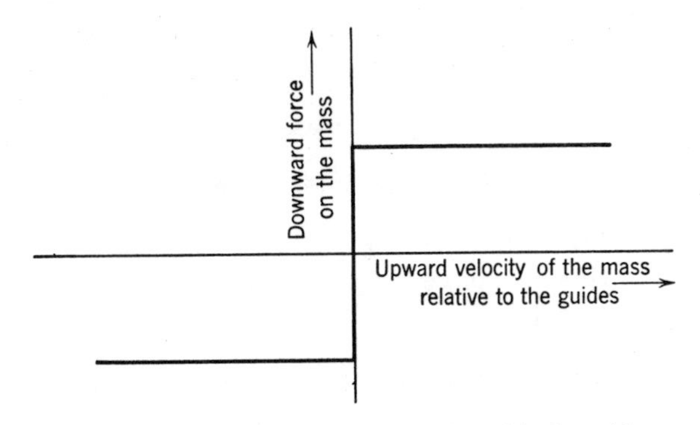

Fig. 4-4. Characteristic of the friction with the guides.

velocity can have any value at all between the positive and negative constant values. We choose the magnitude of the constant value as another parameter.

Having fixed on the factors to enter the mathematical formulation, we are ready to adopt symbols to represent them. In defining the symbols carefully, we expect to discover and clear up doubtful points and ones perhaps entirely overlooked. Proceeding, we write

[Let x = displacement of a point on the mass measured from the position this point occupies when the spring is relaxed, positive downward (feet).]

Notice that we have chosen to measure x positive downward, having discovered no reason favoring one direction over the other. The choice is arbitrary, and really quite unimportant except that, once made, it must be observed with absolute consistency in all that follows.

[t = time measured from the instant the mass is released (seconds).
M = value of the mass (slugs).
g = acceleration of gravity (ft sec^{-2}).]

Here we have in mind expressing the weight as Mg. We might equally well have chosen to introduce a symbol for the weight and to have omitted

g from the list. Or, we might have chosen to use g and the weight and to have omitted M.

[
K = constant of the spring (lb ft^{-1}).
D = constant of the dashpot (lb sec ft^{-1}).
F = magnitude of the friction force while sliding occurs (lb).
]

These, we think, are all the symbols we will need. We have not introduced a symbol for the velocity of the mass even though it is an important variable in the problem. Rather we intend to express this velocity as $\frac{dx}{dt}\cdot$ In so doing, we observe that, by the mathematical definition of a derivative, the positive direction of velocity is the same as the positive direction of displacement,* in this case chosen as downward. Similarly, we need no symbol for acceleration, for we can express it more usefully here by $\frac{d^2x}{dt^2}\cdot$ Again, by the definition of a derivative, the positive direction of the acceleration is the same as that of velocity, which in turn is the same as that of displacement.

At long last, we are almost ready to start writing an equation. One more thing, however, remains to be decided. Newton's law of motion, as we observed in our verbal statement, holds at every instant; but it appears that all instants are not alike. Sometimes the mass is displaced upward and is moving (has velocity) upward; at others it is moving downward while displaced either up or down; or it may be displaced upward and be moving downward. But, as should be clear now, the directions of the separate forces are dependent on the directions of displacement and of velocity. Therefore, what kind of an instant do we select for writing the equation? The answer is that we make an arbitrary choice from among the possibilities and then check to see what differences, if any, there will be for the other possible combinations of directions of displacement and velocity.

For simplicity we choose to consider first an instant when both displacement and velocity are positive; that is, we suppose the mass to be below the reference point and continuing to move further downward. This choice is for convenience and is made arbitrarily without worrying about whether or not the initial displacement is such that both displacement and velocity ever would be positive at once. Whether this condition actually will occur is something we are going to let the mathematical analysis tell us. For the moment, we are concerned only with seeing what Newton's

* If you are not already familiar with this you should satisfy yourself that it is true.

law says must be true if displacement and velocity are both positive at once.

To help us fix the signs in our equation, we draw the diagram, figure 4–5, showing the mass as a free body acted upon by the four forces. The arrows show the directions of the forces on the body when the displacement and the velocity are positive. The spring force Kx acts upward because, when the mass is displaced downward as shown, the spring is stretched and hence pulls upward. The dashpot force $D\,\dfrac{dx}{dt}$ and the friction force F act upward because each acts to oppose the motion (velocity), which is downward. From the diagram it is clear that all the forces except the weight Mg act upward. Hence the resultant downward force is $+Mg - Kx - D\dfrac{dx}{dt} - F$. Consequently the mathematical expression of Newton's law is

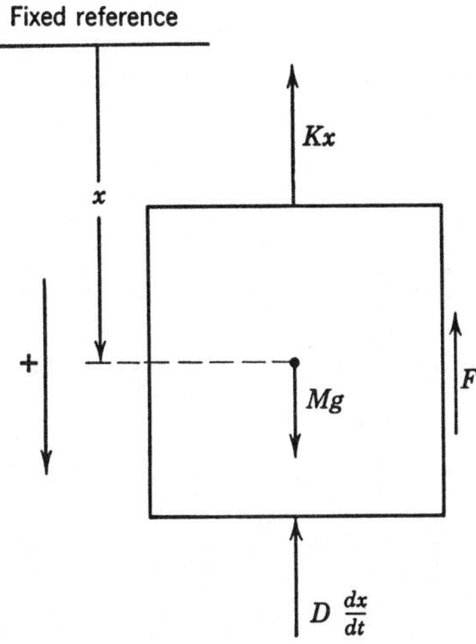

Fig. 4-5. Free-body diagram for an instant when both displacement and velocity are positive.

$$Mg - Kx - D\frac{dx}{dt} - F = M\frac{d^2x}{dt^2} \tag{4–1}$$

Notice that we have not considered the sign of the acceleration $\dfrac{d^2x}{dt^2}$; this equation determines it. Clearly it depends on the magnitude of Mg relative to the other three terms on the left. Incidently, we see a check here, for, in the absence of the spring, the dashpot, and friction, 4–1 becomes the equation for a freely falling body, as it should.

Now we must study 4–1 to see if it applies to instants when x and $\dfrac{dx}{dt}$ are not both positive. To do this, we consider the terms on the left, one at a time.

First, the term Mg representing the weight will always be positive since the force of gravity is downward regardless of the position of the mass or which way it happens to be moving. The term $+Mg$ then stands as it is for every instant in the motion.

Next we have the spring force. This was set up as $-Kx$ on the basis that when the displacement is downward the spring is stretched and therefore exerts an upward or negative force on the mass. Suppose now that x is negative. In this case, the spring will be compressed and thus will push downward on the mass, that is, exert a positive force. The term $-Kx$ is thus correct in this situation too, for, when x is negative, the value of the term will be positive, as it should be. Thus the term $-Kx$ in 4–1 is correct for all instants regardless of the sign of x.

EXERCISE. Show that the term $-D\frac{dx}{dt}$ is also correct in general and not restricted to the case of positive $\frac{dx}{dt}$ for which it was set up.

Finally we come to the friction force term $-F$, and here there is something different. So long as there is motion, the friction force has the magnitude F, but when the motion is downward F acts upward, and when the mass moves upward F acts downward. That is, when $\frac{dx}{dt}$ is positive, the term is $-F$ as in 4–1, but when $\frac{dx}{dt}$ is negative it should be $+F$. But there is nothing to produce automatically the necessary reversal of sign as there is with the preceding two terms. Thus we are forced to conclude that, because of the friction-force term, a separate equation like 4–1 except with the sign of F reversed is needed when $\frac{dx}{dt}$ is negative.

Consequently we find that translation of Newton's law into mathematics leads in this case not to a single equation but to the following pair:

$$\left.\begin{aligned} &\text{For } \frac{dx}{dt} > 0, \qquad Mg - Kx - D\frac{dx}{dt} - F = M\frac{d^2x}{dt^2} \\ &\text{For } \frac{dx}{dt} < 0, \qquad Mg - Kx - D\frac{dx}{dt} + F = M\frac{d^2x}{dt^2} \end{aligned}\right\} \tag{4–2}$$

It is to be emphasized that this complexity comes from the way in which F depends on velocity and the inadequacy of mathematics to express more simply the innocent-looking function shown in figure 4–4. Actually the whole story is not contained even in 4–2 for the case $\frac{dx}{dt} = 0$ is not included.

EXERCISE. Work the case $\frac{dx}{dt} = 0$; it is important as it determines among other things whether or not motion will ever commence.

The occurrence of separate equations in 4–2 complicates the mathematical procedure for solving for x as a function of t, as is desired. Avoidance of the complication introduced by the separate equations is the reason why it is customary in dealing with vibrations to simplify matters by ignoring any constant friction forces that may actually be present. When this is done, it is clear that 4–2 reduces to a single equation throughout.

Exercise. Study this illustration further to see what would happen if the positive direction had been taken up instead of down, and if the origin of x had been taken somewhere else. There is, for example, the possibility of eliminating the term Mg from the equation by the use of a particular origin. Also, find the consequences of some other kinds of forces: for example, a damping force proportional to the square of the velocity, and a non-linear spring whose tensile force is proportional to the cube of its extension.

Our concern in this chapter is the translation of a problem situation into mathematical terms and not with carrying out the mathematical analysis itself, so we will not proceed further with this problem. We have set down in great detail what the whole process of translation into mathematics includes in this case. If the situation were in a field reasonably familiar to him, the practicing engineer working in good professional style probably would put on paper only what we have indicated by brackets, the figures, and the equations. Much of what it has taken whole paragraphs to explain here he might pass over almost intuitively. If, however, the problem is in a field new to him, and such is likely to be the case in professional practice, his argument might need to be as carefully made and as deliberate as this one has been. To keep such a long story straight, most of it would have to be written out.

4–8 Temperature of a Cooling Fin

As another illustration of the process of translating a physical situation into mathematics, we consider the problem of calculating the steady-state radial distribution of temperature in a cooling fin. The fin, shown in figure 4–6, is a thin rectangular piece of metal welded along one edge to the body from which heat is to be dissipated. Air is blown upward in the axial direction along both faces of the fin, parallel to the edge attached to the hot body. So great is the flow that we believe it legitimate at this stage to assume that the temperature of the air is not sensibly increased by the presence of the fin. Also, conditions are such that the temperature of the hot body may be assumed uniform along the line where the fin is attached. Consequently, the temperature of the fin will not vary in the axial direction. Furthermore, the fin is so thin that temperature variation in the direction of its thickness can be neglected.

Exercise. Justify this assumption on the basis of the numerical calculation worked out in connection with equation 2–20, section 2–3.

Thus the only possibility of temperature variation remaining is that with respect to distance radially from the attached edge; and so, by the simplifying assumptions which have been introduced and which are believed to be reasonably close to the truth, the problem has been reduced to that of finding the temperature distribution with respect to one dimension only.

Thinking further about what happens physically, we see that two kinds of heat transfer take place. First there is conduction of heat through the

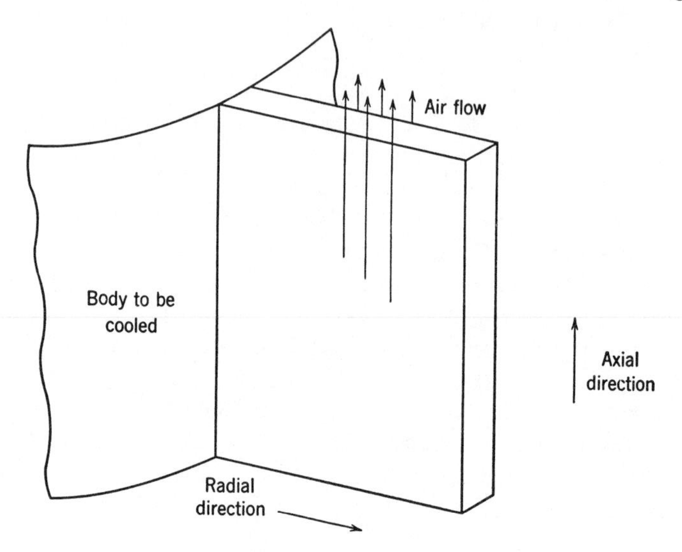

Fig. 4-6. Cooling fin.

metal. This heat is conducted radially outward by virtue of the gradient in the very temperature distribution which we seek. We should be able to deal with it quantitatively by applying Fourier's law of heat conduction. But there is a second kind of heat flow going on too. Heat is carried away by the air from the exposed surfaces of the fin. The laws of forced convection should be applicable here. Radiation will be assumed negligible in this case in comparison to the forced convection.

Somehow relating the flow of heat by conduction to that by convection should lead to a useful solution. From the law of conservation of energy, the heat conducted into the fin at its attached edge must be exactly equal to the total of all the heat conveyed away from the exposed surfaces, because in the steady state there can be no change in the thermal energy stored in the fin. This, however, is not an attractive basis for solution because we foresee that expressing the total of the heat dissipated from the surfaces will place the temperature distribution we seek in an integral relationship, and we think this may lead to difficulty. The idea of applying the conservation of energy, however, is basically sound, and, if we select as a system not the whole fin but a small element of it, we think the result will be useful.

For the system to which to apply the conservation of energy we choose an element of the fin such as that shown in figure 4-7. It is a slice of differential thickness cut out by two planes normal to the lateral surfaces and parallel to the attached edge. Since the dimension in the direction of

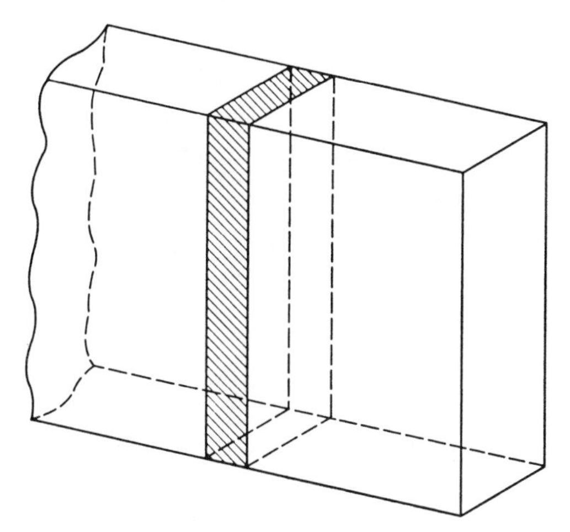

Fig. 4-7. An element of the fin.

the temperature variation is vanishingly small, the temperature can be taken as the same for all parts of the element.

Then, as we are considering a steady state, the principle of conservation of energy tells us that there can be no net flow of heat into the element. Being more specific, we make the following verbal statement of the principle as applied:

The heat conducted per unit time into the element through the metallic interface nearest the attached edge equals the sum of that conducted out through the opposite face and that conveyed away by the air from the exposed surfaces of the element.

To translate this statement of physical fact into a mathematical expression, we have to decide what terms are best for our purpose of determining the distribution of temperature along the fin.

The heat conveyed to the air from the sides of the element is proportional to the difference between the fin temperature and the air temperature. The heat conducted into the element from the adjacent metal is proportional to the temperature gradient, that is, rate of change of fin temperature with respect to distance radially along the fin. Thus appropriate variables will be the temperature of a point on the fin and the distance of the point from some reference.

Expressing the heat transferred by convection will involve also a coefficient of heat transfer and the exposed area of the element. The latter can be expressed in terms of the differential thickness of the element and the dimensions at right angles to this. These dimensions also will enter the expression for heat conducted, as will the thermal conductivity of the metal.

Having in mind now the quantities needed for the mathematical expression we select symbols and make precise definitions as follows:

T = temperature of the fin, a function of distance radially along the fin (°F).
T_a = temperature of the air flowing past the fin, a constant (°F).
x = radial distance along the fin measured positive outward from the attached edge (ft).
h = coefficient of heat transfer by forced convection from the surface to the air (Btu hr^{-1} ft^{-2} $°\text{F}^{-1}$).
k = thermal conductivity of the metal (Btu hr^{-1} ft^{-1} $°\text{F}^{-1}$).
b = thickness of the fin (ft).
w = width of the fin in the axial direction (ft).

Going back to the verbal statement we form expressions for each of the three terms it describes, and as we do so we construct the sketch figure 4–8.

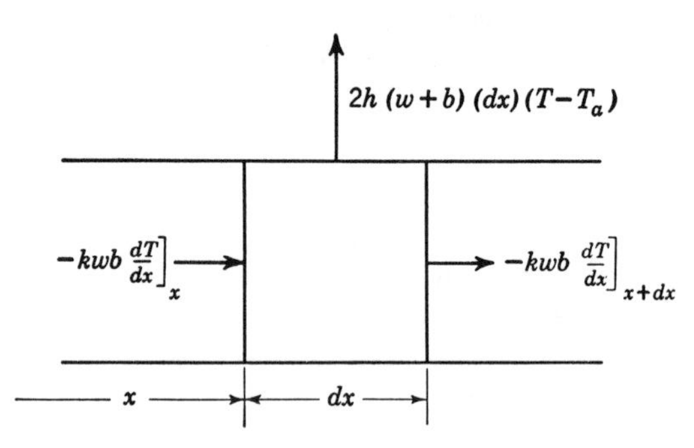

Fig. 4-8. Components of heat balance shown diagrammatically.

The heat conducted into the element through the left-hand metallic face is, by Fourier's law,

$$-kwb\,\frac{dT}{dx}\Big]_x$$

In writing this expression, we proceed as though T were an increasing function of x. If this were true, heat would flow in the direction of decreasing x since heat flows toward lower temperature. But we want an expression for heat flow toward increasing x, that is, into the element from the left; hence the minus sign. The heat conducted out of the element through the right-hand face similarly is

$$-kwb \left.\frac{dT}{dx}\right]_{x+dx}$$

The symbols following the derivatives mean that the gradients are evaluated at the faces of the elements located by x and $x + dx$ respectively.

The heat transferred away by convection from the sides of the element is

$$h(2w + 2b)(dx)(T - T_a)$$

The equation expressing the heat balance is now clear. It is

$$-kwb \left.\frac{dT}{dx}\right]_x = -kwb \left.\frac{dT}{dx}\right]_{x+dx} + 2h(w + b)(dx)(T - T_a) \quad (4\text{–}3)$$

Although 4–3 seems to complete the translation of the physical facts into mathematics, it is not satisfactory. It is true only when the dimension of the element dx is made to approach zero, for only then is it legitimate to assume, as has been done, that all parts of the element are at the same temperature. But if dx approaches zero in 4–3, the first two terms approach equality, or, from another point of view, their difference is a differential quantity which approaches zero. The third term also approaches zero. Hence the equation as it stands fails to disclose anything useful. To overcome this difficulty, we solve 4–3 for the ratio of the differential quantities; thus

$$\frac{\left.\frac{dT}{dx}\right]_{x+dx} - \left.\frac{dT}{dx}\right]_x}{dx} = \frac{2h(w + b)}{kwb}(T - T_a) \quad (4\text{–}4)$$

Now when dx approaches zero, the limit of the left-hand side is recognized as the definition of the derivative of the quantity $\frac{dT}{dx}$ with respect to x, that is, $\frac{d^2T}{dx^2}$. And thus in the limit when the element is made vanishingly thin, 4–4 becomes

$$\frac{d^2T}{dx^2} = \frac{2h(w + b)}{kwb}(T - T_a) \quad (4\text{–}5)$$

A solution to this second-order differential equation is the function T of

x, which is sought. We will not obtain the solution here, however, since our immediate concern is limited to the philosophy of reducing physical situations to mathematical terms. Accordingly let us investigate modifications which might have been made in the foregoing translation to mathematics.

First we observe that a simplification would result by taking the reference for fin temperature as the air temperature instead of the zero of the Fahrenheit scale. Thus letting

θ = temperature rise of a point on the fin above the air temperature in Fahrenheit degrees ($\theta = T - T_a$),

it is clear that instead of 4–5 we would get

$$\frac{d^2\theta}{dx^2} = \frac{2h(w + b)}{kwb}\,\theta \tag{4-6}$$

Solution of this equation is discussed in sections 5–3, 5–4, and 5–5.

Next we notice that, if the thickness of the fin b is small compared to the width w, we might drop b from the sum $(w + b)$ and instead of 4–5 get

$$\frac{d^2T}{dx^2} = \frac{2h}{kb}(T - T_a) \tag{4-7}$$

which is independent of the width w.

Then examining the derivation of 4–5, we notice that, instead of the parameters w and b, we might just as well have used the cross-sectional area of the fin $A = wb$ and the perimeter of its outer surface $P = 2(w + b)$. In these terms the equation would be

$$\frac{d^2T}{dx^2} = \frac{hP}{kA}(T - T_a) \tag{4-8}$$

Furthermore, it is clear that 4–8 is in fact more general than 4–5 in that its derivation need not be restricted to a fin of rectangular cross section. It could, for example, apply to a rod of circular cross section protruding from the hot body. For this case, though, we would need to reexamine the assumption that all temperature gradients in the metal could be neglected except that in the x-direction. Also, the value of the heat-transfer coefficient h would be different than for the fin with plane surfaces.

Equations 4–6, 4–7, and 4–8 are all similar to 4–5 in that they involve temperature as the dependent variable. Precisely the same physical facts, however, can be translated into quite different terms. For example, suppose it is desired to find not the temperature distribution along the fin but rather how the heat conducted radially depends on distance from the edge.

To do this, we introduce in addition to the quantities defined already:

q = heat conducted radially outward in the fin at distance x from the edge (Btu hr^{-1}).

Then, working with precisely the same verbal statement of the conservation of energy as applied to the element, we sketch figure 4–9 and write the following equation:

$$q = q + dq + 2h(w + b)(dx)(T - T_a) \tag{4–9}$$

Notice that, in writing this equation, we proceeded as though q were an increasing function of x although we know physically that in this case it

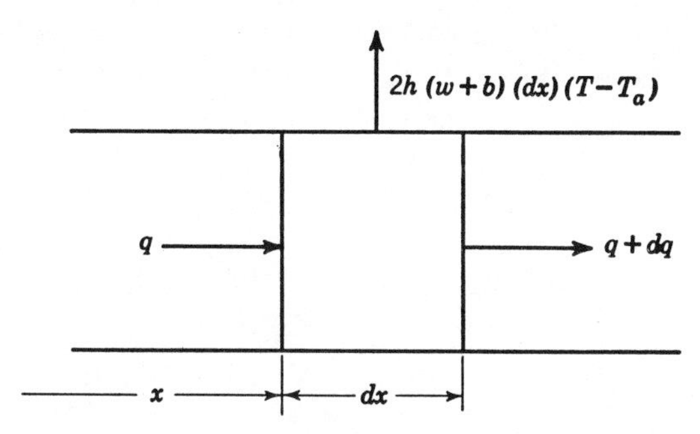

Fig. 4-9. Heat-balance diagram.

decreases with x because of the heat conveyed away from the sides of the fin. Thus, we expect the mathematics to show us that the rate of change of q with respect to x is negative. And this turns out to be so for 4–9 leads at once to

$$\frac{dq}{dx} = -2h(w + b)(T - T_a) \tag{4–10}$$

and the right-hand side is in fact negative if $T > T_a$, as it must be for heat to flow from the fin to the air.

Another relation connecting q and T comes from Fourier's law as in the previous analysis. Thus

$$q = -kwb\frac{dT}{dx} \tag{4–11}$$

If this is substituted in 4–10, we obtain precisely the same result as previously, that is, 4–5.

But now we wish to eliminate T rather than q. To do this we differentiate 4–10 with respect to x, noting that in the process T_a is a constant. Then for $\frac{dT}{dx}$ we substitute the value obtained from 4–11. The result is

$$\frac{d^2q}{dx^2} = \frac{2h(w + b)}{kwb} q \tag{4–12}$$

This second-order differential equation represents a distinctly different mathematical translation of the same physical facts for it is expressed in terms of an entirely different variable, q, instead of T. It is interesting though, and significant, that, although there is an important difference in the nature of the variables q and θ, 4–12 is of precisely the same form as 4–6.

In this example, we have endeavored to show, among other things, that a problem may have more than one mathematical formulation. Starting with a single precise statement of principle there is considerable latitude for exercising discretion in the selection of the terms for the mathematical expression, and the effect on the resulting expression can be much more profound than simply representing a given quantity by a different combination of letters. Furthermore, there is a converse; the same mathematical formulation may be obtained starting from quite different, but consistent, applications of principle.

EXERCISE. Check the last statement above by deriving equation 4–5, starting with a verbal statement of conservation of energy as applied to a finite piece of the fin extending from x to the outer edge. Repeat the exercise using a piece extending from the attached edge to x.

4–9 A Circuit to Simulate Non-Linear Resistance

This illustration differs from the two preceding in that they lead to ordinary differential equations and this one involves no mathematics beyond elementary algebra. Nevertheless, the mental effort required for a solution is probably at least as great.

The purpose here is particularly to show the benefits derivable from orderly choices of symbols and from careful attention to physical symmetry. As these benefits occur throughout the mathematical analysis, the analysis is carried through to a conclusion. The problem situation is as follows:

For use in an electric computing device, we desire a circuit which will simulate a non-linear resistance. Imagine that we have seen a short qualitative description of such a circuit. The circuit is shown in figure 4–10 with a typical current-voltage characteristic. Our problem is to determine whether such a device will be suitable for our use.

The description says that, as the voltage applied to the terminals A-B

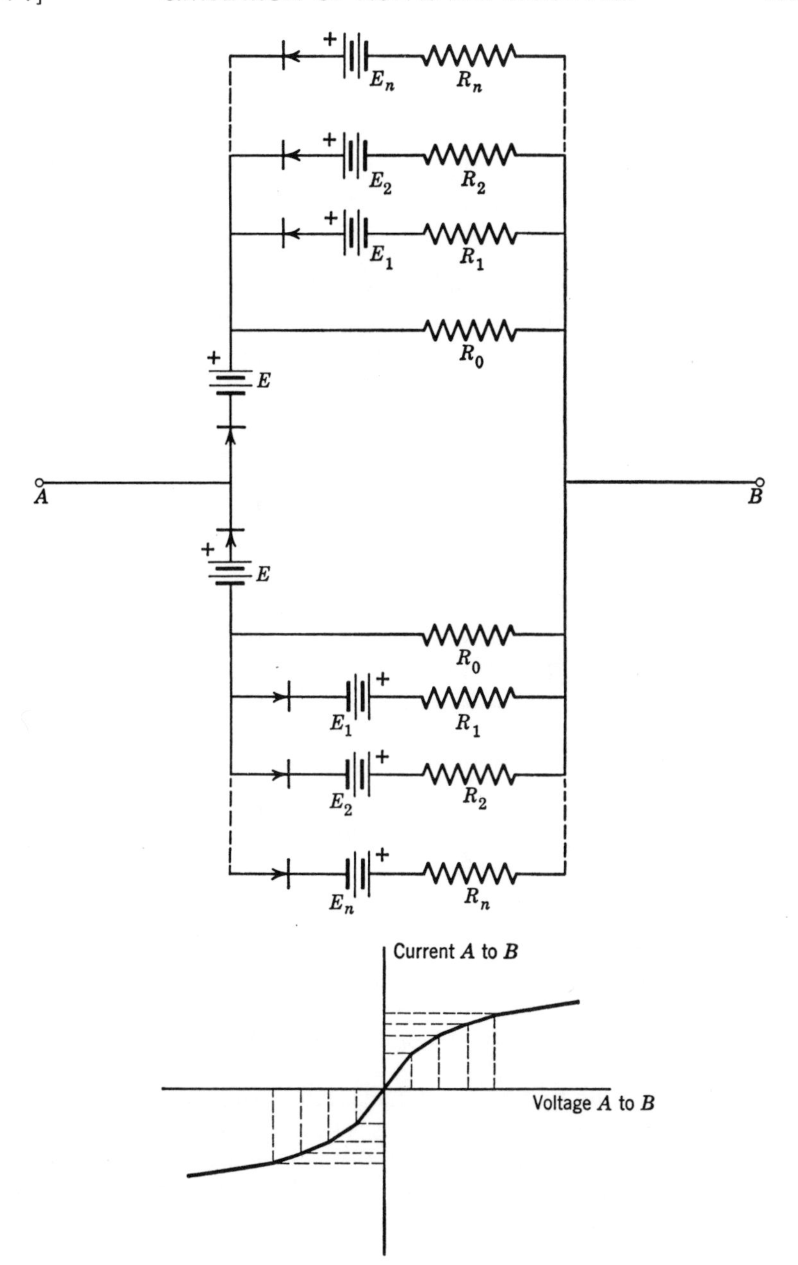

Fig. 4-10. Circuit to simulate non-linear resistance.

is increased, the rectifiers in the separate parallel paths cease to conduct at successively higher applied voltages so that the rate of increase of current with applied voltage diminishes in the stepwise fashion indicated in the curve of figure 4-10. The characteristic can be made smoother by increasing the number of parallel paths, and its form may be altered by adjustment of the component resistors and battery voltages.

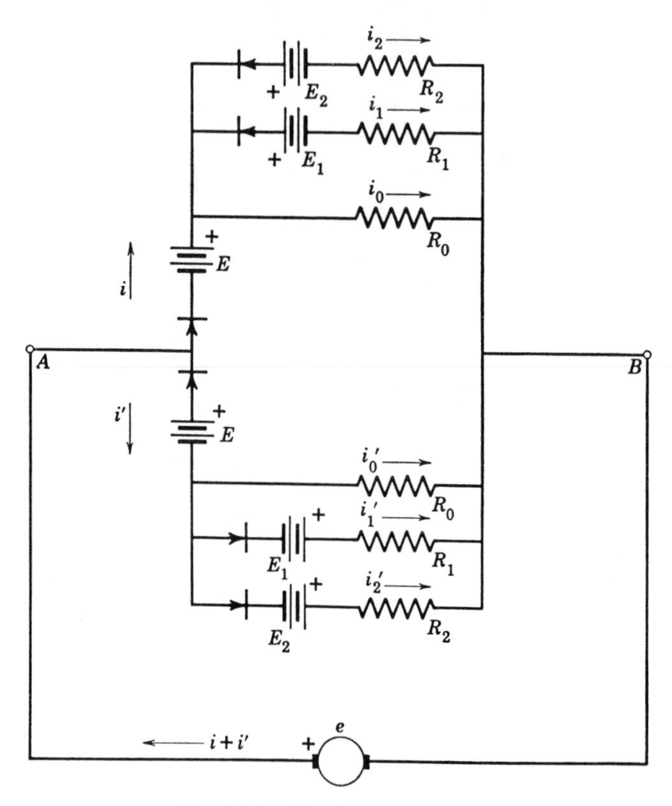

Fig. 4-11. Circuit to be analyzed.

To see whether this type of circuit is adaptable to our purpose we wish to gain a more precise understanding of its behavior. We decide to start by analyzing a simplified version in which the only parallel branches present are those identified by the subscripts 0, 1, and 2. The analysis of this simplified circuit, then, is the subject of the present illustration.

The first problem we set ourselves is to determine how the current through the network from A to B will depend on the voltage applied externally from A to B. To fix ideas, we draw the circuit as in figure 4-11 but without the symbols for currents.

Since we seek a relationship between current and voltage, and since there are a number of paths, it will be appropriate to apply Kirchhoff's voltage and current laws. In doing this, we plan to assume that each rectifier behaves ideally. That is, it conducts current in the direction of its arrow with zero voltage drop but permits no current to flow in the reverse direction regardless of the voltage applied.

Before applying Kirchhoff's laws, we decide on symbols to represent the currents as shown in figure 4–11. The subscripts are the same as those already used to identify the parameters of the respective parallel paths. Primes are used to distinguish the currents in the lower paths from those in the upper ones. The currents in the common parts of the upper and lower portions are designated by i and i' respectively without subscripts. The current we wish to find is evidently the sum $i + i'$.

The positive directions chosen for the currents are indicated by the arrows. The choice is arbitrary; if a current turns out to be negative, it means simply that it flows in the direction opposite to the arrow. In fact it is clear that, because of the rectifiers, i_1, i_2, and i' can never be positive and that i_1', i_2' and i can never be negative. We choose, however, to ignore these obvious facts and to take the arrows in the directions shown because we suspect that this plan may result in a more orderly set of equations.

Applying Kirchhoff's current law, we have then the following equations:

$$i = i_0 + i_1 + i_2 \tag{4–13}$$

$$i' = i_0' + i_1' + i_2' \tag{4–14}$$

Notice that Kirchhoff's current law was already applied implicitly when we identified the current through the external circuit as $i + i'$.

As a preliminary to applying Kirchhoff's voltage law, we complete the external circuit by connecting a generator across the terminals A-B. We denote the terminal voltage of the generator by e with its polarity such as to cause the current $i + i'$ to flow in the arrow direction. The end result we seek is $i + i'$ as a function of e.

Inspection of the network shows that we can choose six loops around which to make six independent applications of Kirchhoff's voltage law. An obvious choice of the six loops would be to start at the top of the diagram with the loop including E_2, R_2i_2, R_1i_1, and E_1; then the loop including E_1, R_1i_1, and R_0i_0, and so on, the sixth loop being the one including e, E, E_2 and R_2i_2'. This systematic selection of loops does not, however, take full advantage of the symmetry which is present in the network. A better plan in this case is to use the six loops each of which includes e and one of the six parallel branches. Each of the resulting six equations will contain a term e, a single Ri term, and either one or two battery-voltage terms. Having worked out one of these equations by physical reasoning, it should be possible to derive the others from it by interchanges of symbols and appropriate changes of signs determined by inspection of the diagram.

Adopting this plan we start by applying Kirchhoff's voltage law to the loop which includes the generator and the upper R_2. We shall express the law by saying that the sum of the voltage drops around the closed path is zero, and we shall take the clockwise direction as positive. We then have the following drops:

1. Through the generator there is an increase in potential; consequently the drop is $-e$.
2. Through the upper common rectifier there is zero drop if i flows, that is, if i is positive.
3. Through the first battery the drop is $-E$.
4. Through the rectifier there is no drop provided i_2 has a negative value.
5. Through the battery there is a drop $+E_2$.
6. Through R_2 the drop is $+R_2 i_2$.

Thus, equating the sum of these drops to zero:

$$-e - E + E_2 + R_2 i_2 = 0 \tag{4-15}$$

provided $i > 0$ and $i_2 < 0$.

Different relations exist for this loop when either rectifier is not conducting. They involve the drops across the rectifiers in the non-conducting direction, and we could write them, but at this stage we think we can reach a useful result by writing equations restricted to the cases where the rectifiers conduct.

Having seen how 4-15 is constructed, it would be easy to write down the corresponding equations for each of the remaining five loops. Rather than do this, however, we see now that it would be still simpler first to solve 4-15 for the current i_2 and then, by using the symmetry of the network, write down directly the corresponding expressions for the other five branch currents.

Thus from 4-15,

$$i_2 = \frac{e + E - E_2}{R_2} \tag{4-16}$$

provided $i > 0$ and $i_2 < 0$.

Again looking ahead we perceive that we can simplify the writing of the expressions which are to come if we characterize each of the resistance elements by its conductance G instead of by its resistance R. Thus,

$$G_2 = \frac{1}{R_2} \tag{4-17}$$

and similary for G_0 and G_1. Then 4-16 becomes

$$i_2 = G_2(e + E - E_2) \tag{4-18}$$

provided $i > 0$ and $e < E_2 - E$.

The second condition, $e < E_2 - E$, follows from this equation and the condition previously observed that $i_2 < 0$.

Now we are in a position to write the expressions for the currents in all the parallel branches simply by an inspection of the diagram in relation to 4-18 and its derivation. Thus,

$$\left.\begin{aligned} i_2 &= G_2(e + E - E_2) \quad \text{if} \quad e < E_2 - E \\ i_1 &= G_1(e + E - E_1) \quad \text{if} \quad e < E_1 - E \\ i_0 &= G_0(e + E) \end{aligned}\right\} \text{ and if } i > 0$$
$$\left.\begin{aligned} i_0' &= G_0(e - E) \\ i_1' &= G_1(e - E + E_1) \quad \text{if} \quad e > -E_1 + E \\ i_2' &= G_2(e - E + E_2) \quad \text{if} \quad e > -E_2 + E \end{aligned}\right\} \text{ and if } i' < 0 \tag{4-19}$$

The second and third of these equations follow from the first by noting that, in the equation for i_1, E_1 replaces E_2 in the circuit, and that in the equation for i_0 there is no battery in the circuit to correspond to E_2 and therefore the corresponding term is simply omitted. The last three equations are derived from the first three by noting that with respect to positive current through the lower branches the polarity of the generator e is the same as in the upper branches but all the polarities of the battery voltages in the lower part of the circuit are reversed in relation to those in the upper part. Consequently the sign of e in the lower equations is the same as that in the upper equations but the signs of E, E_1, and E_2 are reversed.

All the signs of inequality in the conditions for the lower equations in 4-19 are opposite to those for the upper equations. This is because the directions of these signs are determined by the relation of the positive direction of the corresponding current to the conducting direction of its rectifier, and all these relations in the lower half are opposite to those in the upper half. Only one inequality is needed for the third and fourth equations because the loops to which they apply have one less rectifier.

Exercise. Since a physical check of the above argument is clearly desirable, derive the expression for i_1' by applying Kirchhoff's voltage law directly.

Now that we have a good physical check of what has been done so far, we are ready to proceed with our task of finding $i + i'$ as a function of e. We might do this by purely mathematical manipulation of equations 4-13,

4–14, and 4–19 with due regard for the conditions imposed by the rectifiers. We foresee, however, that this may become complicated and confusing unless we first pause to see what can be learned of the physical meaning of the expressions already derived for the currents.

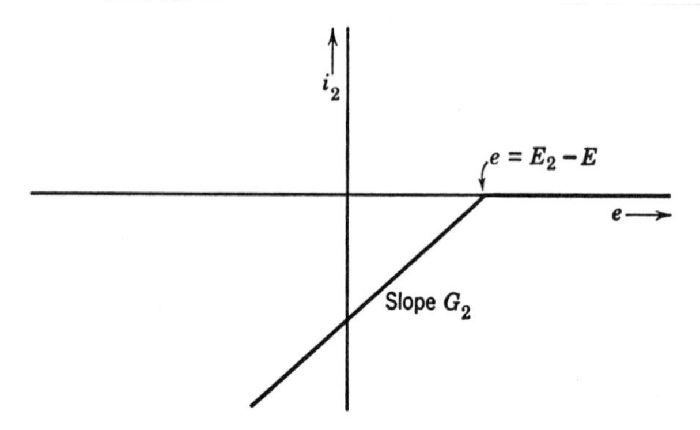

Fig. 4-12. Sketch to show i_2 as given by equations 4-19.

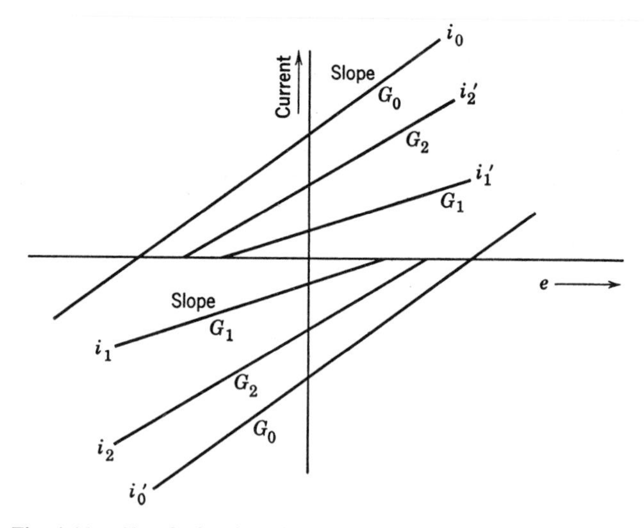

Fig. 4-13. Sketch showing all six of the currents in equations 4-19.

To gain the desired physical understanding, we make the sketch figure 4–12 showing the current i_2 as a function of e as given in equations 4–19. Here we have chosen for the moment to disregard the condition $i > 0$, but we have taken account of the fact that the expression for i_2 holds only when $e < E_2 - E$. When $e > E_2 - E$, it is clear that the rectifier prevents i_2 from becoming positive and consequently to the right of the point $e = E_2 - E$ we have shown i_2 as zero. Evidently the point $e = E_2 - E$

could be anywhere along the e axis depending on the magnitudes of E_2 and E.

Next we sketch figure 4–13 to show all six of the currents but still not taking account of the conditions $i > 0$ and $i' < 0$. For this sketch we have supposed that $E_2 > E_1 > E$ and $G_0 > G_2 > G_1$.

Now, using 4–13, we add i_0, i_1, and i_2 to form i as shown in figure 4–14. In this sketch, we have introduced the condition $i > 0$ by showing the

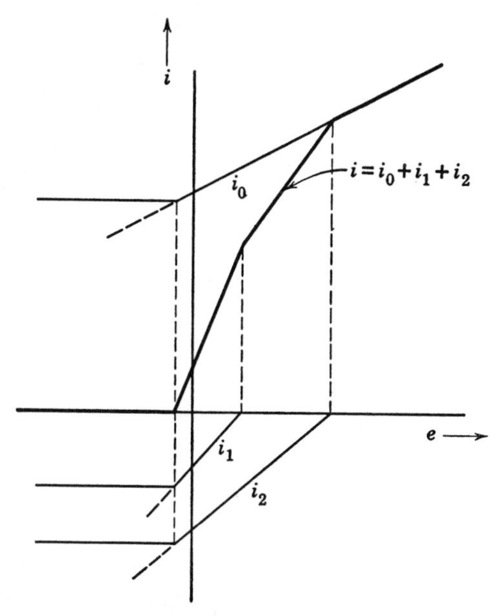

Fig. 4-14. Combination of the branch currents in the upper half of the network.

curve for i as zero to the left of the point where it meets the e axis, thus taking account of the fact that the upper common rectifier permits i to flow only in the positive direction. Also, to the left of the point at which i becomes zero, we have shown its components with constant values.

Exercise. Check the constant values just mentioned by applying Kirchhoff's voltage law to appropriate loops under the condition that the upper common rectifier does not conduct.

The range of e within which the upper part of the network is active may be found analytically by adding the expressions for i_0, i_1, and i_2 in 4–19 to give i and then setting the condition $i > 0$. Thus

$$i = (e + E)(G_0 + G_1 + G_2) - G_2E_2 - G_1E_1 > 0 \tag{4–20}$$

from which

$$e > -E + \frac{G_1E_1 + G_2E_2}{G_0 + G_1 + G_2} \tag{4–21}$$

Evidently a similar expression holds for the range of e in which i' is present. By symmetry it is seen to be

$$e < E - \frac{G_1E_1 + G_2E_2}{G_0 + G_1 + G_2} \tag{4-22}$$

Clearly the ranges of e given by 4–21 and 4–22 may overlap as indicated at the left of figure 4–15, or there may be a dead band between them where the network presents infinite resistance as shown on the right of the figure.

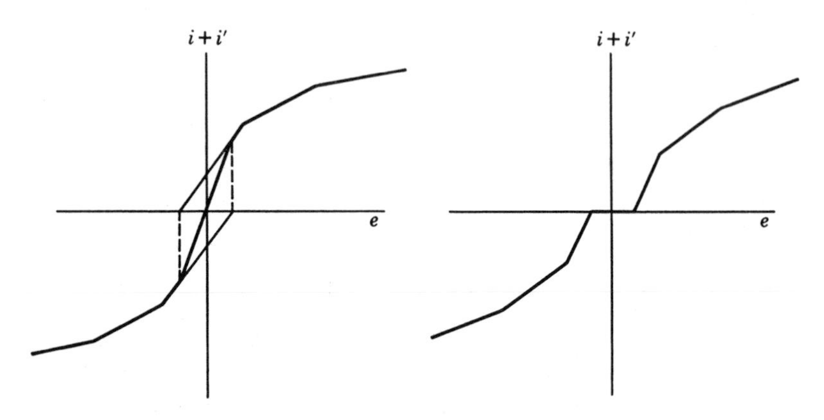

Fig. 4-15. Possible forms of the complete characteristic.

For the remainder of the analysis let us assume that the ends of the ranges of i and i' coincide at $e = 0$. From either 4–21 or 4–22 we see that this will occur if

$$E = \frac{G_1E_1 + G_2E_2}{G_0 + G_1 + G_2} \tag{4-23}$$

We are now in a position to express the terminal current $i + i'$ as a function of e. When e is positive, the terminal current is simply i, since i' is zero under the condition 4–23, which has been assumed. When e is between zero and $E_1 - E$, both i_1 and i_2 exist so that the terminal current is given with the aid of 4–13 and 4–19 as

$$i + i' = (e + E)(G_0 + G_1 + G_2) - E_1G_1 - E_2G_2 \tag{4-24}$$

But E is determined by 4–23, and substituting this value in 4–24 gives

$$i + i' = e(G_0 + G_1 + G_2) \tag{4-25}$$

This is exactly as it would be if the circuit consisted only of the three upper resistors with no batteries and no rectifiers.

When e is negative the terminal current $i + i'$ consists only of i'. By the symmetry of the circuit and of the equations, it may be seen that the expression for i' in the range $-E_1 + E < e < 0$ must be identical with 4–25, and so 4–25 gives the terminal current for the whole range $-E_1 + E < e < E_1 - E$.

Next consider what happens when e exceeds $E_1 - E$. Then i_1 has been cut off and the terminal current is simply $i_0 + i_2$, or, using 4–19,

$$i + i' = e(G_0 + G_2) + E(G_0 + G_2) - E_2G_2 \tag{4–26}$$

If e exceeds $E_2 - E$, all currents are cut off except i_0, and we get for the terminal current

$$i + i' = eG_0 + EG_0 \tag{4–27}$$

Substitution of the values of e for the transition points between 4–25 and 4–26 and between 4–26 and 4–27 shows that the current changes continuously past these points, although of course the slope is discontinuous. This checks with the physical picture already deduced.

Finally, let us collect together the expressions for the function $i + i'$, adding those for the negative ranges of e which have not been worked out but which follow readily from symmetry. We do this in the form of the accompanying table.

Range of e	Terminal current $i + i'$
$e < -E_2 + E$	$eG_0 - EG_0$
$-E_2 + E < e < -E_1 + E$	$e(G_0 + G_2) - E(G_0 + G_2) + E_2G_2$
$-E_1 + E < e < E_1 - E$	$e(G_0 + G_1 + G_2)$
$E_1 - E < e < E_2 - E$	$e(G_0 + G_2) + E(G_0 + G_2) - E_2G_2$
$E_2 - E < e$	$eG_0 + EG_0$
Restrictions	$E = \dfrac{E_1G_1 + E_2G_2}{G_0 + G_1 + G_2}$ $E < E_1 < E_2$

This mathematical expression of the function is interpreted diagrammatically in figure 4–16. This is as far as we will carry the analysis of the circuit, although it should be clear that we have by no means exhausted the possibilities so far as shapes of characteristic are concerned.

In this illustration of the translation of a physical situation into mathematical terms, we have shown how careful attention to physical symmetry can lead to equations whose forms preserve the symmetry. This not only permitted forming the fundamental equations for parts of the network by making changes in equations already derived physically, but also it enabled us to write some of the results without need for repeating algebraic manipulations. Proceeding as we did resulted in an analysis of such form that the way to generalize it to include cases with any number of parallel branches is almost self-evident. It should be clear that this simplification of the work would not have occurred if the symbols, positive directions, and choice of loops for writing Kirchhoff's voltage law had not been chosen with a view to the inherent symmetry which is present. If you are not

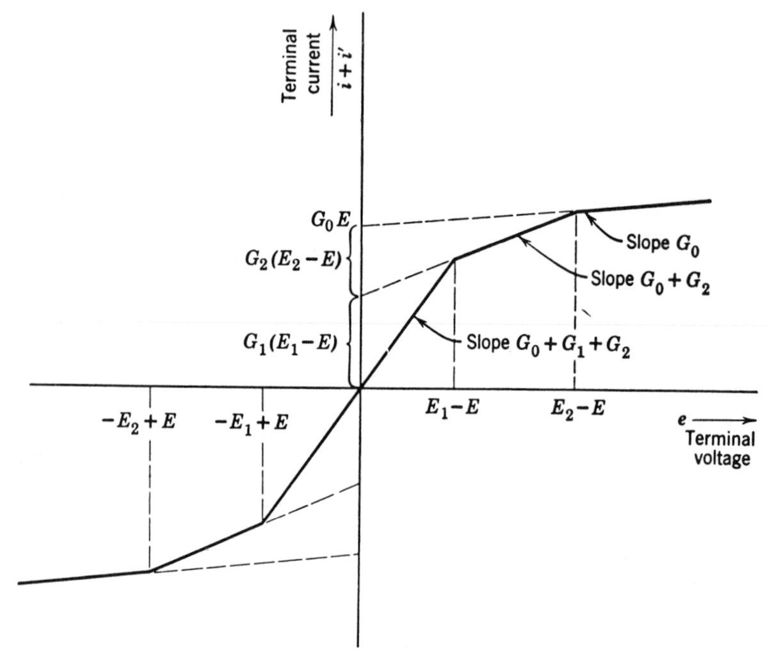

Fig. 4-16. A characteristic of the circuit in figure 4-11.

convinced of this, try analyzing the circuit according to a plan which ignores the symmetry.

In relying as heavily as we did on considerations of symmetry, you must be very careful to make enough physical checks as you go along to be sure that things really are as they seem. Alternatively, you may prefer to do the entire analysis on a direct physical basis and reserve symmetry for checking purposes. Such checking is discussed further in section 6-6.

4-10 Mathematical Formulation: Recapitulation

The chief point we have endeavored to make in this chapter is that you should never try to form a mathematical expression until you really understand clearly and precisely what you wish to express and in terms of what quantities. To make a correct and effective translation into mathematics, the condition to be satisfied as it applies in the situation at hand should be expressed first in clear English. Additionally, before starting to write an equation, you should give careful thought to the quantities which are to enter the formulation and to the symbols to be used, particularly their definitions.

Keep in mind always that mathematics is like a machine into which you puts facts and conditions, and which when you turn the proper cranks gives out the conclusions which must result from the facts and conditions put in. The machine cannot supply the facts nor can it tell if a wrong one is fed in. It may stall if inconsistent facts are supplied, but you cannot be sure that it will.

Although this chapter is devoted specifically to the process of putting ideas into mathematical form, the inverse process of getting ideas out of the mathematical conclusions is of equal importance. Indeed, it is often desirable to interpret physically mathematical expressions which occur before the end of the analysis is reached. Such interpretations are made for purposes of checking or, as in the last illustration, section 4–9, to point the way for further steps in the mathematical work. Interpretation of mathematical expressions is discussed fully in chapter 7. It should be clear already that the interpretation of the mathematical conclusion will be simplified greatly if the original translation into mathematics is done rationally.

CHAPTER FIVE

Solution of Some Ordinary Differential Equations

5–1 Differential Equations in Engineering

Applications of many of the most useful physical laws lead naturally to ordinary differential equations. Such equations may describe conditions at a point in space or at an instant in time; their solutions tell how something varies in space or as a function of time. Differential equations occur so frequently in engineering analysis that a fundamental understanding of how to solve them is an important part of the mathematical background of engineers.

This chapter deals with ordinary differential equations, that is, those containing only ordinary or total derivatives as distinguished from partial derivatives. Ordinary differential equations arise where a quantity depends on a single independent variable, partial differential equations when there are two or more independent variables. In this book we already have met and solved some very simple ordinary differential equations, notably in sections 2–2 and 2–3; some less simple ones which we did not solve were set up in sections 4–7 and 4–8.

This chapter does not presuppose prior knowledge of how to solve differential equations; on the other hand it is not a comprehensive treatment of the subject. The purpose is to help establish a firm physical understanding of some processes by which differential equations are solved and to give working principles for dealing with the most common types.

You must realize at the outset that solving differential equations is inherently less straightforward than the more elementary mathematical operations such as finding a derivative or evaluating an integral. There are numerous ways to attack the subject, ranging from the elegant method of Laplace transforms to the older method which is largely one of trial and error. The latter is the one we shall adopt here; it requires the simplest mathematical tools and remembering the fewest special rules; thus we feel it to be the most practical for the engineer who has not yet become a specialist in a field where learning more sophisticated methods would become worth while.

5-2 A Graphical Solution: Starting Time of a Fan

To begin a study of the steps in solving a differential equation let us consider a problem of predicting from design information how long it will take a fan driven by a motor to come up to speed after the motor is energized. The fan is of the centrifugal type and is to be driven by a three-phase wound-rotor induction motor rated at 10 horsepower. Starting of

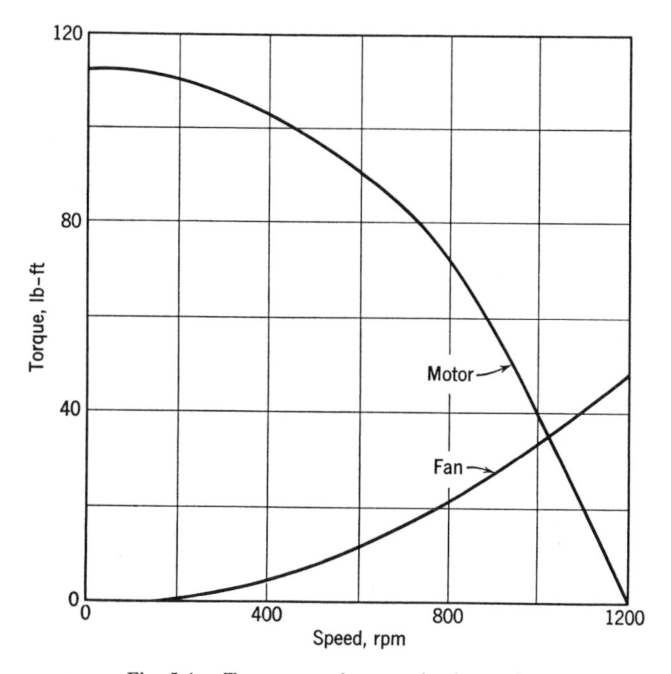

Fig. 5-1. Torque-speed curves for fan and motor.

the motor is to be at full voltage and with resistance connected in the rotor circuit. We wish to find how the speed varies with time during the starting period, the rotor resistance remaining unchanged.

The predicted torque-speed curve for the fan, that is, the relation between the speed and the torque required to drive the fan steadily at that speed with the anticipated discharge conditions, is shown in figure 5-1. Also shown in the figure is the torque-speed curve for the motor. This curve shows the torque which the motor delivers at its coupling as a function of speed. The steady operating speed of the fan and the motor will be at the point of intersection of the motor curve and the fan curve because at this speed the torque supplied by the motor is just equal to that required to keep the fan running. Moreover, operation at this speed will be stable, for at any lower speed there will be an excess of driving torque which will cause the system to accelerate, and conversely at any higher speed there will be a deficiency of driving torque and the system will decelerate, in both cases approaching the intersection point. Thus examination of the curves in figure 5–1 shows that the steady running speed of the fan under these conditions will be 1020 rpm.

In planning how to find the variation of speed with time we begin by considering what happens as the motor is started. In the first place, there

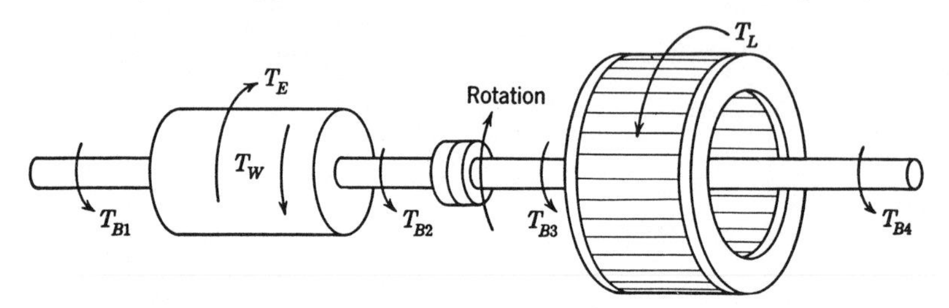

Fig. 5-2. Free-body diagram of the rotor showing the torques.

may be an electrical delay while the current rises from zero to some value before the motor begins to turn. Then, or possibly simultaneously, there is the interval in which the rotating parts are accelerating under the influence of the motor and load torques. Experience leads us to expect the electrical transient in this problem to be much shorter in duration than the mechanical transient, and for simplicity we assume the electrical transient to be negligible. We assume that at any speed the motor produces the torque given by the curve in figure 5–1 even though this curve is for the motor running at constant speeds with steady-state current and in the present situation the speed is changing with time.

Having reduced the problem to a mechanical one, we draw a free-body diagram, figure 5–2, which shows the rotors, shafts, bearings, and coupling, together with the torques that act upon the rotating system. These are the electromagnetic torque of the motor, the load torque of the fan, the bearing torques, and the wind-friction torque exclusive of the fan. According to the law for the angular acceleration of a rigid body, that is, $\Sigma T = I\alpha$, the resultant of the torques of motor, load, bearings, and wind friction is equal to the product of the moment of inertia of the whole system by its angular acceleration. This is the verbal statement which gives the application of the fundamental principle to our particular problem. We expect that it may be possible to manipulate the relation to yield a speed-time curve.

In the execution stage, the verbal statement is first expressed in mathematical language.

Let T_E = electromagnetic torque acting on the rotor, positive when it has the direction of rotation (lb ft).

T_L = load torque of the fan, positive when it has the direction opposite to that of rotation (lb ft).

$T_{B1}, T_{B2}, T_{B3}, T_{B4}$ = bearing friction torques, with the same positive directions and units as T_L.

T_W = wind friction torque exclusive of the fan, with same positive direction and units as T_L.

I = moment of inertia of whole system, (slug ft^2).

ω = angular velocity, positive in the actual direction of rotation, (radians sec^{-1}).

t = time with zero at the instant at which the motor is energized (sec).

Then,

$$T_E - T_{B1} - T_{B2} - T_W - T_L - T_{B3} - T_{B4} = I\frac{d\omega}{dt} \tag{5–1}$$

Some simplification may be made by recognizing that $T_E - T_{B1} - T_{B2} - T_W$ is the torque available at the shaft of the motor when there is no acceleration. This will be designated by T_M; that is, it is the torque that the motor will deliver from the end of the shaft at constant speed and is given by figure 5–1. For this statement to be strictly true, T_W, which was defined as all the wind friction exclusive of the fan, must be the same as the wind friction within the motor. In other words, we neglect any wind friction on the coupling and connecting shaft. Likewise the torque required at the coupling to turn the fan at steady speed is $T_L + T_{B3} + T_{B4}$, designated by T_F. Then

$$T_M - T_F = I\frac{d\omega}{dt} \tag{5–2}$$

Although we might have been tempted to write this equation sooner, it is important to note, that proceeding as we did, we know precisely how and to what extent the frictional torques are accounted for; otherwise there would have been room for doubt; likewise we are sure which moment of inertia is involved.

Equation 5–2 is a differential equation; we come to the problem of how to solve it. First let us summarize what we know about the terms it contains. The motor shaft torque T_M is a function of ω (and for an induction motor in the steady state electrically, as we are assuming here, a function of ω only); it is in fact the function represented by the motor curve in figure 5–1. Therefore we may regard this function of angular velocity, $T_M(\omega)$, as known. Likewise T_F, the torque to drive the fan, is a function of angular velocity, $T_F(\omega)$, and is known from the fan torque-speed curve of figure 5–1. Thus, the entire left-hand member of equation 5–2 is a known function of angular velocity ω; that this function happens to be given by curves instead of by a mathematical formula is not important. The right member of 5–2 contains I, the moment of inertia of the whole rotating system, and this we have estimated from drawings by methods which we will not discuss here to have the value 1.26 slug ft^2. Also in the right member are ω, the dependent variable which is sought, and t, the independent variable.

The equation can be regarded as giving the time rate of change of ω for any value of ω, thus using functional notation

$$\frac{d\omega}{dt} = \frac{T_M(\omega) - T_F(\omega)}{I} \tag{5–3}$$

Our problem is to find a function of t, $\omega(t)$, such that its slope $\frac{d\omega}{dt}$ depends on ω in accordance with the right side of 5–3. The right side of 5–3 is a function of ω which could be plotted against ω by using the data of figure 5–1. Then, if we know the value of ω at some value of t, and we do know the initial condition that $\omega = 0$ at $t = 0$, we can calculate the corresponding value of $\frac{d\omega}{dt}$. This suggests that over a short enough interval of time, Δt, we can find the change of ω with reasonable accuracy by multiplying $d\omega/dt$ by Δt, thus:

$$\Delta\omega = \omega_1 - \omega_0 \doteq \frac{d\omega}{dt}\Delta t = \frac{T_M(\omega_0) - T_F(\omega_0)}{I}\Delta t \tag{5–4}$$

in which ω_0 is the value of ω at the beginning of the interval, ω_1 at the end. Thus, by starting with $\omega_0 = 0$, we could find ω_1 at the end of the small interval Δt. For the next interval Δt, we could proceed similarly except that ω_0 would be replaced by the value $\omega = \omega_1$ just found and a new value, $\omega = \omega_2$, would be found for the end of the second interval. By repeated calculation, we could construct a broken line curve giving ω as a function of t, as sketched in figure 5–3.

How can we tell how small to make the interval Δt? One way would be to make an arbitrary choice, say 0.1 sec, calculate a portion of the curve, then try a smaller value, say 0.05 sec, repeat the calculation, and compare the results of the two calculations. If the results differ significantly, the first interval was too large, and perhaps the second is, also; in this case, we should try a still smaller interval. If the results are in good agreement, the interval is small enough, possibly unnecessarily small (to our regret, since the calculation is tedious), but at least it is on the safe side.

We now have one way of solving the differential equation. Our procedure illustrates very clearly the kind of relation that a differential equation represents, and that to get a particular solution we have to know ω at some value of t, called an initial condition. Observe that the same differential equation would hold if the problem were somewhat different, for

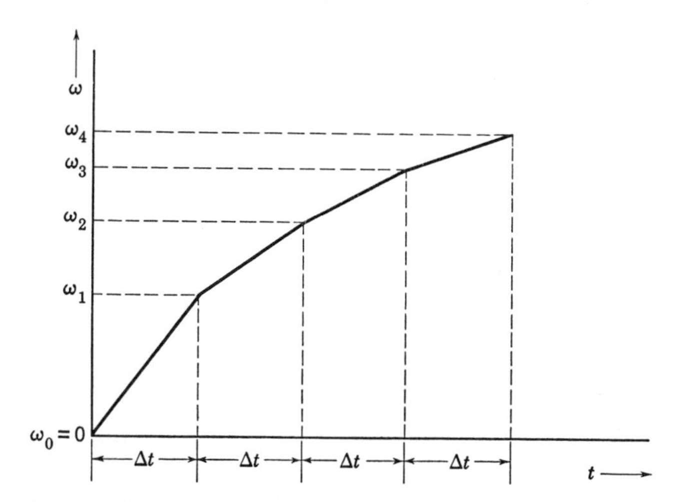

Fig. 5-3. Illustrating a step-by-step process for solving equation 5-3.

example, to find the speed-time curve following a sudden change in rotor resistance to the value corresponding to figure 5–1 from some other value with the motor already running. Although the differential equation would be the same, the initial condition would be different and a different curve would be found for the solution.

Although we have a straightforward method for solving the differential equation, we are not pleased with it because of the apparent tediousness of ascertaining that the intervals Δt are small enough. So we try another approach to see whether anything can be gained by separating the variables ω and t, thus:

$$dt = I \frac{d\omega}{T_M(\omega) - T_F(\omega)} \tag{5-5}$$

We observe that t comes from dt by integration and try

$$t = \int_0^t dt = I \int_0^\omega \frac{d\omega}{T_M(\omega) - T_F(\omega)} \tag{5-6}$$

where, since ω is to correspond to t, the upper limit on the right side is ω and the lower limit is zero provided $\omega = 0$ at $t = 0$. Thus the initial condition enters this solution too! If we can integrate the right side, we have a solution. It should certainly be possible to integrate it graphically or numerically, for $T_M(\omega)$ and $T_F(\omega)$ are known, hence we should be able to draw a curve of $\dfrac{1}{T_M(\omega) - T_F(\omega)}$ as a function of ω, as sketched in figure 5-4. According to equation 5-6, the time required for the fan to come up

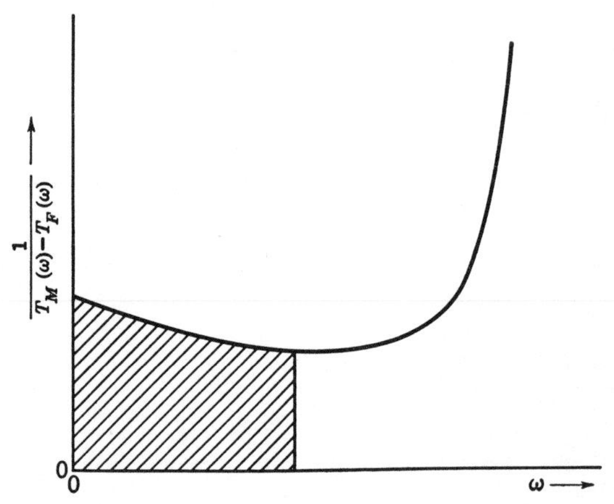

Fig. 5-4. Rough sketch of the function to be integrated in equation 5-6.

to any speed ω is the area under the curve up to the vertical line through ω, multiplied by I. Thus, we have a plan for solving 5-2 which appears to be less tedious and perhaps more accurate than the step-by-step procedure worked out first.

To carry out the integration we replace the curve which was sketched in figure 5-4 by one plotted accurately on rectangular coordinate paper, figure 5-5, and which for simplicity is in terms of speed N in rpm instead of ω. Then we replace $d\omega$ in 5-6 by $\dfrac{2\pi}{60}\,dN$ and obtain

$$t = \frac{2\pi}{60} I \int_0^N \frac{dN}{T_M(N) - T_F(N)} \tag{5-7}$$

where t is still the time in seconds.

The area can now be found by counting squares or with the aid of a planimeter. The authors' experiences lead them to feel that, unless a large number of areas are to be determined, it does not pay to find a planimeter and learn or relearn the technique of using it; the same may be said of formula methods such as Simpson's rule. Counting squares is fully as accurate and less costly in time than the more sophisticated methods.

Counting squares, we find, for example, the time to reach 100 rpm. The area under the curve up to 100 rpm we estimate to be 1.8 squares. The

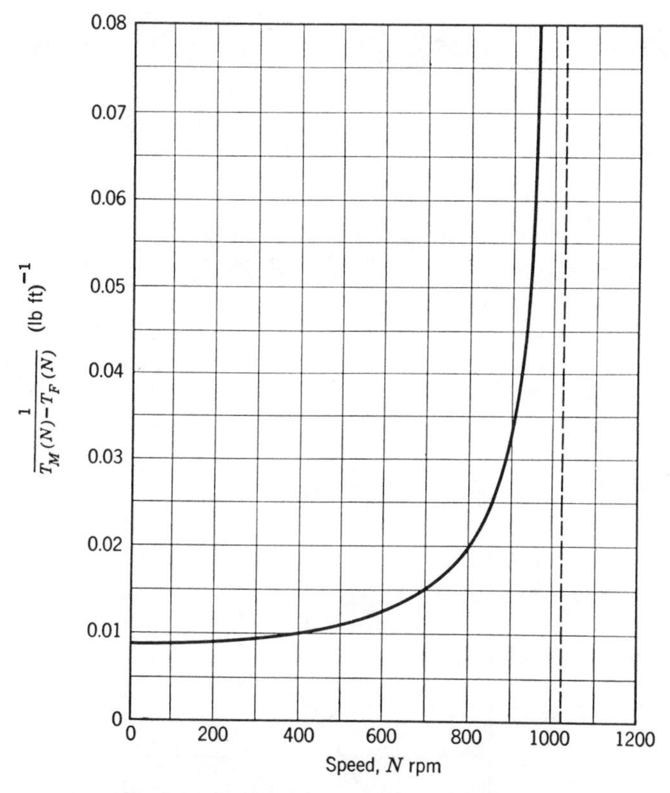

Fig. 5-5. Plot of the integrand in equation 5-7.

value of a square is dependent on the scales to which the curve is plotted and in this case is $100 \times 0.005 = 0.5$. The integral up to 100 rpm is then $1.8 \times 0.5 = 0.9$, and the time to reach 100 rpm is the product of this number by the value of I $(=1.26)$ and $2\pi/60$ or $0.9 \times 1.26 \times 2\pi/60 = 0.12$ seconds. Next, we count the squares between 100 and 200 rpm, add them to those already counted, and get the time to reach 200 rpm as before. Proceeding in this way, we construct the accompanying table and draw the speed-time curve, figure 5-6, and so attain the goal we set for ourselves.

Speed N (rpm)	Squares for last increment of N	Total squares up to N	t (sec) to reach N = (squares) $\times$ 0.5 $\times \frac{2\pi}{60} \times$ 1.26
0	0	0	0
100	1.8	1.8	0.12
200	1.8	3.6	0.24
300	1.9	5.5	0.36
400	2.0	7.5	0.50
500	2.15	9.65	0.64
600	2.35	12.0	0.79
700	2.8	14.8	0.98
800	3.5	18.3	1.21
900	5.2	23.5	1.55
950	4.3	27.8	1.83

Before using figure 5–6, the result of our analysis, we make some overall checks. (The arithmetic, of course, was checked carefully as it was performed.) In the first place, based on experience with machines of this general size and character, 2 sec to reach 90 per cent of full speed seems to be of the correct order of magnitude. Had our result been a few milliseconds or several minutes, we would have had good grounds for suspecting a gross error of some sort.

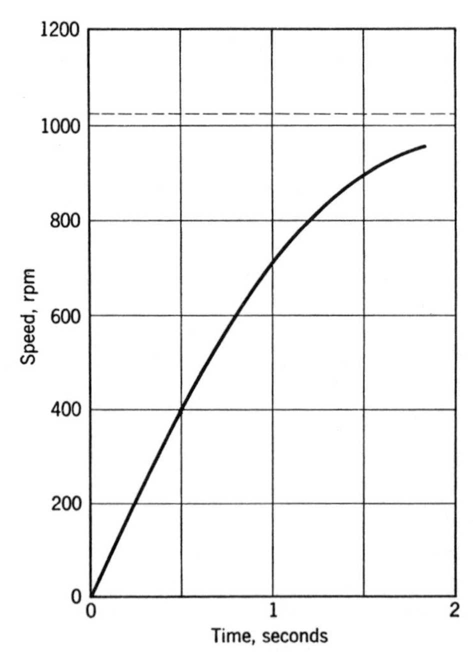

Fig. 5-6. Speed-time curve calculated by taking areas under the curve of figure 5-5.

Now we make a more specific check. Examining the original data in figure 5–1, we observe that the resultant torque available for acceleration up to 500 rpm is never far from 100 lb ft. If it were exactly constant at this value, T_a, the time to reach 500 rpm would be

$$t = \frac{I\omega}{T_a} = \frac{1.26 \times 2\pi \times 500}{100 \times 60} = 0.66 \text{ sec} \tag{5-8}$$

and this agrees closely with the value calculated in our table. Further examination of figure 5–1 shows that the torque available for acceleration above 500 rpm is always much less than 100 lb ft. and therefore the time to accelerate the rest of the way to full speed should be much more than 0.66 sec, which also agrees with our analysis.

The preceding argument suggests that the whole analysis could have been made more easily by simply averaging the torque available for acceleration with respect to speed and then making a calculation like 5–8. It is true that in problems like this there is always some way of taking an average that will give a correct answer, but the correct average is often not that found by taking the arithmetic mean. Here, for instance, it is clear that although the simple average of torque with respect to speed is good enough for a rough check it is not really accurate; the correct kind of average is found by averaging the reciprocal of net torque and then taking the reciprocal of the result. Thus in general the use of averages is a questionable procedure until the problem has been solved without them in order to find out what kind of an average to use.

Let us see now what we can learn from this illustration that may be generally useful in solving differential equations. In the first place, basing our reasoning on physical pictures of the relationship given by the differential equation, we were able to plan two entirely different methods for solution, both perfectly straightforward. For the first method we thought of the desired result $\omega(t)$ as a graph of ω versus t and of the differential equation as defining the slope of this curve. Then the plan was to step our way along the curve in a series of short straight-line segments, each having the slope given by the differential equation for the value of ω at that point. To start this process we had to know the value of ω at some one value of t; and we did know the initial value. The second method we planned, and this is the one we adopted, was possible in this case because the variables in the differential equation were separable so that the time t could be expressed as an integral with respect to ω of a function of ω. Here too, we reasoned in terms of physical pictures, thinking of the integral as an area under a curve; and in this case, since the data were known in curve form, we performed the integration by actually finding an area under a curve plotted to scale. In this second method, as in the first, we found it necessary to know the value of ω at one value of t in order to get started.

Thus in general we can see that the differential equation by itself did not suffice to determine the solution to our particular problem; something else was needed, the fact that the fan starts from rest, that is, $\omega = 0$ at $t = 0$. In other words the differential equation can be satisfied by any one of a whole family of solutions—curves of ω versus t—each starting at a different initial value of ω. To find the particular solution we want, we must determine the appropriate initial conditions from physical relationships which are not contained in the differential equation. Not only is this conclusion valid in this instance, it is also true in general of the solutions of all differential equations. We can see also that for this first-order differential equation, one in which the highest derivative is the first as in this illustration, only one condition was needed to determine a particular solution, and this is true of all first-order differential equations.

5-3 A Second-Order Differential Equation: Temperature of a Fin

To gain further insight into the solution of differential equations let us go on with the determination of the temperature distribution in a cooling fin under conditions of thermal equilibrium which we commenced in section 4-8. The essence of the situation we treated there follows:

As shown in figure 4-6 a thin rectangular fin is attached along one edge to a body to be cooled. Heat flows radially outward along the fin and is then transferred to the air by forced convection. We wish to know the temperature of the fin as a function of distance radially outward from the attached edge after steady-state conditions are reached; that is, after the temperature at any point no longer changes with time. The situation has been simplified by assuming the temperature along the attached edge of the fin to be uniform and the air temperature to be uniform and unaffected by the heat picked up from the fin. Also, temperature variation within the fin in the direction of its thickness has been neglected.

Several translations of this situation into mathematical terms were made, each leading to a second-order differential equation. We wish to study equation 4-6, which for convenience is repeated here:

$$\frac{d^2\theta}{dx^2} = \frac{2h(w + b)}{kwb}\,\theta \tag{5-9}$$

In this equation

θ = temperature rise of the fin above the air temperature, a function of distance radially along the fin (°F).

x = radial distance along the fin measured positive outward from the attached edge (ft).
h = coefficient of heat transfer by forced convection from the surface to the air (Btu hr^{-1} ft^{-2} °F^{-1}).
k = thermal conductivity of the metal (Btu hr^{-1} ft^{-1} °F^{-1}).
b = thickness of the fin (ft).
w = width of the fin in the axial direction (ft).

In considering how to solve 5–9 let us think of the result we seek as a curve of θ plotted against x to rectangular coordinates, and see how we might use the information contained in the differential equation to construct this curve in step-by-step fashion somewhat as described in section 5–2 for equation 5–3. For this purpose we will suppose that we have available numerical values of h, k, b, and w. To solve 5–3 by the step-by-step procedure we showed by means of 5–4 how to find approximately the change $\Delta\omega$ in the function $\omega(t)$ for a small change Δt in the independent variable t. Reasoning for the present case in similar fashion we see that we can use 5–9 to get not the change in the function $\Delta\theta$ but the change in its slope, $\Delta\left(\frac{d\theta}{dx}\right)$, corresponding to a small change Δx in the independent variable x. To start constructing the curve we need to know, as in section 5–2, a value of the function, and here, in addition, the value of its slope at the same point. Supposing, then, that we know the values of θ and $\frac{d\theta}{dx}$ at $x = 0$ we could advance in small steps Δx, constructing two curves simultaneously, one of $\frac{d\theta}{dx}$ against x and the other of θ against x, the two being related at every point by the differential equation 5–9.

EXERCISE. Describe the above process in more detail using sketched curves to illustrate the steps.

To construct the solution in this way requires that two conditions be known, both θ and $\frac{d\theta}{dx}$, at some one value of x. To find the two values we must return to the physical reality of the problem. That further consideration of the physical situation should be necessary is not surprising, for the differential equation expresses the application of a physical law (conservation of energy) merely to an element of the fin and takes no account of physical conditions at the ends of the fin and the effect of the terminations on fin temperature. It is significant that two conditions are needed to solve the second-order equation of this illustration, whereas in the preceding section one sufficed for the first-order equation.

We may suppose that the temperature of the fin at the edge attached to the body to be cooled is known. Then the value of θ at $x = 0$, call it θ_0, is this temperature expressed as a rise above the air temperature which also is presumed to be known. Now what about the value of $\frac{d\theta}{dx}$ at $x = 0$? Reference to the derivation of 4–3 in section 4–8 shows that by Fourier's law $\frac{d\theta}{dx}$ at $x = 0$ is proportional to the heat conducted into the fin from the hot body. Suppose for the moment that this heat flow, and hence $\frac{d\theta}{dx}$ at $x = 0$, is known; indeed, we might be designing the fin to carry away a specified flow of heat. Thus we would have a physical basis for starting the step-by-step construction of the curves of θ and $\frac{d\theta}{dx}$ as functions of x; and we could expect everything to go smoothly until we reached the far end of the fin. At the far end the curves would determine a value of fin temperature and, through $\frac{d\theta}{dx}$, a value of heat flow past the end of the fin. But where does this heat go?

Suppose, as implied by figure 4–6, that the far end of the fin is cut off squarely and exposed to cooling air flowing at the same velocity as it flows past the other surfaces. The heat transferred from the end by convection to the air would be equal to the product of the heat transfer coefficient h, the surface area of the end wb, and the temperature rise of the end above air temperature, or, in symbols,

$$hwb\theta\Big]_{x=L}$$

where L is the radial length of the fin. But this heat flow must be identical with the conduction of heat radially along the fin just short of $x = L$, and this by Fourier's law is expressed in symbols as

$$-kwb\,\frac{d\theta}{dx}\Big]_{x=L}$$

Equating these two expressions for the heat flow through the end of the fin gives

$$h\theta\Big]_{x=L} = -k\,\frac{d\theta}{dx}\Big]_{x=L} \tag{5–10}$$

And this is a physical condition which must be satisfied by a solution to equation 5–9 if the fin terminates as we have assumed. If the fin ended in some other way, for instance, by attachment to another hot body of known temperature, a condition other than 5–10 would apply at $x = L$, but in any case the condition to be met would be based on physical facts not included in the differential equation.

Starting with given values of θ and $\frac{d\theta}{dx}$ at $x = 0$ and carrying the step-by-step construction of the curves to $x = L$ it would be only by chance that θ and $\frac{d\theta}{dx}$ at $x = L$ would turn out to have the relationship required physically and expressed by 5–10. This is because the length L is specified. On the other hand, if L is to have such a value that a specified amount of heat is to be carried away by the fin, then the step-by-step construction could be carried out until such a distance is reached that 5–10 is satisfied. In our case, however, we are assuming L fixed, and so we conclude that we have put too many conditions on the solution; and that since 5–10 clearly must be satisfied we cannot specify both θ and $\frac{d\theta}{dx}$ at $x = 0$. If one of these is known then it should be possible to find what the other must be. We will suppose that at $x = 0$ it is θ that we know, or in symbolic form the condition to be satisfied at this boundary of the fin is

$$\theta\Big]_{x=0} = \theta_0 \tag{5–11}$$

The problem now may be restated as that of finding a function $\theta(x)$ which satisfies the differential equation 5–9 and the two boundary conditions 5–11 and 5–10 at $x = 0$ and $x = L$ respectively. Our step-by-step method is still a possibility for doing this. It would involve assuming a value of $\frac{d\theta}{dx}$ at $x = 0$, proceeding as already described to $x = L$, testing to see if 5–10 is satisfied, and, if it is not, starting over with another assumed value of $\frac{d\theta}{dx}$ at $x = 0$, and so on. Although this procedure is perfectly feasible provided numerical values of the parameters are known, we propose to adopt it only if we cannot find some less laborious method. From what we have done it should be clear, however, that no matter what the method for finding a solution our conclusions about the conditions to be satisfied at the boundaries $x = 0$ and $x = L$ remain valid.

Looking for a more direct and easier method of solution we return to 5–9 and examine it anew. This equation tells us that the second derivative of the function $\theta(x)$ which we seek is equal to the function itself multiplied by the factor $2h(w + b)/kwb$, and this from physical considerations we know to be positive and, in our problem, constant. We ask ourselves, what function differentiated twice is the same as the function itself except for multiplication by a positive constant? A possibility would be the exponential

$$\theta = \epsilon^{mx} \tag{5-12}$$

for if we differentiate it twice with respect to x we get

$$\frac{d^2\theta}{dx^2} = m^2\epsilon^{mx} = m^2\theta$$

By comparison of this with 5-9 we see that 5-12 satisfies equation 5-9 provided

$$m^2 = \frac{2h(w + b)}{kwb} \tag{5-13}$$

or m may have either of the two values

$$m = \pm\sqrt{\frac{2h(w + b)}{kwb}} \tag{5-14}$$

Thus we have found two functions either of which satisfies 5-9. They are

$$\theta = \epsilon^{+\alpha x} \tag{5-15}$$

and

$$\theta = \epsilon^{-\alpha x} \tag{5-16}$$

where, for brevity, α represents the positive root in 5-14, that is,

$$\alpha = +\sqrt{\frac{2h(w + b)}{kwb}} \tag{5-17}$$

Reconsidering what we have done we see that if the exponential function 5-12 is multiplied by a constant, 5-9 still is satisfied. Consequently we have two more possible solutions:

$$\theta = C_1\epsilon^{+\alpha x} \tag{5-18}$$

and

$$\theta = C_2\epsilon^{-\alpha x} \tag{5-19}$$

Each of these satisfies the differential equation regardless of the values of C_1 and C_2; moreover, they include 5-15 and 5-16 as special cases in which the constants equal one.

Although 5-18 and 5-19 each satisfies the differential equation, neither one in itself permits simultaneous fitting of the two conditions 5-10 and 5-11 because in either case there is only one constant, C_1 or C_2, that can be adjusted. It appears, however, from what has been done that a solution with the degree of flexibility necessary to fit two boundary conditions can be had by adding 5–18 and 5–19; thus

$$\theta = C_1\epsilon^{\alpha x} + C_2\epsilon^{-\alpha x} \tag{5–20}$$

EXERCISE. Confirm the fact that 5–20 satisfies the differential equation 5–9, no matter what values are assigned to C_1 and C_2.

Equation 5–20 is a general solution; it satisfies the differential equation no matter what values the constants C_1 and C_2 may have, and these two constants permit it to be adjusted to fit two boundary conditions as we have found to be necessary to determine the solution for any particular situation.

To find the values of C_1 and C_2 for the particular problem at hand, we substitute the general solution 5–20 into the boundary conditions 5–11 and 5–10 and so obtain the following two relationships between C_1 and C_2:

$$C_1 + C_2 = \theta_0 \tag{5–21}$$

$$h(C_1\epsilon^{\alpha L} + C_2\epsilon^{-\alpha L}) = -k(\alpha C_1\epsilon^{\alpha L} - \alpha C_2\epsilon^{-\alpha L}) \tag{5–22}$$

Solving these simultaneously for C_1 and C_2 yields

$$C_1 = \frac{(k\alpha - h)\epsilon^{-\alpha L}\theta_0}{(k\alpha + h)\epsilon^{\alpha L} + (k\alpha - h)\epsilon^{-\alpha L}} \tag{5–23}$$

$$C_2 = \frac{(k\alpha + h)\epsilon^{\alpha L}\theta_0}{(k\alpha + h)\epsilon^{\alpha L} + (h\alpha - h)\epsilon^{-\alpha L}} \tag{5–24}$$

and upon putting these values in 5–20,

$$\theta = \frac{(k\alpha + h)\epsilon^{\alpha(L-x)} + (k\alpha - h)\epsilon^{-\alpha(L-x)}}{(k\alpha + h)\epsilon^{\alpha L} + (k\alpha - h)\epsilon^{-\alpha L}}\,\theta_0 \tag{5–25}$$

If we have made no manipulative errors 5–25 is the particular solution which we have been seeking, for it is so constructed that it should satisfy all the physical conditions which have been imposed. These physical conditions are expressed by the differential equation 5–9 and the particular conditions 5–10 and 5–11 which we found must hold at the boundaries $x = L$ and $x = 0$.

EXERCISE. Check to see that 5–25 is correct dimensionally and that it satisfies 5–9, 5–10, and 5–11.

Interpretation of 5–25 will be made easier if its terms are grouped differently:

$$\theta = \frac{k\alpha[\epsilon^{\alpha(L-x)} + \epsilon^{-\alpha(L-x)}] + h[\epsilon^{\alpha(L-x)} - \epsilon^{-\alpha(L-x)}]}{k\alpha[\epsilon^{\alpha L} + \epsilon^{-\alpha L}] + h[\epsilon^{\alpha L} - \epsilon^{-\alpha L}]}\,\theta_0 \tag{5–26}$$

Also it will be helpful to have the rate of change of θ with respect to x as found by differentiating 5–26:

$$\frac{d\theta}{dx} = \frac{-k\alpha^2[\epsilon^{\alpha(L-x)} - \epsilon^{-\alpha(L-x)}] - h\alpha[\epsilon^{\alpha(L-x)} + \epsilon^{-\alpha(L-x)}]}{k\alpha[\epsilon^{\alpha L} + \epsilon^{-\alpha L}] + h[\epsilon^{\alpha L} - \epsilon^{-\alpha L}]}\,\theta_0 \qquad (5\text{–}27)$$

In general we can say that $\epsilon^A \geq \epsilon^{-A}$ if $A \geq 0$. From this it follows that the denominators of 5–26 and 5–27 must be positive, and for all values of

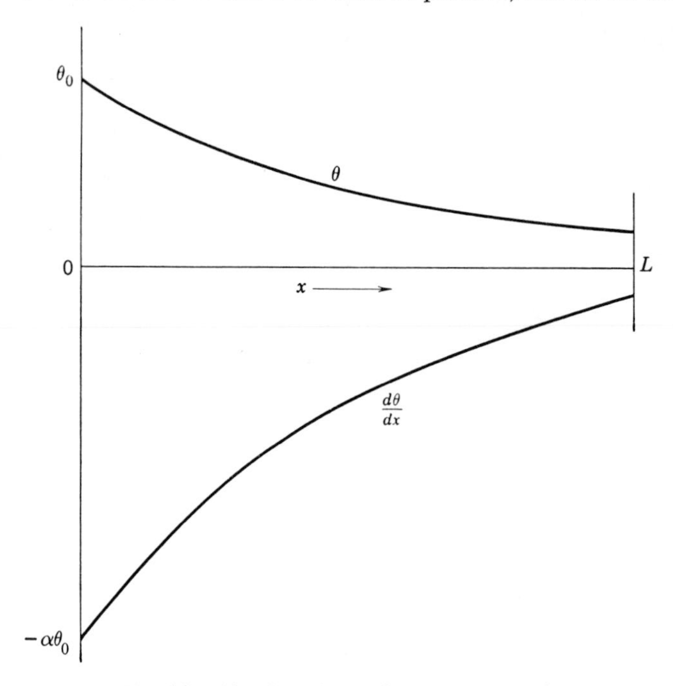

Fig. 5-7. Sketches of equations 5-26 and 5-27.

x from 0 to L the numerator of 5–26 must be positive and that of 5–27 negative. Therefore if the body to which the fin is attached is hotter than the air (θ_0 positive) θ will be positive and $\dfrac{d\theta}{dx}$ will be negative throughout the length of the fin. And this finding from the mathematical expressions is in agreement with the result of purely physical reasoning. It is worthy of note that the whole of the foregoing analysis applies equally well if θ_0 is negative, as would be so if the fin were used to transfer heat from hot air to a cooler body.

The general forms of the results 5–26 and 5–27 can now be sketched as in figure 5–7, where it is assumed that θ_0 has a positive value. These are the curves which also would have been found for a numerical case had we pursued the step-by-step method of solution.

In the study of this problem we have shown that the differential equation alone is not enough to determine the temperature distribution in the cooling fin. By physical reasoning we have found that certain conditions at the ends or boundaries of the fin also need to be known, just as, in the problem on the starting of the fan, the speed at the first instant had to be known. But in the earlier problem involving a first-order differential equation one such condition sufficed, whereas with the second-order equation of the present problem two conditions are necessary. The two conditions used were the value of the function, that is, the temperature, at one boundary of the fin, and a relation between the function and its slope at the other boundary. With different physical connections at the ends of the fin a different pair of boundary conditions would have applied, but the differential equation would have been the same.

Next a function satisfying the differential equation and the two boundary conditions was determined by trial. While this function might have been built up as a curve if numerical values of the parameters had been available, we found it possible in this case to construct a suitable function expressible in mathematical form. This solution was general in the sense that it might have been fitted to boundary conditions other than the pair for the particular problem at hand. The possibility of finding other such general solutions is investigated in the next sections.

5–4 Another Solution for the Fin: Hyperbolic Functions

The form of equation 5–26, the result found in section 5–3 for the temperature distribution in the cooling fin, suggests expression in terms of hyperbolic functions instead of exponentials.

The hyperbolic sine is defined by

$$\sinh ax = \frac{\epsilon^{ax} - \epsilon^{-ax}}{2} \tag{5–28}$$

and the hyperbolic cosine by

$$\cosh ax = \frac{\epsilon^{ax} + \epsilon^{-ax}}{2} \tag{5–29}$$

These functions are sketched in figure 5–8. Using these definitions 5–26 can be written as

$$\theta = \frac{k\alpha \cosh \alpha(L - x) + h \sinh \alpha(L - x)}{k\alpha \cosh \alpha L + h \sinh \alpha L}\, \theta_0 \tag{5–30}$$

which the typesetter will prefer to 5–26 and which is convenient for numerical calculation if a table of hyperbolic functions is at hand. Thus we have a solution to the differential equation in appearance quite unlike the one previously found but which, of course, is exactly equivalent in magnitude at every point.

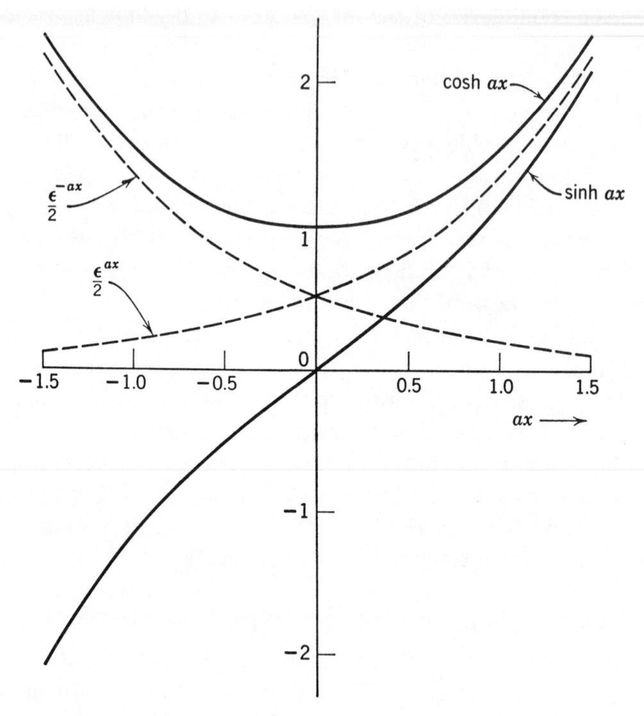

Fig. 5-8. Sketches of the hyperbolic sine and the hyperbolic cosine.

Now let us investigate the possibility of working with hyperbolic functions from the beginning of the solution. The formulas for differentiation of the hyperbolic sine and hyperbolic cosine are easily found from 5–28 and 5–29 to be

$$\frac{d}{dx} \sinh ax = a \cosh ax \tag{5–31}$$

$$\frac{d}{dx} \cosh ax = a \sinh ax \tag{5–32}$$

The second derivative of either of these hyperbolic functions evidently is the function itself multiplied by a^2. This suggests at once that as a solution to 5–9 we try

$$\theta = C \sinh ax \tag{5–33}$$

Upon substituting this in 5–9 we find that the differential equation is satisfied if

$$a = \pm\sqrt{\frac{2h(w+b)}{kwb}} = \pm\alpha \tag{5-34}$$

Since there are two values of a, we might be tempted to try as the whole solution

$$\theta = C_3 \sinh \alpha x + C_4 \sinh(-\alpha x) \tag{5-35}$$

But from 5–28 it is evident that $\sinh(-\alpha x) = -\sinh \alpha x$, and so our tentative solution collapses to

$$\theta = (C_3 - C_4) \sinh \alpha x = C_a \sinh \alpha x \tag{5-36}$$

That is, the two constants enter the equation in such a way that they may be replaced by a single constant C_a, and we can adjust our solution to fit only one of the two boundary conditions. This was not so with C_1 and C_2 in 5–20 (unless $\alpha = 0$, an exceptional circumstance in which 5–20 breaks down as a solution). Hence 5–36 may be part but not all of a solution capable of adjustment to two boundary conditions.

By similar reasoning we find as another part of a possible solution

$$\theta = C_b \cosh \alpha x \tag{5-37}$$

where the constant C_b can have any value whatever.

Like 5–36, 5–37 has but one constant and is not general enough to permit adjustment to fit the two boundary conditions which we have found necessary to determine a particular temperature distribution in the fin. To get a solution having sufficient generality for this we add 5–36 and 5–37:

$$\theta = C_a \sinh \alpha x + C_b \cosh \alpha x \tag{5-38}$$

EXERCISE. Show that 5–38 satisfies the differential equation, then evaluate C_a and C_b so as to satisfy the boundary conditions 5–10 and 5–11, and finally show that the particular solution thus found is identical with 5–20, and hence with 5–26.

Where there is some sort of physical symmetry the general solution 5–38 in terms of hyperbolic functions may result in greater simplicity and ease of numerical computation than the solution 5–20 in terms of exponential functions.

EXERCISE. Find the temperature distribution supposing that the ends of the fin each terminate in hot bodies maintained at equal temperatures. Take the origin $x = 0$ midway between the ends of the fin.

5–5 Solution in Terms of Power Series

The two solutions to the fin problem which have been established respectively in terms of exponential functions and hyperbolic functions were found because insight led us to suspect that these functions would work. There remains the possibility that other functions may exist which also would do the job. To investigate this we shall try a very general solution, a power series in x for the temperature distribution $\theta(x)$.

$$\theta = a_0 + a_1x + a_2x^2 + a_3x^3 + a_4x^4 + \cdots \tag{5-39}$$

The a's are constants which we shall endeavor to determine so that the differential equation 5-9 is satisfied by 5-39. To do this we substitute 5-39 into 5-9, obtaining

$$2a_2 + 2 \cdot 3a_3x + 3 \cdot 4a_4x^2 + 4 \cdot 5a_5x^3 + \cdots = \frac{2h(w + b)}{kwb}[a_0 + a_1x + a_2x^2 + a_3x^3 + \cdots] \tag{5-40}$$

Replacing the coefficient of the series on the right with α^2, already defined by 5-17, and collecting like powers of x, we get

$$(2a_2 - \alpha^2a_0) + (2 \cdot 3a_3 - \alpha^2a_1)x + (3 \cdot 4a_4 - \alpha^2a_2)x^2 + (4 \cdot 5a_5 - \alpha^2a_3)x^3 + \cdots = 0 \tag{5-41}$$

The differential equation is true at every value of x from 0 to L, and if the function 5-39 is to be suitable it must satisfy 5-9 at all values of x. Consequently 5-41 must be true at every value of x. From this it follows that the first term of 5-41 and each coefficient of x must be zero so that the sum of all the terms will be zero no matter what the value of x. Thus we find the following set of relations among the a's:

$$\begin{aligned} 2a_2 - \alpha^2a_0 &= 0 \\ 2 \cdot 3a_3 - \alpha^2a_1 &= 0 \\ 3 \cdot 4a_4 - \alpha^2a_2 &= 0 \\ 4 \cdot 5a_5 - \alpha^2a_3 &= 0 \\ &\vdots \end{aligned}$$

From these we find that all the a's with even subscripts can be expressed in terms of any one of them, for instance, in terms of a_0; thus

$$\begin{aligned} a_2 &= \frac{\alpha^2a_0}{2} = \frac{\alpha^2a_0}{2!} \\ a_4 &= \frac{\alpha^2a_2}{3 \cdot 4} = \frac{\alpha^4a_0}{4!} \\ a_6 &= \frac{\alpha^2a_4}{5 \cdot 6} = \frac{\alpha^6a_0}{6!} \\ &\vdots \end{aligned} \tag{5-42}$$

Similarly the a's with odd subscripts may be expressed in terms of a_1:

$$
\begin{aligned}
a_3 &= \frac{\alpha^2 a_1}{2 \cdot 3} = \frac{\alpha^2 a_1}{3!} \\
a_5 &= \frac{\alpha^2 a_3}{4 \cdot 5} = \frac{\alpha^4 a_1}{5!} \\
a_7 &= \frac{\alpha^2 a_5}{6 \cdot 7} = \frac{\alpha^6 a_1}{7!} \\
&\;\;\vdots \qquad\quad \vdots \qquad\quad \vdots
\end{aligned}
\tag{5-43}
$$

Apparently two of the a's, a_0 and a_1 as we have arranged matters, cannot be determined from the differential equation.

Now let us incorporate the relations 5–42 and 5–43 into the series 5–39 and see how it looks.

$$
\theta = a_0\left[1 + \frac{\alpha^2 x^2}{2!} + \frac{\alpha^4 x^4}{4!} + \frac{\alpha^6 x^6}{6!} + \cdots\right] + a_1\left[x + \frac{\alpha^2 x^3}{3!} + \frac{\alpha^4 x^5}{5!} + \cdots\right] \tag{5-44}
$$

Search through a list of power-series expansions reveals that the first series in 5–44 is identical with cosh αx, and that if the second series is multiplied by α it becomes sinh αx. Thus 5–44 is identically equivalent to

$$
\theta = a_0 \cosh \alpha x + \frac{a_1}{\alpha} \sinh \alpha x \tag{5-45}
$$

But so far as satisfying the differential equation is concerned, a_0 and a_1 may have any values whatever; they must be determined by other conditions, and since they are independent of one another it will take two such conditions to determine them. Thus a_0 and a_1/α play exactly the same role as do C_b and C_a respectively in 5–38, which indeed is identical with 5–45. Thus this latest effort leads us to the same result already found from two other approaches.

There is, however, a significant difference between this last approach and the two former ones. In the earlier approaches we had established, by physical reasoning, that a particular solution to the differential equation would need to be determined by two boundary conditions; therefore our aim was to construct a solution general enough to permit fitting to two conditions. In the last approach, on the other hand, we started by assuming the most general type of solution we could think of—a power series with coefficients undetermined—and found that even after this function was adjusted to satisfy the differential equation it necessarily contained two independent constants which could have any values, and consequently that to find a particular solution from it would require two boundary conditions.

5-6 The General Solution of an Ordinary Differential Equation

For the two differential equations investigated so far, those for the starting time of a fan and for the temperature distribution in a cooling fin, we found the possibility of an infinite number of particular solutions, each one corresponding to particular initial or boundary conditions but all satisfying the differential equation. In the fan problem the speed at the first instant might have any given value, and corresponding to each such initial speed would be a different particular solution to the differential equation. With the cooling fin we saw the possibility of different sets of conditions at the boundaries, each set giving rise to a particular solution of the differential equation. Also, in solving this problem we used the concept of a *general solution*, that is, one which can be adjusted through proper evaluation of constants contained in it to fit any physically consistent set of boundary conditions and so reduce to a particular solution.

The adjustable constants in the general solution are called *arbitrary constants* because they can be given any values whatever and still the differential equation will be satisfied. With the second-order differential equation of the fin problem we saw that there had to be two arbitrary constants to permit fitting the two boundary conditions found by physical reasoning to be necessary. The two boundary conditions are entirely independent of each other, but both must be fitted; consequently the arbitrary constants must occur in the general solution in such a way as to permit this independent fitting of conditions. For this reason the constants are further characterized as *independent* arbitrary constants.

In showing that the general solution to our second-order differential equation must contain two independent arbitrary constants we discovered one instance of an important truth which is generalized in the mathematical theory of differential equations as follows:

The general solution to any ordinary differential equation of n*th order (the order of the highest derivative) must contain* n *independent arbitrary constants.*

That this truth is already familiar is realized if the solving of a differential equation can be considered as a series of integrations. To remove the nth derivative and those of lower order there must be n integrations. Each integration introduces a new constant of integration; hence there will be n constants of integration in the solution, and these are precisely what we have been calling the independent arbitrary constants. In the fan problem we solved the differential equation by a simple integration because the variables were separable. We might have used indefinite instead of definite integrals in 5–6, obtaining

$$t = I \int \frac{1}{T_M(\omega) - T_F(\omega)} \, d\omega$$

and then t would be some function of ω, say $F(\omega)$, the derivative of which is equal to the integrand $[T_M(\omega) - T_F(\omega)]^{-1}$, plus any constant C or $t = F(\omega) + C$, and C is the arbitrary constant that appears in this method of solution. That there is but one constant corresponds to the fact that the differential equation is of first order. In the cooling fin problem we did not separate the variables and so did not obtain the solution by simple integration; nevertheless what we were doing was integrating, and this is what happens in many problems as we shall see in later sections.

After finding one general solution to the differential equation for the cooling fin we tried in sections 5–4 and 5–5 to find other general solutions. We succeeded in finding two other forms, 5–38 and 5–44, which certainly differ in appearance from the original solution, 5–20, but which nevertheless are identically equivalent to it. Thus try as we would in this instance we could find no more than one general solution. Here we have evidence of another generalization of mathematics, namely, that a solution having the requisite number of independent arbitrary constants *is* the general solution, and that once such a solution is found there is no need to look further.

From a practical point of view we can say that to solve an ordinary differential equation arising in a particular problem we must establish by physical reasoning the boundary or initial conditions. Then by some process, systematic or otherwise, we find a solution which satisfies the differential equation and the boundary or initial conditions. It is shown in the theory of differential equations that such a solution is unique. A value of the generalization on page 196 is that it tells us how many boundary or initial conditions need to be established, something which otherwise is not always easy to see.

5–7 Differential Equation for Free Vibration without Damping

To illustrate the use of the foregoing generalizations and to study further the solution of differential equations, consider the idealized mechanical system shown in figure 5–9. It consists of a mass of M slugs constrained to horizontal motion and connected to a spring having a constant of K lb ft^{-1}.

Assuming that the forces of friction and damping are negligible we wish to determine the motion of the mass after some initial disturbance from its position of equilibrium. With these simplifications the only force acting on M is that due to the extension or compression of the spring. By Newton's law this force must equal the mass M times its acceleration. In symbols,

$$-Kx = M\frac{d^2x}{dt^2} \tag{5-46}$$

where t is time in seconds and x is the coordinate in feet, positive to the right, locating M with respect to its position when the spring is unstressed. Equation 5-46 may be checked with earlier work by noting that it could

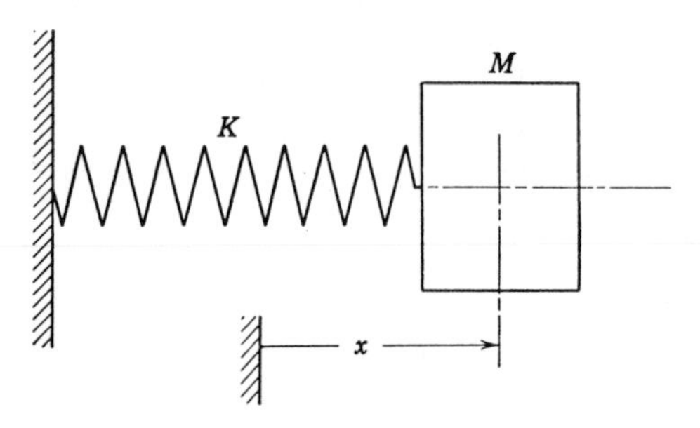

Fig. 5-9. Mass and spring system constrained to horizontal motion.

have been derived from equation 4-1 by setting the forces of gravity, damping, and friction equal to zero.

Our problem is to find the displacement x as a function of time t after M is set in motion; that is, we must solve the second-order differential equation 5-46. To do this we may start by making full use of physical insight. From experience we know that such a system will vibrate indefinitely if the mass is displaced from equilibrium and then released; that is to say, we expect M to describe harmonic motion. Therefore we believe the solution will have sinusoidal form. Reasoning in a different way we might have observed that 5-46 says the solution must be a function whose second derivative is of the same form as the function itself except for a constant multiplier and a reversal of sign. Search among the commonly encountered functions such as C_0, C_1t, C_2t^2, $C_3{}^{mt}$, $C_4 \cos \omega t$ and $C_5 \sin \omega t$ where m, ω, and the C's are constants would then show us that both the sine and the cosine have the desired property. So on either basis we are led to try as a solution $C \cos \beta t$ or $C \sin \beta t$ where β is the angular frequency and has yet to be determined. But either of these functions by itself would have only one arbitrary constant, and by the generalizations made in section 5–6 we must find a solution for this second-order equation having two independent arbitrary constants. Thus we try

$$x = C_1 \cos \beta t + C_2 \sin \beta t \tag{5–47}$$

and test by substituting it in 5–46. We find that 5–46 is satisfied if

$$\beta = +\sqrt{\frac{K}{M}} \tag{5–48}$$

EXERCISE. Considering the properties of the sine and the cosine, justify dropping the negative value of the root in 5–48.

The quantity $\sqrt{\frac{K}{M}}$ is known as the *natural angular frequency* and, since t in this case is measured in seconds, is in radians per second. It is the frequency which characterizes the free vibration of the system.

Thus we have found as the general solution to the differential equation 5–46,

$$x = C_1 \cos \sqrt{\frac{K}{M}}\, t + C_2 \sin \sqrt{\frac{K}{M}}\, t \tag{5–49}$$

and our next task is to evaluate the arbitrary constants C_1 and C_2. To do this two independent physical conditions are necessary, and to decide what they must be we consider how the system is set in motion, that is, we seek initial conditions. In our statement of the problem we did not include the manner of starting the motion. It could be started by moving the mass to one side and then releasing it from rest, that is, with initial velocity zero. Another way would be to impart an initial velocity with the displacement x initially zero, perhaps by giving M a hammer blow. In general the motion might be started by giving M both an initial displacement x_0 and an initial velocity v_0, and this is what we shall assume, taking as initial conditions:

$$x\Big]_{t=0} = x_0 \tag{5–50}$$

$$\frac{dx}{dt}\Big]_{t=0} = v_0 \tag{5–51}$$

It is worthy of note in passing that to start motion of the system comprising the mass and spring requires an initial supply of energy from outside since during the motion there are no external forces to do work on the system. To displace the mass initially requires stretching or compressing the spring, and this involves the storage of strain energy in the spring. Giving the mass an initial velocity is to impart to it kinetic energy. In writing 5–50 and 5–51 we are supposing initial supplies of energy in each of these forms.

Substituting conditions 5–50 and 5–51 into 5–49 we find that C_1 and C_2 respectively are x_0 and $v_0\sqrt{\dfrac{M}{K}}$, and hence that the particular solution we seek is

$$x = x_0 \cos \sqrt{\frac{K}{M}}\, t + v_0 \sqrt{\frac{M}{K}} \sin \sqrt{\frac{K}{M}}\, t \tag{5–52}$$

This result is interpreted by sketched plots in figure 5–10 where the motion is shown for three different initial velocities but a single initial displacement.

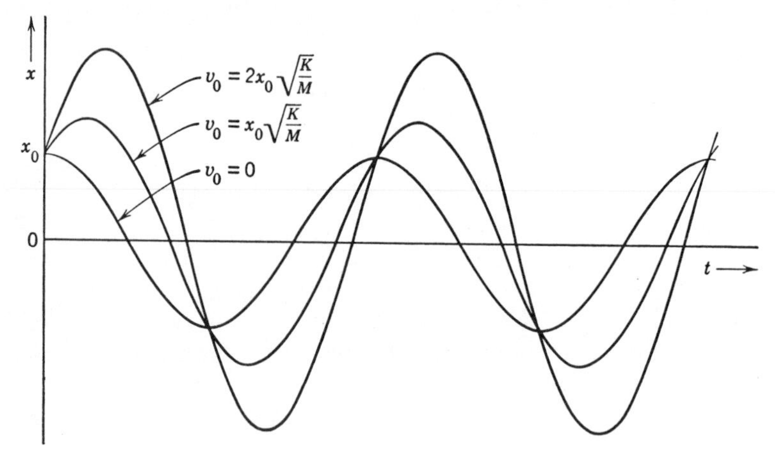

Fig. 5-10. Sketches of equation 5-52 showing the motion of the system of figure 5-9.

The forms of the curves in figure 5–10 suggest that the sine and cosine terms in 5–52 may be combined advantageously into a single sine or cosine function of time displaced along the t-axis by a phase angle. Indeed, as alternatives to the general solution 5–49 we could have used either

$$x = A_1 \cos\left(\sqrt{\frac{K}{M}}\, t + \phi_1\right) \tag{5–53}$$

or

$$x = A_2 \sin\left(\sqrt{\frac{K}{M}}\, t + \phi_2\right) \tag{5–54}$$

Each of these satisfies the differential equation 5–46 and each has the requisite number of arbitrary constants, A_1 and ϕ_1, or A_2 and ϕ_2.

Exercise. Choose either 5–53 or 5–54, check to see that it satisfies 5–46, then use 5–50 and 5–51 to evaluate the arbitrary constants, and finally show that the particular solution thus found is identically equivalent to 5–52.

We have solved 5–46 as we set out to do, but our procedure involved starting with the fact derived through insight that the solution would have to be a cosine or sine function. Suppose that we had not thought of these functions; suppose that we did not even notice the essential difference in the forms of 5–9 and 5–46, that is, the minus sign in 5–46, and had hastily concluded that what worked with 5–9 would also work with 5–46. To solve 5–9 we tried the exponential form 5–12. Proceeding similarly here, let us attempt to solve 5–46 by trying

$$x = C\epsilon^{mt} \tag{5-55}$$

Substituting this into differential equation 5–46 and dividing out the common factor $C\epsilon^{mt}$, which is certainly not zero, we get

$$-K = Mm^2 \tag{5-56}$$

from which it follows that 5–55 is a solution provided that

$$m = \pm\sqrt{-\frac{K}{M}}$$

That is, the two values of m are imaginaries and may be written

$$m = \pm j\sqrt{\frac{K}{M}} \tag{5-57}$$

where
$$j \equiv \sqrt{-1} \tag{5-58}$$*

Thus we have a solution with two arbitrary constants

$$x = C_a\epsilon^{j\sqrt{K/M}t} + C_b\epsilon^{-j\sqrt{K/M}t} \tag{5-59}$$

but which, at least superficially, does not resemble the solutions in sinusoidal form already found. There are, however, two relationships between exponentials with imaginary exponents and trigonometric functions. These, named Euler's relations, are

$$\epsilon^{j\theta} = \cos\theta + j\sin\theta \tag{5-60}$$

$$\epsilon^{-j\theta} = \cos\theta - j\sin\theta \tag{5-61}$$

Actually these may be regarded as a single relationship since the second can be obtained by replacing θ in the first by $-\theta$.

* Mathematicians and engineers other than electrical conventionally use i instead of j to represent $\sqrt{-1}$. Electrical engineers have adopted j to avoid confusion with the symbol they customarily use for current. The authors use j since many of the problems considered here involve current.

EXERCISE. Verify either of the Euler relations by reference to expansions of each of the terms in a power series.

Euler's relations may be interpreted geometrically if we think of a complex number as a vector resolved into two components at right angles to each other: the real component and the imaginary component. Representing $\cos\theta$ along the axis of real numbers, as in figure 5-11, and $\sin\theta$

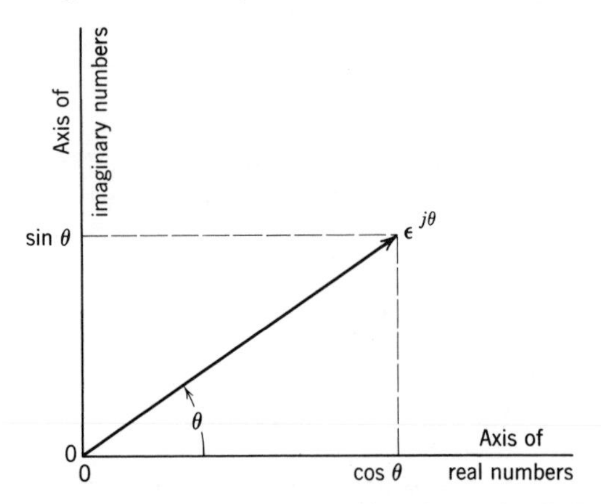

Fig. 5-11. Geometric interpretation of equation 5-60.

along the axis of imaginary numbers we see that $\epsilon^{j\theta}$ is a vector of unit length making an angle θ with the axis of reals. Similarly $\epsilon^{-j\theta}$ is a unit vector at a negative angle θ.

EXERCISE. Make a geometric interpretation of $A^{j\beta t}$ where A and β are constants and t is time. Consider the behavior of the real and imaginary components as t increases from zero.

Using Euler's relations in 5-59 to replace the exponentials with trigonometric functions we find at once that

$$x = (C_a + C_b)\cos\sqrt{\frac{K}{M}}\,t + j(C_a - C_b)\sin\sqrt{\frac{K}{M}}\,t \tag{5-62}$$

Then comparing this with the former general solution 5-49 we see that the two are equivalent if

$$C_a + C_b = C_1 \tag{5-63}$$

and

$$i(C_a - C_b) = C_2 \tag{5-64}$$

Solving these simultaneously we find

$$C_a = \tfrac{1}{2}(C_1 - jC_2)$$

$$C_b = \tfrac{1}{2}(C_1 + jC_2)$$

Thus if C_1 and C_2 represent real numbers, as they must in a physical problem, then C_a and C_b are complex numbers. And then we observe the interesting fact that, although each of the two terms of the solution 5–59 is complex, the sum is real.

We are now in a position to generalize to a certain extent by saying that equations of the form of either 5–9 or 5–46 may be solved by assuming an exponential form such as 5–12 or 5–55. If the exponents come out imaginary then Euler's relations 5–60 and 5–61 are used to convert to trigonometric functions as a convenience in interpretation. It may be convenient also even when the exponents are real to convert to another form of solution, the hyperbolic functions as indicated in section 5–4.

EXERCISE. Solve 5–28 and 5–29 simultaneously for ϵ^{ax} and ϵ^{-ax} and observe the parallel with Euler's relations 5–60 and 5–61.

5–8 Damped Free Vibration

The preceding section dealt with the vibration of a spring and mass system which was idealized by ignoring any frictional forces which in practice would be present at least to some small degree. Now we wish to see how the solution to the problem is affected by the inclusion of one kind of friction, a damping force proportional to the first power of the velocity of the mass. Such damping may approximate the effect of air friction or, more exactly, the effect of a dashpot in which the friction comes from forcing a viscous fluid through small openings. The system to be studied is represented in figure 5–12. We wish to find the motion of the mass M, that is, x as a function of time t, after an initial disturbance which is either a displacement or a velocity or both.

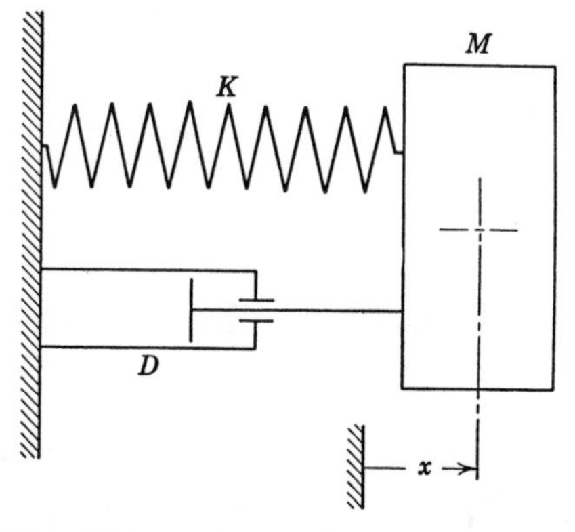

Fig. 5-12. Mass, spring, and dashpot system constrained to horizontal motion.

We shall start by applying Newton's law to M exactly as we did in section 5–7 except that now there is an additional force acting on the mass, one proportional to the velocity of M and directed always so as to oppose this velocity. Thus we have

$$-D\frac{dx}{dt} - Kx = M\frac{d^2x}{dt^2} \tag{5-65}$$

where D is the dashpot constant or damping coefficient in lb sec ft^{-1} and the other symbols are as defined in the first paragraph of section 5–7.

To solve 5–65 for x as a function of t we begin by seeing what physical insight can tell us. If the force of damping is quite small, as, for instance, due to air friction alone, we would expect M to vibrate after displacement from equilibrium somewhat as we found in section 5–7 except that because of the friction we would expect the vibrations to die down gradually and that ultimately M would come to rest. Thus the solution might include sine or cosine functions of time, but less simply than in the solution for no damping; something in the expression would have to make the amplitude decrease with time. On the other hand if the damping force were made very large, perhaps by filling the dashpot with a very viscous fluid such as molasses, we would not expect M to vibrate but rather to return slowly, perhaps *very* slowly, to its position where the spring is unstressed. In such a case we would hardly expect the solution to involve sine or cosine functions of time; exponential functions decaying slowly with time would seem far more reasonable. So with little damping the solution conceivably could include sines and cosines in some way, but with large damping the solution is much more likely to have the character of decaying exponential functions.

Now in the latter part of section 5–7 we saw that exponential functions can be reduced to sine and cosine functions if the exponents are imaginary instead of real numbers. With this, then, as a lead we try as a solution to 5–65 the function

$$x = C\epsilon^{mt} \tag{5–66}$$

hoping that it may satisfy the differential equation and include all the possible solutions.

Let us systematize the work of substituting 5–66 into the differential equation by first arranging the terms of 5–65 in descending order of derivatives; thus

$$M\frac{d^2x}{dt^2} + D\frac{dx}{dt} + Kx = 0 \tag{5–67}$$

Then upon substituting 5–66

$$Mm^2C\epsilon^{mt} + DmC\epsilon^{mt} + KC\epsilon^{mt} = 0$$

or

$$C\epsilon^{mt}(Mm^2 + Dm + K) = 0$$

Thus if 5–66 is to satisfy the differential equation, either $C = 0$, which is not a useful solution, or ϵ^{mt} would have to be zero, which it is not for any value of t greater than zero, or

$$Mm^2 + Dm + K = 0 \tag{5–68}$$

which appears to be possible, if m has the right value or values. To determine what these values might be we solve the quadratic and find two values of m:

$$m = -\frac{D}{2M} \pm \sqrt{\left(\frac{D}{2M}\right)^2 - \frac{K}{M}} \tag{5–69}$$

and we notice in passing that if $D = 0$ this result reduces as it should to 5–57.

Corresponding to the two values of m in 5–69 we have found two functions of the form of 5–66 which satisfy the differential equation, and by taking their sum we can provide for two independent arbitrary constants. By the generalizations made in section 5–6 we now have the general solution to the second-order differential equation 5–65. It is

$$x = C_a \epsilon^{\left[-\frac{D}{2M} + \sqrt{\left(\frac{D}{2M}\right)^2 - \frac{K}{M}}\right]t} + C_b \epsilon^{\left[-\frac{D}{2M} - \sqrt{\left(\frac{D}{2M}\right)^2 - \frac{K}{M}}\right]t} \tag{5–70}$$

Now we must consider what 5–70 means physically. First we would like to know if it fulfills our prediction that the motion would be vibratory when the damping is very small. If the damping is absent entirely, $D = 0$, the exponents become imaginary, and 5–70 reduces identically to 5–59, which we already have seen can be expressed in terms of sinusoidal functions. If the damping is not zero but small enough that $(D/2M)^2 < K/M$ the exponents in 5–70 will be complex numbers. The value of damping which makes the radicals zero is $D = 2\sqrt{KM}$ and is called *critical damping*. Let us try to interpret 5–70 in the range of less than critical damping, that is, in the range of D:

$$0 < D < 2\sqrt{KM} \tag{5–71}$$

To do this we factor -1 out of the quantity under the radical in the exponents and then represent $\sqrt{-1}$ by j as in section 5–7. The result is

$$x = C_a \epsilon^{\left[-\frac{D}{2M} + j\sqrt{\frac{K}{M} - \left(\frac{D}{2M}\right)^2}\right]t} + C_b \epsilon^{\left[-\frac{D}{2M} - j\sqrt{\frac{K}{M} - \left(\frac{D}{2M}\right)^2}\right]t} \tag{5–72}$$

Now we can take out the factor $\epsilon^{-\frac{Dt}{2M}}$; thus

$$x = \epsilon^{-\frac{Dt}{2M}} \left[C_a \epsilon^{j\sqrt{\frac{K}{M} - \left(\frac{D}{2M}\right)^2}\,t} + C_b \epsilon^{-j\sqrt{\frac{K}{M} - \left(\frac{D}{2M}\right)^2}\,t} \right]$$

Next the quantity in parenthesis can be put in terms of sinusoidal functions through the use of Euler's relations 5–60 and 5–61 exactly as was done in converting 5–59 to 5–62. Doing this we find

$$x = \epsilon^{-\frac{Dt}{2M}} \left[C_1 \cos \sqrt{\frac{K}{M} - \left(\frac{D}{2M}\right)^2}\, t + C_2 \sin \sqrt{\frac{K}{M} - \left(\frac{D}{2M}\right)^2}\, t \right] \qquad (5\text{–}73)$$

where C_1 and C_2 are new arbitrary constants related to the original ones by 5–63 and 5–64.

To evaluate C_1 and C_2 so as to obtain a particular solution let us suppose that the motion is started by giving the mass M an initial displacement x_0 and an initial velocity v_0. That is, we take as initial conditions

$$x\Big]_{t=0} = x_0 \qquad (5\text{–}74)$$

$$\frac{dx}{dt}\Big]_{t=0} = v_0 \qquad (5\text{–}75)$$

For convenience in carrying out the algebra let us define the following quantities:

$$\alpha = \frac{D}{2M} \qquad (5\text{–}76)$$

$$\beta = \sqrt{\frac{K}{M} - \left(\frac{D}{2M}\right)^2} \qquad (5\text{–}77)$$

Then 5–73 may be written

$$x = \epsilon^{-\alpha t}(C_1 \cos \beta t + C_2 \sin \beta t) \qquad (5\text{–}78)$$

and the velocity as

$$\frac{dx}{dt} = \epsilon^{-\alpha t}[(-\alpha C_1 + \beta C_2) \cos \beta t - (\alpha C_2 + \beta C_1) \sin \beta t] \qquad (5\text{–}79)$$

Upon substituting the initial conditions 5–74 and 5–75 in 5–78 and 5–79, the following values for C_1 and C_2 are found:

$$C_1 = x_0 \qquad (5\text{–}80)$$

$$C_2 = \frac{v_0 + \alpha x_0}{\beta} \qquad (5\text{–}81)$$

and thus the expression for the motion following these initial conditions becomes

$$x = \epsilon^{-\alpha t} \left[x_0 \cos \beta t + \frac{v_0 + \alpha x_0}{\beta} \sin \beta t \right] \qquad (5\text{–}82)$$

Now let us restore the values of α and β so as not to lose sight of physical meanings

$$x = \epsilon^{-\frac{D}{2M}t}\left[x_0 \cos\sqrt{\frac{K}{M} - \left(\frac{D}{2M}\right)^2}\, t + \frac{v_0 + \frac{D}{2M}x_0}{\sqrt{\frac{K}{M} - \left(\frac{D}{2M}\right)^2}} \sin\sqrt{\frac{K}{M} - \left(\frac{D}{2M}\right)^2}\, t\right] \tag{5-83}$$

This result is sketched in figure 5-13 for the case of release from rest ($v_0 = 0$) as the curves for $D = 0.5(2\sqrt{KM})$ and for $D = 0.1(2\sqrt{KM})$,

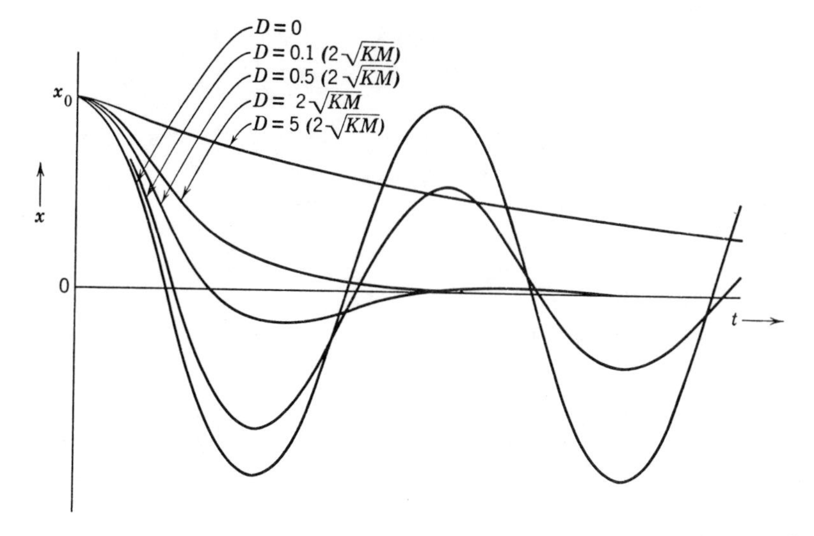

Fig. 5-13. Motion of the system of figure 5-12 started from an initial displacement for various values of damping.

that is, damping equal to one-half and one-tenth the critical value respectively, and for $D = 0$. The latter curve is the same as the one in figure 5-10 for zero initial velocity. Figure 5-13 includes also some curves resulting from the analysis given in the remainder of this section.

Having investigated the solution for the case when $0 \leq D < 2\sqrt{KM}$ let us return to the general solution 5-69 and see what happens when D is above the critical value; that is, we wish to study the effect of damping in the range

$$D > 2\sqrt{KM} \tag{5-84}$$

In this range the quantity under the square-root signs in 5-70 is positive and consequently the exponents are real quantities. Moreover, both are negative no matter how large D is made. Thus in this range of large damping the solution confirms our guess that it would be in terms of negative exponential functions.

To investigate 5-70 in more detail in the overdamped case, let us assume the same initial conditions 5-74 and 5-75 as for the underdamped case already studied. If we evaluate the arbitrary constants C_a and C_b for these initial conditions, there results:

$$x = \frac{1}{2\sqrt{\left(\frac{D}{2M}\right)^2 - \frac{K}{M}}}\left[\left(v_0 + \frac{D}{2M}x_0 + \sqrt{\left(\frac{D}{2M}\right)^2 - \frac{K}{M}}\,x_0\right)\epsilon^{-\left[\frac{D}{2M} - \sqrt{\left(\frac{D}{2M}\right)^2 - \frac{K}{M}}\right]t} - \left(v_0 + \frac{D}{2M}x_0 - \sqrt{\left(\frac{D}{2M}\right)^2 - \frac{K}{M}}\,x_0\right)\epsilon^{-\left[\frac{D}{2M} + \sqrt{\left(\frac{D}{2M}\right)^2 - \frac{K}{M}}\right]t}\right] \tag{5-85}$$

This result is illustrated in figure 5-13 by the curve for $D = 5(2\sqrt{KM})$, that is, for damping equal to five times the critical value. Equation 5-85 may be put in terms of hyperbolic functions with a result which presents an interesting parallel with 5-83 for the underdamped case. Thus

$$x = \epsilon^{-\frac{Dt}{2M}}\left[x_0 \cosh\sqrt{\left(\frac{D}{2M}\right)^2 - \frac{K}{M}}\,t + \frac{v_0 + \frac{D}{2M}x_0}{\sqrt{\left(\frac{D}{2M}\right)^2 - \frac{K}{M}}}\sinh\sqrt{\left(\frac{D}{2M}\right)^2 - \frac{K}{M}}\,t\right] \tag{5-86}$$

In finding expressions for the solutions in the two ranges $D < 2\sqrt{KM}$ and $D > 2\sqrt{KM}$ we passed over the special case where $D = 2\sqrt{KM}$, that is, the case of critical damping. Putting the critical value of D in 5-69 gives only one value of m and thus leads to the solution

$$x = C\epsilon^{-\sqrt{\frac{K}{M}}t} \tag{5-87}$$

which contains only one arbitrary constant, C. Therefore 5-87 is not the general solution because it does not contain *two independent* arbitrary constants. We must seek the general solution for the critical case in some more subtle way.

Let us try a modified form of the solution that worked before, namely, the product of a constant and an exponential function as in 5–66. One modification that occurs to us is that of keeping the exponential part of the solution (which works but gives us only one arbitrary constant) but replacing the constant by a function of time. Thus we think of

$$x = f(t)\epsilon^{mt}$$

What function of time shall we try? In a previous case, section 5–5, we had some success in representing a general function by a power series, so we might try

$$f(t) = a_0 + a_1 t + a_2 t^2 + a_3 t^3 + a_4 t^4 + \cdots$$

or

$$x = (a_0 + a_1 t + a_2 t^2 + a_3 t^3 + \cdots)\epsilon^{mt} \tag{5–88}$$

where we hope to be able to adjust m, and if necessary the a's, to make the solution satisfy the equation. We compare the new trial solution 5–88 with 5–87, which does satisfy the equation, and are encouraged by the fact that 5–87 is similar to the first product of 5–88, $a_0\epsilon^{mt}$. It might even happen that the desired solution of 5–67 is the first two products of 5–88,

$$x = (a_0 + a_1 t)\epsilon^{mt} \tag{5–89}$$

where a_0 and a_1 would be the two independent arbitrary constants.

To test 5–88 we substitute it in differential equation 5–67 and carry along, say, three terms, one more than may be needed:

$$\begin{aligned} M[(m^2a_0 + 2ma_1 + 2a_2) + (m^2a_1 + 4a_2m)t + m^2a_2t^2]\epsilon^{mt} \\ + D[(ma_0 + a_1) + (ma_1 + 2a_2)t + ma_2t^2]\epsilon^{mt} \\ + K(a_0 + a_1t + a_2t^2)\epsilon^{mt} = 0 \end{aligned}$$

and the question is, can a_0, a_1, a_2, and m be chosen to satisfy this equation? We rearrange and collect terms according to variation with time, and, by the a's:

$$\begin{aligned} \epsilon^{mt}\{(m^2M + mD + K)a_0 + (2mM + D)a_1 + 2Ma_2 \\ + t[(m^2M + mD + K)a_1 + 2(2mM + D)a_2] \\ + t^2(m^2M + mD + K)a_2\} = 0 \end{aligned} \tag{5–90}$$

The factor ϵ^{mt} cannot be zero; moreover, if the left side of the equation is to be zero independently of t, each line must separately be zero. This may be accomplished if

$$m^2M + mD + K = 0 \tag{5–91}$$

$$2mM + D = 0 \tag{5–92}$$

$$a_2 = 0 \tag{5–93}$$

and, moreover, we do not see any other useful way to satisfy the equation ($a_2 = a_1 = a_0 = 0$ also satisfies it but corresponds to $x = 0$, or physically, to the fact that the system may be at rest; but this does not interest us).

The first of the conditions, 5–91, is precisely the same equation encountered before as 5–68. In the case under consideration, where $D = 2\sqrt{KM}$, both 5–91 and 5–92 are satisfied by the same value of m, namely, $m = -\sqrt{\frac{K}{M}}$. It is interesting and fortunate that this is true, that different values of m are not required to satisfy the two conditions. With $a_2 = 0$, which is the third condition, 5–93, the trial solution reduces to 5–89 and so this actually does satisfy the differential equation in the case of critical damping. Moreover, from 5–90 it does not matter what values are assigned to a_0 and a_1; these are the arbitrary constants. From 5–89 it is clear that they are independent. Hence the solution for the critically damped case is

$$x = (a_0 + a_1 t)\epsilon^{-\sqrt{\frac{K}{M}}t} \tag{5-94}$$

Exercise. Put $D = 2\sqrt{KM}$ in differential equation 5–67 and check by substitution that it is satisfied by solution 5–94.

Choosing the same initial conditions 5–74 and 5–75 as for the other cases and evaluating the arbitrary constants a_0 and a_1 in 5–94, we obtain

$$x = \epsilon^{-\sqrt{\frac{K}{M}}t}\left[x_0 + \left(v_0 + \sqrt{\frac{K}{M}}\,x_0\right)t\right] \tag{5-95}$$

which is represented for the case $v_0 = 0$ in figure 5–13 by the sketched curve for $D = 2\sqrt{KM}$.

Summing up, we have solved our problem, which was to find the effect of damping on the free vibration of a simple spring-and-mass system. To interpret the solution to the differential equation describing this situation, we found it convenient to introduce the concept of critical damping and to consider the three cases where the damping actually present is less than, greater than, and equal to critical damping. We see that the critically damped case separates the cases where the motion is oscillatory (underdamped) from those which are non-oscillatory (overdamped). These results are discussed further in section 7–8.

The differential equation we have studied here is important because it is an example of a type which occurs often in engineering analysis, and a number of points we have encountered in solving it have much wider application. These matters are generalized in section 5–9.

5-9 Homogeneous Linear Differential Equations with Constant Coefficients

The differential equation solved in the preceding section is

$$M \frac{d^2x}{dt^2} + D \frac{dx}{dt} + Kx = 0 \tag{5-96}$$

Its solution was found by a procedure which applies to a whole class of equations: homogeneous linear differential equations with constant coefficients. The meanings of the descriptive terms which characterize such equations are discussed in the next three paragraphs.

Equation 5-96 is said to be *homogeneous* in x and its derivatives because each term contains either x or one of its derivatives to the same power, the first in this case. One way to make 5-96 non-homogeneous would be to replace the zero on the right with a constant or function of t; we deal with such equations in sections 5-10 through 5-13.

Equation 5-96 is *linear*, meaning that the dependent variable x and its derivatives appear only to the first power. It would be non-linear if, for instance, the third term were kx^3 instead of Kx. We already have met and solved a non-linear differential equation, 5-3, for the starting of the motor-driven fan. That equation is non-linear because of the non-linear dependence of $T_M(\omega)$ and $T_F(\omega)$ on the dependent variable ω.

Finally, 5-96 has *constant coefficients*, that is, M, D, and K are constants. The following differential equation is both homogeneous in y and its derivatives and linear, but its coefficients are not all constants; that of the first derivative is $1/x$.

$$\frac{d^2y}{dx^2} + \frac{1}{x}\frac{dy}{dx} + ky = 0 \tag{5-97}$$

This equation, it may be noted in passing, arises, among other places, in the analysis of the vibration of a circular membrane. Its solution comes out in terms of Bessel functions which, although outside the scope of this book, are inherently no more difficult to work with than the familiar exponential, trigonometric, and hyperbolic functions.

A general nth order homogeneous linear differential equation with constant coefficients may be written

$$a_n \frac{d^ny}{dt^n} + a_{n-1} \frac{d^{n-1}y}{dt^{n-1}} + \cdots + a_0y = 0 \tag{5-98}$$

in which y and t are the dependent and independent variables respectively, and the coefficients $a_n, a_{n-1} \cdots a_0$ are constants. It should be clear that in the preceding sections we have solved two other equations besides 5–96 which are of the same type as 5–98; they are 5–9 for the cooling fin and 5–46 for the undamped mass-and-spring system.

To solve 5–98 we assume the solution $y = C\epsilon^{mt}$, which upon substitution results in

$$a_n m^n + a_{n-1} m^{n-1} + \cdots + a_0 = 0 \tag{5-99}$$

This is called the *characteristic equation.* In these terms 5–13 is the characteristic equation for the differential equation 5–9; similarly 5–56 is that for 5–46; and 5–68 that for 5–67. Evidently the characteristic equation can be written down by inspection of the differential equation without need for the formality of actually substituting the exponential. A differential equation of nth order clearly will have a characteristic equation of nth degree, and this, of course, will have n roots; that is, it will be satisfied by n values of m.

If n is no higher than 2, finding the values of m is easy; if n is higher than 2 the task is likely to be formidable. For solving cubic and higher-degree algebraic equations various methods are available,* but they will not be discussed here. That solving the differential equation becomes a problem in solving an algebraic equation is not a consequence of the particular procedure we are describing here. All analytical methods for solving 5–98 require at some point finding the roots of the characteristic equation 5–99.

Having found the n roots of 5–99 ($m = m_1, m_2, \cdots m_n$) the general solution of 5–98 may be written as

$$y = C_1\epsilon^{m_1 t} + C_2\epsilon^{m_2 t} + \cdots + C_n\epsilon^{m_n t} \tag{5-100}$$

except when some of the values of m are equal to each other, the case of repeated roots. If for example, the first four roots are equal and written as m_r, then the first four terms of 5–100 *must* be replaced by

$$(C_1 + C_2 t + C_3 t^2 + C_4 t^3)\epsilon^{m_r t} \tag{5-101}$$

We encountered an example of this exception in the case of critical damping near the end of section 5–8.

Some roots of 5–99 may be complex. If so, they always will occur as conjugate pairs; that is, the real parts will be equal, and the imaginary parts also equal but of opposite sign. For instance, suppose this is the

* See, for example, Rosenbach and Whitman, *College Algebra*, Third Edition, Ginn, Boston, 1949, pp. 325–332, which also contains other references; and Doherty and Keller, *Mathematics of Modern Engineering*, Vol. I, Wiley, New York, 1936, pp. 94–130, with other references on p. 301.

case with m_1 and m_2 and that they are given by

$$m_1 = \alpha + j\beta \tag{5-102}$$

$$m_2 = \alpha - j\beta \tag{5-103}$$

It is then convenient to replace the part of the solution 5–100 corresponding to m_1 and m_2 by

$$\epsilon^{\alpha t}(C_a \cos \beta t + C_b \sin \beta t) \tag{5-104}$$

Instances of this have been met in sections 5–7 and 5–8.

Also, it may be convenient to employ still other forms of the general solution, as we have seen in the preceding sections. For instance, the cosine and sine terms in 5–104 may be replaced by a single cosine or sine with phase angle, and sometimes the negative exponential solutions can be replaced to advantage by hyperbolic sines and cosines.

5–10 A Non-Homogeneous Equation: Integrating Circuit

The circuit in figure 5–14 containing inductance of L henries and resistance of R ohms in series is sometimes used in automatic control or measurement devices to perform electrically the operation of integrating a function with respect to time. Under certain conditions the current i or the voltage drop Ri across the resistance is a faithful representation of the time integral of the voltage e applied to the circuit.

To see this, apply Kirchhoff's voltage law around the circuit

$$e = L\frac{di}{dt} + Ri \tag{5-105}$$

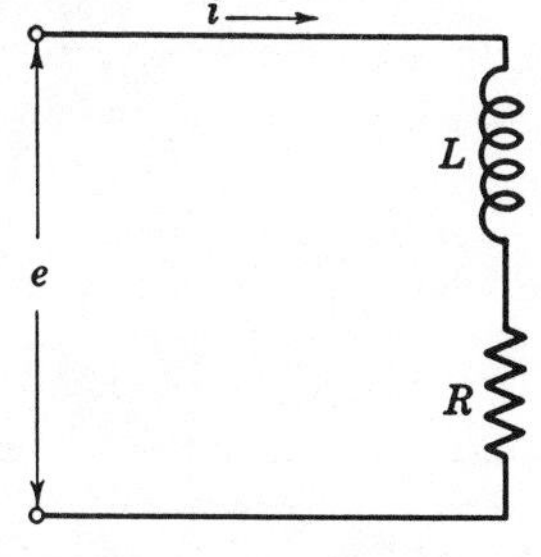

Fig. 5-14. An integrating circuit.

If conditions are such that the voltage drop Ri can be neglected in comparison to the voltage across the inductance, then 5–105 can be replaced by

$$e = L\frac{di}{dt} \tag{5-106}$$

from which

$$\int_{t_0}^{t_1} e\,dt = L(i_1 - i_0) \tag{5-107}$$

where t_0 and t_1 are the times at the beginning and end of the interval considered and i_0 and i_1 are the corresponding values of current. It may be convenient to start the integration by closing the circuit at an instant designated as $t = 0$, and then the current is a measure of the integral up to any time t; thus

$$i = \frac{1}{L} \int_0^t e\, dt \tag{5-108}$$

The problem before us is to see to what extent the accuracy of the integration is affected by the presence of resistance. Our plan will be to assume some simple time functions for e and in each such case to solve the differential equation 5-105 for i and then to compare the result with 5-108.

Suppose, first, that the voltage e is a constant designated by E. Then the equation to be solved is

$$L \frac{di}{dt} + Ri = E \tag{5-109}$$

Although this equation resembles those discussed in section 5-9 in that it is linear and has constant coefficients, it is not quite of the type discussed there because the term E on the right makes it non-homogeneous in the dependent variable i and its derivatives.

To find a solution, let us begin by guessing. The right-hand member is a constant. If we were to assume i constant, the first term would be zero and the second term would have to equal the right-hand member, that is, E. Thus we find as a possible solution

$$i = \frac{E}{R}. \tag{5-110}$$

There is no doubt that this satisfies 5-109, but it hardly seems reasonable that it could be the whole solution, for it is merely what would result if the circuit comprised only R and the voltage source; surely the solution should depend also on the inductance L. Furthermore, this solution contains no arbitrary constant. By the generalization of section 5-6, which applies to all ordinary differential equations, the general solution in this case should have one arbitrary constant since the equation is of first order.

If the solution 5-110 had something added to it which contained an arbitrary constant, and which when substituted in the terms on the left of the differential equation gave zero, we should have what we want. In other words, we might find the solution to

$$L \frac{di}{dt} + Ri = 0 \tag{5-111}$$

and add it to 5-110 to get the general solution to 5-109. Now we recognize 5-111 as a homogeneous linear differential equation with constant coefficients; it is of the type to which the generalizations of section 5–9 apply. To solve it we write the characteristic equation

$$Lm + R = 0 \tag{5–112}$$

from which

$$m = -\frac{R}{L} \tag{5–113}$$

and the solution is

$$i = C\epsilon^{-Rt/L} \tag{5–114}$$

where C is an arbitrary constant.

Thus the general solution to 5–109 is

$$i = \frac{E}{R} + C\epsilon^{-Rt/L} \tag{5–115}$$

since it satisfies the differential equation and has the single arbitrary constant C which is required.

EXERCISE. Substitute 5–115 into 5–109 to be sure of the preceding statement. Also check 5–115 independently by separation of the variables in 5–109 and integrating.

To evaluate C we have the condition of zero current at the instant $t = 0$ when the circuit is closed. From this condition we find $C = -E/R$ and consequently the particular solution of interest to us is

$$i = \frac{E}{R}(1 - \epsilon^{-Rt/L}) \tag{5–116}$$

To test the performance of the circuit as an integrating device for this case where $e = E$, a constant, we evaluate 5–108 and obtain the true value of the integral:

$$i = \frac{1}{L}\int_0^t E\,dt = \frac{Et}{L} \tag{5–117}$$

The two results are sketched together for comparison in figure 5–15. Here we see that the integration performed by the circuit, 5–116, agrees with the accurate result, 5–117, only when t is small compared to the time constant of the circuit L/R.

To investigate the range of agreement of the two results in a more precise way, expand the exponential in 5–116 in a power series; thus

$$i = \frac{E}{R}\left[1 - 1 + \frac{Rt}{L} - \frac{1}{2!}\left(\frac{Rt}{L}\right)^2 + \frac{1}{3!}\left(\frac{Rt}{L}\right)^3 - \cdots\right]$$

$$= \frac{Et}{L}\left[1 - \frac{1}{2}\frac{Rt}{L} + \frac{1}{6}\left(\frac{Rt}{L}\right)^2 - \cdots\right]$$

From this we conclude that 5–116 will be correct to about 1 per cent if Rt/L is less than 0.02. In other words, the circuit will integrate a constant correctly to within 1 per cent if the time constant of the circuit L/R is at least fifty times the duration of the constant voltage E.

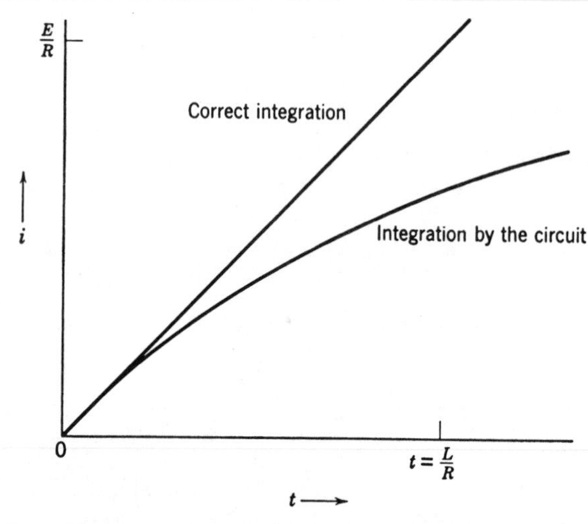

Fig. 5-15. Performance of the integrating circuit; integration of a constant.

Now let us investigate the performance of the integrating circuit for the time function $e = kt$, that is, a voltage which increases linearly with time at a rate k volts per second. The differential equation we must solve is

$$L\frac{di}{dt} + Ri = kt \tag{5–118}$$

Basing our method on what we have learned by solving 5–109, let us think of the solution to this equation as the sum of two parts:

1. A part which contains the arbitrary constant, and which when substituted in the left of 5–118 will give zero. This part clearly will be identical with 5–114.
2. A part which when substituted in the left of 5–118 will satisfy the equation; that is, will give what appears on the right, namely, kt.

To find this second part we are tempted to try

$$i = At \tag{5–119}$$

where A is a coefficient to be determined. Putting this in the left of the differential equation 5–118 we get

$$LA + RAt$$

If $A = k/R$ the second term becomes kt, which is what we want, and gives also the constant term kL/R, which is disconcerting. But this constant term can be canceled out by adding $-Lk/R^2$ to the trial solution 5-119. So the part of the solution we seek is

$$i = \frac{kt}{R} - \frac{kL}{R^2} \tag{5-120}$$

and the general solution to 5-118 is

$$i = C\epsilon^{-Rt/L} + \frac{k}{R}\left(t - \frac{L}{R}\right) \tag{5-121}$$

EXERCISE. Check to see that 5-121 is indeed the general solution of 5-118.

Assuming for the initial condition $i = 0$ at $t = 0$, we evaluate C in 5-121 and find the particular solution.

$$i = \frac{k}{R}\left[\frac{L}{R}(\epsilon^{-Rt/L} - 1) + t\right] \tag{5-122}$$

EXERCISE. Determine the conditions under which 5-122 agrees with 5-108, that is, the conditions under which the circuit will integrate faithfully a linearly increasing function of time.

As our concluding test of the integrating circuit consider its behavior with the voltage $E \cos \omega t$ applied. For this case the differential equation is

$$L\frac{di}{dt} + Ri = E \cos \omega t \tag{5-123}$$

To solve it let us again proceed as with 5-118, that is, by considering the solution as the sum of two parts; a part containing the arbitrary constant and resulting in zero when substituted in the left member; and a part which when substituted gives $E \cos \omega t$.

This latter part we suspect must contain a term which is something times $\cos \omega t$. But if we substitute this in the first term, a term in $\sin \omega t$ will appear which will have no counterpart in the other terms, and so the solution also must contain something to balance this. It appears, then, that this part of the solution may be the sum of a cosine and a sine. Consequently we are led to try

$$i = A \cos \omega t + B \sin \omega t \tag{5-124}$$

where A and B are constants as yet undetermined. To find their values and whether 5-124 in truth is the right guess as to form, we substitute in the differential equation

$$-\omega LA \sin \omega t + \omega LB \cos \omega t + RA \cos \omega t + RB \sin \omega t = E \cos \omega t \quad (5\text{–}125)$$

Collecting similar terms in the left member,

$$(-\omega LA + RB) \sin \omega t + (RA + \omega LB) \cos \omega t = E \cos \omega t \quad (5\text{–}126)$$

Now if 5–124 is a solution, the left member of 5–126 must reduce identically to the right member; that is, on the left the coefficient of the cosine must be E and the coefficient of the sine must be zero. This gives two conditions to be satisfied by A and B, namely,

$$RA + \omega LB = E \quad (5\text{–}127)$$

$$-\omega LA + RB = 0 \quad (5\text{–}128)$$

From these we find

$$A = \frac{RE}{R^2 + \omega^2 L^2} \quad (5\text{–}129)$$

$$B = \frac{\omega LE}{R^2 + \omega^2 L^2} \quad (5\text{–}130)$$

and thus the general solution to 5–123 is

$$i = C\epsilon^{-Rt/L} + \frac{E}{R^2 + \omega^2 L^2} (R \cos \omega t + \omega L \sin \omega t) \quad (5\text{–}131)$$

Exercise. Show by substitution that 5–131 is indeed the general solution of 5–123

Equation 5–131 can be put in form more convenient for some purposes as follows. First expand the coefficients thus:

$$i = C\epsilon^{-Rt/L} + \frac{E}{\sqrt{R^2 + \omega^2 L^2}} \left(\frac{R}{\sqrt{R^2 + \omega^2 L^2}} \cos \omega t + \frac{\omega L}{\sqrt{R^2 + \omega^2 L^2}} \sin \omega t \right)$$

This form of the equation suggests another one:

$$i = C\epsilon^{-Rt/L} + \frac{E}{\sqrt{R^2 + \omega^2 L^2}} (\cos \phi \cos \omega t + \sin \phi \sin \omega t)$$

which by the trigonometric formula for the cosine of the difference of two angles is equivalent to

$$i = C\epsilon^{-Rt/L} + \frac{E}{\sqrt{R^2 + \omega^2 L^2}} \cos (\omega t - \phi) \quad (5\text{–}132)$$

In these forms

$$\cos \phi = \frac{R}{\sqrt{R^2 + \omega^2 L^2}} \quad (5\text{–}133)$$

$$\sin \phi = \frac{\omega L}{\sqrt{R^2 + \omega^2 L^2}} \tag{5-134}$$

$$\tan \phi = \frac{\omega L}{R} \tag{5-135}$$

and these relationships are shown in the construction of figure 5-16.

To carry on with the study of the integrating circuit we should evaluate C and then compare this result with that of 5-108. We shall not do this, however, because our real concern here is learning to solve differential equations, and we are now ready to state some generalizations about the method we have been employing in this section.

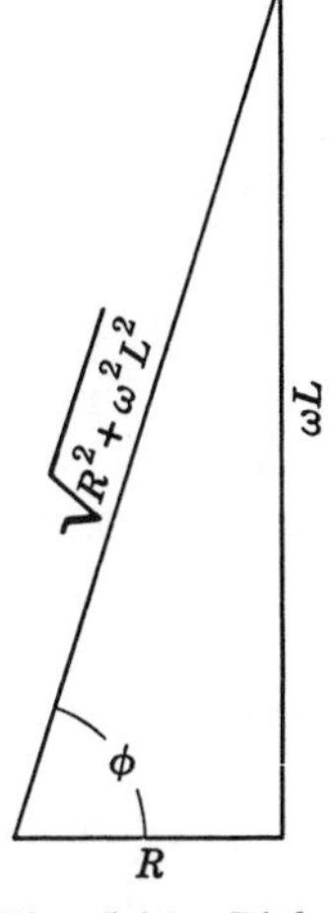

Fig. 5-16. Right triangle for defining the phase angle ϕ in equation 5-132.

5-11 Non-Homogeneous Linear Differential Equations: Generalizations

In section 5-10 the non-homogeneous linear differential equations 5-109, 5-118, and 5-123 were solved by treating the solution as the sum of two parts: a part associated with the non-homogeneous term on the right, and a part found by solving the homogeneous equation 5-111. This technique is applicable to the solving of all linear differential equations and will now be generalized.

A general nth order non-homogeneous linear differential equation is

$$b_n \frac{d^n y}{dt^n} + b_{n-1} \frac{d^{n-1} y}{dt^{n-1}} + \cdots + b_0 y = F(t) \tag{5-136}$$

where t and y are the independent and dependent variables, respectively, the coefficients $b_n, b_{n-1} \cdots b_0$ are either constants or functions of t, and the right-hand member is a function of t. The general solution of this equation is the sum of two parts called respectively the *particular integral* and the *complementary function.*

The particular integral includes none of the n arbitrary constants, which of course the general solution must contain, and when it is substituted in 5-136 the terms on the left reduce identically to $F(t)$, the function on the right.

The complementary function contains all the n arbitrary constants. It is found as the general solution to the homogeneous equation

$$b_n \frac{d^n y}{dt^n} + b_{n-1} \frac{d^{n-1} y}{dt^{n-1}} + \cdots + b_0 y = 0 \tag{5-137}$$

which is formed from 5-136 by replacing the function on the right, $F(t)$, by zero.

The two parts of the general solution are illustrated by 5–131, which is the solution of 5–123. The first term is the complementary function; it is the solution of the homogeneous equation 5–111, and contains the arbitrary constant C. The remainder of the expression constitutes the particular integral. It is important to notice that the arbitrary constants must not be evaluated until after the whole general solution, particular integral plus complementary function, is assembled; the values of the constants will depend on the particular integral as well as on the initial or boundary conditions.

Notice that, as time t increases, the complementary function $C\epsilon^{-Rt/L}$ in 5–131 becomes smaller and smaller, so that for practical purposes it ultimately disappears altogether. The particular integral, on the other hand, continues to constitute a sinusoidal current of constant amplitude no matter how long time runs on. This sort of thing happens frequently in circuit and dynamics problems where time is the independent variable; the complementary function dies away with time while the particular integral persists. For this reason engineers usually call the complementary function in such cases the *transient solution* and the particular integral the *steady state.* Under these circumstances it is common to call the function $F(t)$ on the right of 5–136 the *driving function* or sometimes more specifically the driving voltage or driving force, as appropriate; the driving function is responsible for there being a steady state other than zero.

There are many situations where the transient practically disappears in a matter of seconds or even microseconds, and interest may be only in what happens at greater values of time. For instance, in the circuit of 5–14, not necessarily now being considered as an integrating circuit, with a voltage $e = E \cos \omega t$ applied, one may want to know the current only a considerable time after the switch is closed, that is, the steady-state current. This, of course, may be found as the particular integral without need for finding the complementary function at all. With higher-order differential equations finding the transient solution may be exceedingly difficult, whereas the steady state comes out with relative ease.

Returning to a further examination of the solutions obtained in section 5–10, compare each of the three driving functions and the resulting particular integrals as follows:

Driving Function	*Particular Integral*
E	$\dfrac{E}{R}$
kt	$\dfrac{k}{R}\left(t - \dfrac{L}{R}\right)$
$E \cos \omega t$	$\dfrac{E}{R^2 + \omega^2 L^2}(R \cos \omega t + \omega L \sin \omega t)$

Now observe that in each case there is a kind of direct proportionality between the particular integral and its driving function; for instance, in the second case, if k is doubled, giving a driving voltage twice as large at each instant, then the particular integral at each instant is also doubled. The magnitude of the particular integral in each of these cases is related in this linear fashion to the driving function. This behavior is a special case of a more general property of the solutions of linear differential equations which is expressed by the *principle of superposition.*

In mathematical terms the principle of superposition states that if $F(t)$ in 5–136 is the sum of several terms the particular integral may be found as the sum of the particular integrals for each term of $F(t)$ as if it only were present. In physical terms, for a system in which the independent variable is time, the principle is that the steady-state response to a number of driving functions acting together is the same as the sum of the responses to each driving function acting alone. The principle applies only to linear systems, that is, systems whose behavior can be expressed by linear equations.

EXERCISE. Use the principle of superposition and the foregoing table to solve

$$L\frac{di}{dt} + Ri = E + kt$$

and show by substitution in the differential equation that the result so obtained is in fact the solution.

The generalizations of this section pertaining to the solution of the nonhomogeneous linear differential equation 5–136 are true regardless whether the coefficients $b_n, b_{n-1} \cdots b_0$ are constants or functions of t. But our further discussion of ways to find solutions to such equations is restricted to the case of constant coefficients.

Finding the complementary function is the same as finding the general solution of the related homogeneous equation, and for linear differential equations with constant coefficients this has been generalized in section 5–9. Finding particular integrals is discussed in sections 5–12 and 5–13.

5–12 Finding the Particular Integral: Method of Undetermined Coefficients

For finding the particular integral of 5–136 in the case where the coefficients $b_n, b_{n-1} \cdots b_0$ are constants we shall generalize the method employed in the problem of section 5–10. This is the method of undetermined coefficients. It consists first in predicting the form of the particular integral but with the values of the coefficients of the several terms left undetermined. Then as a second step the coefficients are found. Thus in solving 5–123 we predicted that the particular integral would have the form of 5–124 in which A and B are the undetermined coefficients. The coefficients are determined by substituting the predicted solution into the differential equation and then applying the condition that the left side must reduce identically to the right side, as illustrated by equations 5–125 through 5–130.

A system for predicting the kinds of terms in the particular integral will be described, which works if the right-hand side of 5–136, $F(t)$, is a well-behaved function of the kind likely to be met in engineering practice.

Suppose, first, that $F(t)$ has but one term. Set down in a list the functional form (t-dependent part) of this term and the functional form of each of the t-derivatives of the term. It will be found in the cases likely to be encountered that the list will have either a finite number of forms terminating in a constant or it will have a finite number of forms which recur over and over as differentiation is continued. If no function in the list appears in the complementary function the particular integral to try is a sum of terms like each different form in the list, each term having an undetermined coefficient.

To illustrate, suppose that

$$F(t) = kt^3 \tag{5–138}$$

then the list would be

$$t^3,\ t^2,\ t,\ \text{Constant}$$

and the particular integral corresponding to the driving function $F(t)$ is of the form

$$y = At^3 + Bt^2 + Ct + D \tag{5–139}$$

where A, B, C, and D are the undetermined coefficients.

As another example, encountered already in 5–123, suppose that

$$F(t) = k \cos \omega t \tag{5–140}$$

The list would be

$$\cos \omega t,\ \sin \omega t$$

since further differentiation results merely in recurrence of these two forms. In this case the form of the particular integral would be

$$y = A \cos \omega t + B \sin \omega t \tag{5–141}$$

and is, of course, like 5–124, already deduced for this case by other reasoning.

To evaluate the coefficients, the predicted form of the particular integral, 5–139 or 5–141 for instance, is substituted in the differential equation, and the condition is applied that the result must be an identity, or in other words that the result must be true for all values of the independent variable t. If the predicted form is correct, enough conditions will be found to determine all the hitherto undetermined coefficients. Not only does this operation serve to determine the coefficients, but also it checks the correctness of the predicted form. If there is a mistake in the prediction either too many or too few conditions to determine the coefficients will be found; it will become obvious that something is wrong; and to this extent the method is foolproof.

EXERCISE. See what happens if the term B sin ωt is omitted from 5–124.

EXERCISE. Try out the method of undetermined coefficients by finding the particular integral of 5–105 for the case where the voltage applied is a pulse of the form $e = Et\epsilon^{-bt}$.

In the two foregoing illustrations it has been assumed that no term in the list of functions composed of $F(t)$ and its derivatives appears in the complementary function. If a term in the list is the same as one appearing in the complementary function, multiply all the terms by the lowest power of t which makes all of them different from any in the complementary function and use these instead of the original ones.

EXERCISE. Find the general solution to 5–105 when $e = E\epsilon^{-\alpha t}$ and it happens that $\alpha = R/L$.

If the right-hand member of 5–136, $F(t)$, contains more than one term, the principle of superposition may be used to advantage. First, the particular integrals are found for each term of $F(t)$ as if it only were present, using the procedure just described. Then the particular integral for all the terms of $F(t)$ present together is the sum of the particular integrals for the separate terms. In physical terms the response of the system to each part of the driving function is found and then these responses are added to get the actual response to the whole driving function. Concluding with a note of caution, we emphasize that the above method is legitimate only because 5–136 is a linear equation.

5–13 Complex-Number Method for Sinusoidal Driving Functions

Our purpose in this section is to show the connection between what we have been discussing and the complex-number method commonly used in the steady-state analysis of alternating-current circuits. To do this we shall find the steady-state current in the circuit of figure 5–14 for a voltage $e = E \cos \omega t$. That is, we wish to find the particular integral of 5–123, but we shall use a method quite different from that employed in section 5–10.

The method we shall use depends on Euler's relations 5–60 and 5–61. By the first of these we can write

$$E\epsilon^{j\omega t} = E\cos\omega t + jE\sin\omega t \tag{5-142}$$

The driving voltage we are interested in is $E \cos \omega t$, which is the real part of $E\epsilon^{j\omega t}$. Now by the principle of superposition as given in section 5–11, the steady-state current caused by the driving voltage $E\epsilon^{j\omega t}$, is the sum of the steady-state currents due to $E \cos \omega t$ and $jE \sin \omega t$ acting separately. Our plan is to use the converse of this. We shall find the steady-state current due to $E\epsilon^{j\omega t}$ and then take only the real part of the result, that is, the part due to $E \cos \omega t$, and this should be the same as would result if $E \cos \omega t$ acted alone. The advantage of this procedure is that exponential functions are easier to manipulate than sinusoidal functions.

Thus we wish to find the real part of the particular integral of

$$L\frac{di'}{dt} + Ri' = E\epsilon^{j\omega t} \tag{5-143}$$

Here we have replaced the dependent variable i by a fictitious complex current i' the real part of which is i.

By the method of undetermined coefficients, section 5–12, the form of the particular integral should be

$$i' = I\epsilon^{j\omega t} \tag{5-144}$$

where I is the undetermined coefficient. To determine it and to see that 5–144 satisfies the differential equation, we substitute 5–144 into 5–143 and obtain

$$j\omega LI\epsilon^{j\omega t} + RI\epsilon^{j\omega t} = E\epsilon^{j\omega t}$$

from which

$$I = \frac{E}{R + j\omega L} \tag{5-145}$$

To interpret this complex number we first consider the denominator $R + j\omega L$, known as the impedance of the circuit. Using the triangle construction of figure 5–16 and relations 5–133 and 5–134, we can express $R + j\omega L$ in terms of an angle ϕ; thus

$$R + j\omega L = \sqrt{R^2 + \omega^2 L^2}\,(\cos\phi + j\sin\phi)$$

Next, using Euler's relation 5–60, we may replace the parenthesis so that

$$R + j\omega L = \sqrt{R^2 + \omega^2 L^2}\,\epsilon^{j\phi}$$

Substituting this in 5–145 we obtain

$$I = \frac{E}{\sqrt{R^2 + \omega^2 L^2}}\,\epsilon^{-j\phi} \tag{5-146}$$

But

$$i' = I\epsilon^{j\omega t} = \frac{E}{\sqrt{R^2 + \omega^2 L^2}} \epsilon^{j(\omega t - \phi)} \tag{5–147}$$

and, using Euler's relation 5–60 again,

$$i' = \frac{E}{\sqrt{R^2 + \omega^2 L^2}} [\cos(\omega t - \phi) + j \sin(\omega t - \phi)] \tag{5–148}$$

The real part of i' is the result we wished to find. It is

$$\text{Real part of } i' = i = \frac{E}{\sqrt{R^2 + \omega^2 L^2}} \cos(\omega t - \phi) \tag{5–149}$$

and this we observe is identical, as it should be, with the steady-state part of the former result 5–132.

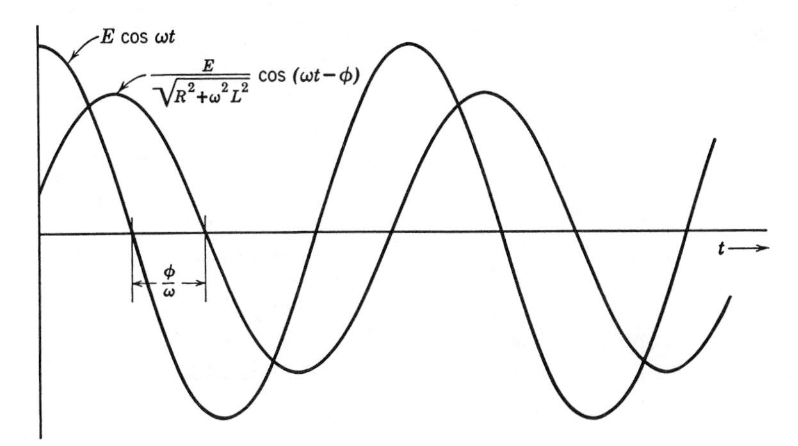

Fig. 5-17. Voltage and current plotted against time.

The current as given by 5–149 is interpreted in figure 5–17 as a function of time. It is a cosine wave which lags the cosine wave of voltage by the angle ϕ. We may, however, make a graphical interpretation at an earlier stage, and this is what is done in the analysis of alternating currents. The voltage applied, $E \cos \omega t$, is by 5–142 the real part of $E\epsilon^{j\omega t}$. Now $E\epsilon^{j\omega t}$ may be represented as a vector in the plane of complex numbers, as shown in figure 5–18. The length of the vector is E, and it makes an angle ωt with the axis of reals. Since ωt increases linearly with time, the vector rotates counterclockwise at a uniform angular velocity. As it does so its projection on the axis of reals generates the function $E \cos \omega t$. By 5–147 and 5–148 the current resulting from the voltage $E\epsilon^{j\omega t}$ may also be represented by a vector rotating in the same direction at the same angular velocity ω as the voltage vector but lagging behind by the angle ϕ. The magnitude of the current vector is obtained from the magnitude of the voltage vector by dividing by the factor $\sqrt{R^2 + \omega^2 L^2}$, which is known as the magnitude of the a-c impedance. The instantaneous value of the actual current which flows as a result of the voltage $E \cos \omega t$ is, of course, the instantaneous projection of the current vector on the axis of reals.

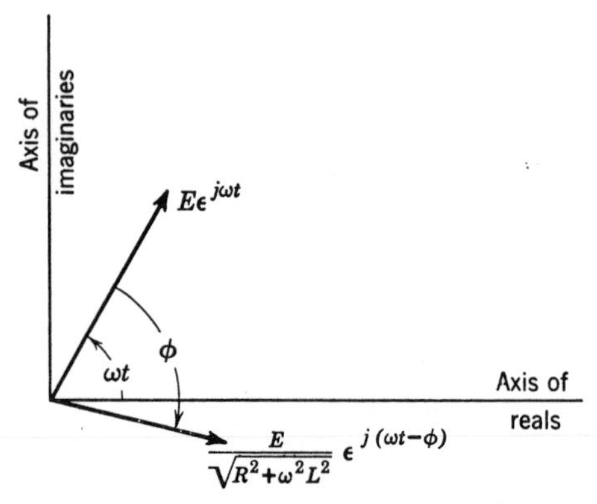

Fig. 5-18. Vector representation of the time functions in figure 5-17.

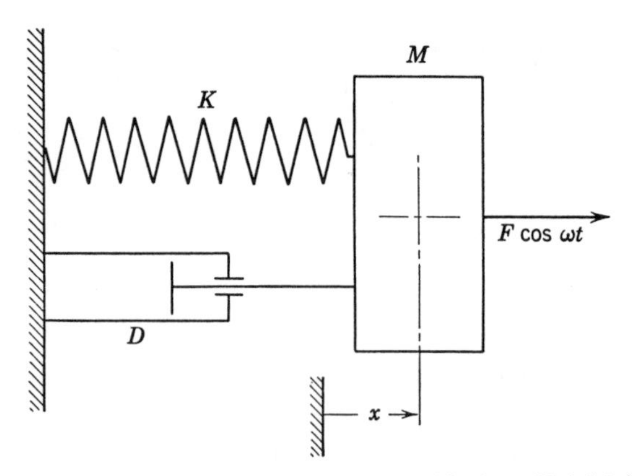

Fig. 5-19. Mass, spring, and dashpot system with sinusoidal driving force.

The complex-number method is by no means restricted to a-c circuits. Clearly it applies for the steady-state response of any linear system to a sinusoidal driving force. For example, consider the system represented in figure 5–19, which consists of a mass M constrained to horizontal motion and connected to a spring and a dashpot whose constants respectively are K and D. The mass is acted upon by an alternating force $F \cos \omega t$, and we wish to find the steady-state motion.

We begin by applying Newton's law of motion to the mass, and in terms of the displacement x of the mass from a fixed reference we get

$$-Kx - D\frac{dx}{dt} + F\cos\omega t = M\frac{d^2x}{dt^2} \tag{5-150}$$

by reasoning similar to that used in establishing equation 4-1.

Rearranging the terms of 5-150 we recognize a non-homogeneous linear differential equation with constant coefficients:

$$M\frac{d^2x}{dt^2} + D\frac{dx}{dt} + Kx = F\cos\omega t \tag{5-151}$$

We are concerned here only with the steady-state motion, that is, the motion after the transient has died away. Consequently we seek only the particular integral and we shall use the complex-number method. In other words, we want the real part of the particular integral of

$$M\frac{d^2x'}{dt^2} + D\frac{dx'}{dt} + Kx' = F\epsilon^{j\omega t} \tag{5-152}$$

where x' is a new, complex variable the real part of which is x. Using the method of undetermined coefficients we see that the particular integral should be of the form

$$x' = X\epsilon^{j\omega t} \tag{5-153}$$

To check this and to find the undetermined coefficient X we substitute in the differential equation, obtaining

$$(-\omega^2M + j\omega D + K)X\epsilon^{j\omega t} = F\epsilon^{j\omega t}$$

From this

$$X = \frac{F}{(K - \omega^2M) + j\omega D} \tag{5-154}$$

Then, as in the preceding example, we think of the denominator of 5-154 in terms of a right-triangle construction; and, using Euler's relation, write

$$(K - \omega^2M) + j\omega D = \sqrt{(K - \omega^2M)^2 + \omega^2D^2}\,\epsilon^{j\phi} \tag{5-155}$$

where

$$\phi = \tan^{-1}\frac{\omega D}{K - \omega^2M} \tag{5-156}$$

Then, substituting 5-155 in 5-154,

$$X = \frac{F}{\sqrt{(K - \omega^2M)^2 + \omega^2D^2}}\,\epsilon^{-j\phi} \tag{5-157}$$

and thus

$$x' = X\epsilon^{j\omega t} = \frac{F}{\sqrt{(K - \omega^2M)^2 + \omega^2D^2}}\,\epsilon^{j(\omega t - \phi)} \tag{5-158}$$

We may conclude the mathematical analysis of the problem at this point with the following interpretation. Equation 5–158 represents a vector of length $F/[(K - \omega^2 M)^2 + \omega^2 D^2]^{1/2}$, which rotates at a constant angular velocity ω, lagging behind the force vector by the angle ϕ given by 5–156. Whether the angle of lag is positive or negative depends, of course, on the relative magnitudes of K and $\omega^2 M$, a matter we shall not go into here. The steady-state motion of M due to the actual force $F \cos \omega t$ is represented by the projection of the rotating vector on the axis of reals, that is,

$$\text{Real part of } x' = x = \frac{F}{\sqrt{(K - \omega^2 M)^2 + \omega^2 D^2}} \cos (\omega t - \phi) \qquad (5\text{–}159)$$

The quantity $[(K - \omega^2 M)^2 + \omega^2 D^2]^{1/2}$ is sometimes called the displacement impedance. A further interpretation of 5–159 is made in section 7–8.

Exercise. To check the foregoing analysis and to see the advantages of the complex-number method, obtain the steady-state solution of 5–151 directly by the method of undetermined coefficients.

5–14 Concluding Remarks

In this chapter we have been particularly concerned with solving linear differential equations with constant coefficients and we have stated and illustrated certain generalizations which are important in this connection. Some aspects of our treatment, however, have not been restricted to linear equations with constant coefficients. In the approaches leading to the various generalizations we followed procedures which have a wide range of usefulness. For instance, in the early part of the chapter we discussed numerical step-by-step solutions, and such methods most certainly are not restricted to linear equations. We used power-series expansions to find solutions to differential equations, thus illustrating a very generally useful method having wide application. Finally, the method of reasoned trial-and-error by which we attacked many of the problems finds very frequent usefulness in the mathematics of engineering analysis.

CHAPTER SIX

Checking

6–1 Role of Checking

Engineering work not thoroughly checked for correctness has little value. In the practice of the engineering profession there are no answer books, nor can you check with a classmate, for, although engineers often work in groups, the group is usually a team with each individual alone responsible for his part of the whole job. Not only is it a point of professional pride to produce correct results, but also a reputation for reliability is an extremely weighty factor in determining advancement. Thus you must learn to check your own work surely and effectively so as to develop the justified self-confidence that goes with professional competence.

The ability to apply checks for reliability is equally important to the man who becomes a supervisor of engineering work. Clearly, he cannot check every task by doing it himself, but, since his own reputation depends upon the accuracy of the work done for him by others, he invariably checks it as thoroughly as he can. Obviously you yourself can use checks to convince your supervisor of the soundness of your results, and conversely there are few situations more embarrassing than having your superior catch a manipulative error that you could have corrected previously.

We cannot help observing that somehow many students get an exaggerated idea of the importance of speed, perhaps because of academic emphasis on short cuts and because of the quantity factor in examinations. In practice, reliability turns out to be much more important than speed.

By far the most effective way to check an analysis is by means of a properly designed experiment. But experiments are likely to be both expensive and time consuming, and therefore the engineer must usually establish the need for the experiments by careful analysis before authorizing or recommending them. Moreover, the design of an experiment to give the information desired may itself involve difficult and complicated analysis. Consequently you will be forced to rely to a major extent on means other than experimental for checking your analytical work.

Good professional method implies not only checking an end result in every reasonable way but also checking frequently as the work progresses so that the analysis is accurate at every stage of its development. This tends to reduce waste of time and effort caused by undetected errors early in the analysis, and helps establish the correctness of the end result because good overall checks sometimes are very difficult to find, and then major reliance must be placed on the work being correct in every detail.

There is a wide range of kinds of checking. At one extreme are checks of the manipulations as, for instance, checks of arithmetic and dimensional checks of equations, or, if experiments are involved, the calibration of instruments. At the other extreme are overall checks so thorough as to constitute critical analyses of all that has been done and which are hardly distinguishable from the learning stage of the professional method where you endeavor to find everything useful from your work that you can. Indeed, thoroughly searching checks are always likely to contribute heavily to the learning stage, and in fact the two often merge.

The remarks above pertain to the checking of any problem whether the analysis is mathematical or experimental, or quite non-quantitative as, for instance, in the problem on the relay-heater production test, section 2–5. The remainder of this chapter, however, mainly treats of methods for checking mathematical work.

6–2 Dimensional Checking

Dimensional checking of equations, which has been used extensively in the previous examples, is testing to see whether each term of the equation has the same units. If the units are the same the equation is said to be dimensionally homogeneous. An extension of this technique is of great value in treating problems that cannot be dealt with by straightforward mathematical methods and in using models to predict the behavior of prototypes.

Opportunity for checking by dimensions may occur with the very first equation in a mathematical analysis and is likely to recur almost line by line thereafter until the final result is reached. A dimensional check not only serves as a sensitive detector of the omission and misplacing of factors such as tend to occur with distressing frequency when most of us do algebra and calculus, but also it may reveal errors in the original set-up. Although an equation which checks dimensionally still may be incorrect, failure to meet this test is positive indication of error. This method of checking is so easy and so effective that it should be learned thoroughly and practiced so regularly that its use becomes habitual.

In making a dimensional check, you simply investigate to see whether each term of an equation is the same kind of quantity; for example:

$$5 \text{ apples} + 6 \text{ apples} = 11 \text{ apples}$$

is dimensionally homogeneous, but

$$5 \text{ apples} + 1 \text{ cow} = 6 \text{ apples}$$

not only is dimensionally inhomogeneous but is obvious nonsense; indeed, dimensional inhomogeneity in an equation implies nonsense. However, most of the equations we seek to check are not as easy to see through as the above; a more representative example would be

$$a \text{ (apples per basket) } A \text{ (baskets)} + b \text{ (apples per tree) } B \text{ (trees)} = C \text{ (apples)}$$

in which, by combining units, it is clear that each term represents apples.

To see whether all terms of an equation measure the same kind of thing, four useful procedures are described below.

1. BY NAMES OF UNITS. This procedure was illustrated in the checks of equations 2–5 and 2–21. To carry it out, the equation is written with the symbol for each quantity followed by the name of the units in which that quantity is expressed; for instance, a velocity might be written as v (ft sec^{-1}). Then unit name parts are combined with similar ones in other factors of the same term as though they were algebraic factors; for example v (ft sec^{-1}) ρ (slug ft^{-3}) would become $v\rho$ (slug ft^{-2}sec^{-1}). If the result of this is to give all terms of the equation identical name combinations, the equation is dimensionally homogeneous. Sometimes to complete the process, some known relationship between units must be introduced as in the check of equation 2–6.

2. BY DIMENSIONS. In this procedure, replace each factor in the equation by its dimensional formula in terms of arbitrary primary quantities and then proceed as in the unit-name method except that dimensions rather than unit names are collected or canceled. Procedure 1 above has an advantage over this one in that it checks the consistency of the system of units used as well as the dimensional homogeneity. Dimensional formulas are derived by considering the fundamental nature of the quantity; for example, density is mass per unit volume or mass divided by length cubed and hence its dimensional formula is $[ML^{-3}]$, where M represents the dimension of mass and L the dimension of length. Similarly, it might be desirable to have a dimensional formula for electrical resistance in terms of the dimensions voltage $[V]$, charge $[Q]$, and time $[T]$; it is $[VQ^{-1}T]$, as may be deduced from Ohm's law and the definition of current as charge passing through a circuit per unit time. The dimensions selected for this purpose are called primary quantities. If too many primary quantities happen to be chosen in a particular case, it will be necessary to eliminate one or more of them by using known physical relationships. For instance, in a problem in dynamics if all four of the dimensions, force

$[F]$, mass $[M]$, length $[L]$, and time $[T]$ are used as primary quantities, it will be found necessary to eliminate one of them. This is done very easily by using Newton's law $f = ma$; it shows that force $[F]$ has the dimensional formula $[MLT^{-2}]$, or, if you prefer, that mass $[M]$ has the formula $[FL^{-1}T^2]$.

EXERCISE. Check the following equations by dimensions: 1–5, 2–5, 2–21, and 2–67.

3. BY INSPECTION. This is a short-cut version of the method of dimensions, 2, above and is appropriate where the natures of the terms are so apparent that writing dimensional formulas is unnecessary. Equation 2–67 was checked in this way simply by observing that each term is a stress.

4. BY COMPARISON. If the terms of an equation have the same forms as those of another equation which is available for comparison and is already known with certainty to be correct, the formality of a dimensional check is obviously unnecessary. For example, 2–70 so far as dimensions are concerned is identical with 2–67, which already had been found dimensionally homogeneous; accordingly, by comparison with 2–67, we observe that 2–70 is also correct dimensionally.

In testing for dimensional homogeneity, note that trigonometric functions, exponential functions, Bessel functions, and so forth, are pure numbers and hence dimensionless. Also, the arguments of such functions are dimensionless. The dimensional checking of equations including exponential functions was illustrated for 2–15 and 2–25. Logarithms and their arguments are dimensionless too, but sometimes you may find a logarithm whose argument has dimensions; in such a case another logarithm should be present also with dimensional argument and the two should combine to give a logarithm whose argument has no dimensions. For example, equation 2–23 contains two logarithms whose arguments have the dimensions (Btu hr^{-1}), but in the next equation the two combine to form a single logarithm with dimensionless argument.

6–3 Checking by Experience

Perhaps the simplest overall check is to see whether a result is reasonable in the light of experience. Even before you have acquired the specialized experience that comes from practice in a particular field of engineering, experiences with everyday things and common sense serve admirably for checking the general reasonableness of results.

In checking a formula look to see whether making each of its factors bigger or smaller causes the predicted quantity to vary in the direction common sense says should be right. Common sense is amazingly dependable for this purpose, but of course not infallible.

Again, in checking numerical magnitudes, ordinary experiences can be very reassuring. Typical illustrations of this are given in the non-fogging mirror problem, section 2–3.

Development of capacity for making checks of this sort is one of the many reasons for trying to learn the maximum from experience.

6–4 Limiting-Case Checks: Parting Velocity of Relay Contacts

Skill in making limiting-case checks is a most valuable asset. It enables you to establish the correctness of at least parts of a result with absolute certainty, and sometimes even makes possible a complete confirmation.

To make a limiting-case check, a factor in the general result is assigned a special value, for example, zero or infinity, for which the correct result is either known already or can be found easily by an independent solution. If the general result fails to reduce to the one known for the simple case, it must be incorrect. If the general result does reduce to that known for the limiting case, this shows that the general result may be correct but does not make it certain. By repeating the process successively for different limiting cases, the margin of uncertainty in the general result being tested can be narrowed, perhaps to zero. Numerous simple illustrations of the method have already been made; see, for example, the checks of 2–15 and 2–25; and if you did not do the exercise following 2–55 we suggest you do it now.

To illustrate limiting-case checking more fully, consider the following problem. It concerns the small electric relay shown in figure 6–1, designed for especially fast opening of its contacts so as to reduce sparking. The contacts are held closed by means of the electromagnet at the top. When current in the coil is interrupted, the holding force of the magnet drops almost instantly to zero, and the two members carrying the contacts are then moved upward by the two springs. Since the lower element is of small mass and the lower spring considerably stiffer than the upper, the contacts remain firmly together until the lower element comes against its stop. The contacts then separate suddenly with a high relative velocity. In investigating the phenomena of current interruption by this relay, we wish to be able to calculate the relative velocity with which the contacts separate at the instant of parting.

Considering the situation, we see that, if the two contacts really do remain in contact until the lower stop is reached, and if this stop brings the lower contact to rest without deflecting, then the velocity we seek is simply the velocity of the two parts moving as one, just as the lower stop is reached.

In planning how to find this velocity, we decide to make the following additional simplifying assumptions:

1. The holding force drops instantly to zero, and there are no residual magnetic forces.
2. All frictional forces are negligible.
3. The masses of the springs are negligible.

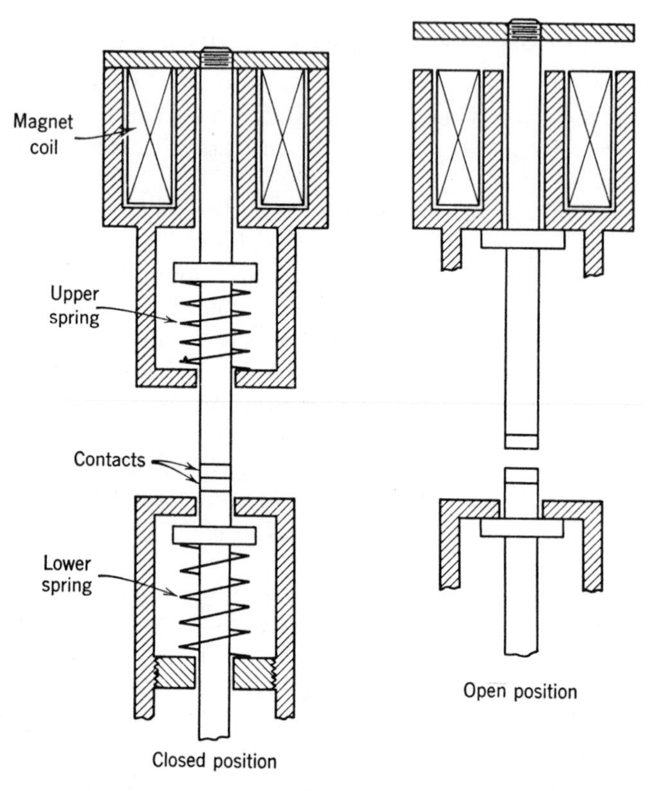

Fig. 6-1. Relay designed for fast parting of contacts.

Then application of the principle of conservation of energy will yield a relation involving the desired velocity and easily measurable quantities. First applying the principle in words:

The total energy stored in the springs in the closed position must equal the sum of:

1. The kinetic energy of the moving parts at the instant the lower stop is reached.
2. The energy remaining in the springs at that instant.
3. The work done against gravity in raising the moving parts from the closed position to the position of parting.

In mathematical symbols:

$$\frac{1}{2}K_UC_U^2 + \frac{1}{2}K_LC_L^2 = \frac{M_U + M_L}{2}v^2 + \frac{1}{2}K_U(C_U - s)^2 + \frac{1}{2}K_L(C_L - s)^2 + (M_U + M_L)gs \quad (6\text{–}1)$$

where K_U and K_L = spring constants of the upper and lower springs respectively (lb in.$^{-1}$).

C_U and C_L = respectively, shortenings from their free lengths of the upper and lower springs in the closed position (in.).

M_U and M_L = masses of the upper and lower moving parts (lb sec^2 in.$^{-1}$).

v = velocity at parting (in. sec^{-1}).

g = acceleration of gravity (in. sec^{-2}).

s = travel of the lower contact from the closed position to its stop (in.).

With a little algebra, we obtain

$$v = \left[\frac{s[(K_U(2C_U - s) + K_L(2C_L - s) - 2g(M_U + M_L)]}{M_U + M_L}\right]^{1/2} \quad (6\text{–}2)$$

Exercise. Check equation 6–2 dimensionally.

Now consider some limiting-case checks of equation 6–2.

1. Suppose the lower stop is set so that $s = 0$. Then neither spring would release any stored energy before the instant for which v has been determined; consequently, v would be zero. Putting $s = 0$ in 6–2 gives $v = 0$, and thus we have a check. It may seem that this check is so obvious as to be within the realm of common sense, as discussed in the preceding section; perhaps so, but the important thing is that we have a check, not what we call it or whose common sense we take for a criterion!

2. Suppose the masses of the moving parts are zero. Then, since there would be no weight in this instance either, the spring forces alone would act and would accelerate the parts to infinite velocity immediately. Putting $M_U + M_L = 0$ in 6–2 gives $v = \infty$, and we have another check.

3. If there were no springs and if the relay were turned upside down so that gravity acted to open the contacts we would have the case of a body falling freely from rest. The correct result for this case is one of those miscellaneous formulas with which most of our brains are cluttered; it is $v = \sqrt{2gs}$. Putting K_U and K_L equal to zero (equivalent to removing the

springs) and replacing g by $-g$ to reverse the effect of gravity makes 6-2 give the correct result.

4. For the last case, suppose that both springs are relaxed (have their free lengths) at the point of contact parting and that we neglect gravity. In this case the energy originally stored, corresponding to the springs being compressed an amount s, would be converted entirely to kinetic energy. The following calculation gives the result that should obtain for thir simplified case:

$$\tfrac{1}{2}(K_U + K_L)s^2 = \tfrac{1}{2}(M_U + M_L)v^2$$

$$v = s\sqrt{\frac{K_U + K_L}{M_U + M_L}} \tag{6-3}$$

This case occurs if $C_U = s$, $C_L = s$, and $g = 0$. Putting these in 6-2 gives the same result as the simple independent calculation.

We will be content for the present with these four limiting-case checks of this result; perhaps you can think of others. It should be apparent that ingenuity is important in devising checks of this kind.

6-5 Evaluation of Indeterminate Forms: Dynamic Braking of a Motor

A complication frequently incidental to a limiting-case check is the occurrence of an indeterminate form which must be evaluated before the check can be completed. The following problem illustrates this.

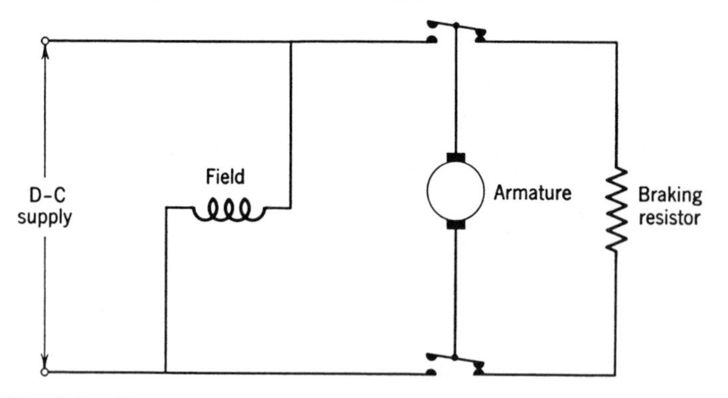

Fig. 6-2. Circuit for dynamic braking; contactor in position for braking.

In making an application of dynamic braking to a d-c shunt motor which is positively connected to a rotating machine, we are faced with the problem of finding the effect of the magnitude of the braking resistance on the number of revolutions turned by the armature during braking. The circuit is shown schematically in figure 6-2. When it is desired to stop the motor, the contactor is caused to disconnect the armature from the supply and connect it instead across a resistor, the field being left excited as for running. The motor in effect thus becomes a generator converting the kinetic energy of the rotating parts to electric energy, which in turn is converted in the resistor to thermal energy and is finally dissipated as heat.

In considering a way to find the angle turned by the armature as it comes to rest, we decide after some thought to make the following simplifications:

1. Neglect the effect of inductance in the armature circuit.
2. Neglect the effect of armature reaction so that the magnetic flux can be assumed constant, independent of armature current.
3. Assume the frictional torques acting are constant, independent of speed (thus torques due to windage and eddy currents will not be accounted for accurately). The frictional torque of the connected load will be referred to the armature shaft.
4. Refer the moment of inertia of the connected load to the motor shaft.

We also assume that the armature current does not become so large as to cause some failure such as a flash-over of the commutator.

As a plan of attack, we decide to apply the law for rotation $T = I\alpha$ to the armature and connected rotating parts considered as a single equivalent rigid body. There are two torques acting: the constant frictional torque and the electromagnetic torque, directly proportional under the conditions assumed to the armature current. The resultant of these torques must equal the moment of inertia of the armature and connected parts times the angular acceleration. Thus

$$-T - K_T i = J \frac{d\omega}{dt} \tag{6–4}$$

where T = constant frictional torque (lb ft).
K_T = torque developed per unit of armature current (lb ft amp^{-1}).
i = armature current at any instant, positive in the direction corresponding to generator action (amp).
J = combined moment of inertia referred to the motor shaft (lb ft sec^2).
ω = instantaneous angular velocity taken positive in the original direction of rotation (rad sec^{-1}).
t = time (sec).

Equation 6–4 is restricted to the case of positive ω. If somehow the motor were to reverse in direction, an equation similar to 6–4 but with T plus would have to be used as was found in the problem discussed in section 4–7.

To solve 6–4, we first obtain another relation between i and ω by applying Kirchhoff's voltage law to the circuit through the armature and the braking resistor. Neglecting inductance, we can say that the voltage rise generated in the armature, proportional to angular velocity, equals the voltage drop through the entire resistance of the circuit. Thus

$$K_E\omega = Ri \tag{6–5}$$

where K_E = voltage generated per unit of angular velocity (volt sec rad^{-1}).
R = sum of the resistances of armature, braking resistor, and their connections (ohms).

We are seeking a relation involving angle turned rather than current; so we use 6–5 to eliminate the current i from 6–4 and obtain

$$-T - \frac{K_E K_T \omega}{R} = J\frac{d\omega}{dt} \tag{6–6}$$

Since we have defined our problem as finding only the angle turned and we are not concerned with the time, we choose to eliminate t from 6–6. We do this by noting that the differential of angle turned $d\theta$ is $\omega\, dt$ or $dt = d\theta/\omega$. Substituting this in 6–6,

$$-T - \frac{K_E K_T}{R}\,\omega = J\omega\frac{d\omega}{d\theta} \tag{6–7}$$

Next we separate variables and integrate between limits:

$$\int_0^{\theta_s} d\theta = -J\int_{\omega_0}^{0} \frac{\omega\, d\omega}{T + \dfrac{K_E K_T}{R}\,\omega} \tag{6–8}*$$

Here θ_s is the entire angle turned during stopping and ω_0 is the initial angular speed. Then we use a table of integrals, substitute the limits, and obtain

$$\theta_s = \frac{JR^2}{K_E{}^2 K_T{}^2}\left[\frac{K_E K_T \omega_0}{R} - T\ln\left(1 + \frac{K_E K_T \omega_0}{RT}\right)\right] \tag{6–9}†$$

EXERCISE. Check equation 6–9 by solving equation 6–6 for ω in terms of t and then from this result finding the angle. Indeed, as part of the solution to the whole problem, it might be very desirable to know the time it takes for the motor to stop.

Equation 6–9 is the result we wish to check by limiting cases.

* Another derivation of this equation is given in section 6–7.
† This equation is interpreted in sections 7–5 and 7–6.

1. First suppose there is no inertia. Then the motor should stop instantly, that is, θ_s should be zero. Our result is in accord with this prediction, as may be seen by setting $J = 0$. The result is correct also for the converse case where the inertia is infinitely large.

2. As a second case, suppose that the frictional torque T is zero. To find the correct result for this case we return to differential equation 6–7 and put $T = 0$. Then ω may be divided out, and upon integrating between limits we find at once that

$$\theta_s = \frac{JR\omega_0}{K_E K_T} \tag{6–10}$$

We realize that if 6–9 gives this when $T = 0$ we will be checking only the integration of 6–7 for the general case and not the correctness of 6–7 itself. However, even though a more inclusive check is desirable, checking just the integration is worth while.

Putting $T = 0$ in 6–9 gives for the second term in brackets zero times infinity, an indeterminate form, so we must endeavor to find its value by some process more subtle than mere substitution of $T = 0$. Suppose T is given a value small enough that the second term in the logarithm becomes so large that the term 1 can be ignored. Then suppose T is decreased to 0.1 of this value. This adds only 2.3 (= ln 10) to the former value of the logarithm, but divides the coefficient of the logarithm by 10. If T is made to approach zero by successive divisions by 10, it thus is easily seen that the second term of 6–9 will approach zero. Consequently the limit of 6–9 as T approaches zero is

$$\theta_s = \frac{JR^2}{K_E{}^2 K_T{}^2} \cdot \frac{K_E K_T \omega_0}{R} = \frac{JR\omega_0}{K_E K_T} \tag{6–11}$$

which agrees with 6–10.

3. As a final limiting case, suppose there is no electromagnetic torque acting, but only the constant friction torque T. Then we note that the original kinetic energy of the rotating parts would be dissipated as work against the constant torque, so that the correct result for this case is simply

$$\theta_s = \frac{J\omega_0{}^2}{2T} \tag{6–12}$$

This case could be brought about either by making the magnetic flux zero, which would cause both K_E and K_T to be zero, or by opening the circuit, that is, making R infinite. Since these factors occur in the single quantity $K_E K_T/R$, making this quantity zero should make 6–9 reduce to

6–12. Maybe it does, but if so the fact is not immediately apparent, for substituting $K_E K_T/R = 0$ in 6–9 gives the indeterminate form infinity times zero.

Since 6–9 misbehaves when $K_E K_T/R$ is set equal to zero, we decide to see what happens as this factor becomes smaller and smaller, that is, as it approaches zero, and we decide to do this first by numerical calculation. To simplify the calculation, we plan to work not with $K_E K_T/R$ but with $K_E K_T \omega_0/RT$, which appears in the logarithm. By a simple rearrangement we can put 6–9 in terms of this factor; thus

$$\theta_s = \frac{\omega_0{}^2 J}{T}\left[\frac{\dfrac{K_E K_T \omega_0}{RT} - \ln\left(1 + \dfrac{K_E K_T \omega_0}{RT}\right)}{\left(\dfrac{K_E K_T \omega_0}{RT}\right)^2}\right] \tag{6–13}$$

Now we observe that when $K_E K_T \omega_0/RT = 0$, the quantity in brackets becomes 0/0, and by comparison with 6–12 that this indeterminate form must equal ½ if 6–9 is correct.

Our problem, then, is to investigate the behavior of the function

$$y = \frac{u - \ln(1 + u)}{u^2} = \frac{N(u)}{D(u)} \tag{6–14}$$

as u approaches zero. We do this with the aid of a slide rule and construct the accompanying table.

u	$\ln(1 + u)$	$N(u) = u - \ln(1 + u)$	$D(u) = u^2$	$y = \dfrac{N(u)}{D(u)}$
1	0.693	0.307	1	0.307
0.5	0.406	0.094	0.25	0.376
0.2	0.1823	0.0177	0.04	0.443
0.1	0.0953	0.0047	0.01	0.47
0.05	0.0488	0.0012	0.0025	0.48
0.02	0.0198	0.0002	0.0004	0.50
0.01	0.00995	0.00005	0.0001	0.50

Values of $N(u)$ and $D(u)$ are sketched in figure 6–3. We see at once from the table that $y = N(u)/D(u)$ does indeed seem to approach ½ as u is made smaller, although we cannot be sure that this is exactly so because the calculation becomes less and less precise toward the bottom of the table and the last value of y shown could be anything from 0.4 to 0.6.

To be more certain of our limiting check, we try another method of evaluating the indeterminate form 6–14, in which we expand $\ln(1 + u)$ in a power series in u in the neighborhood of $u = 0$. This series and others likely to be needed may be found in mathematical tables or can be derived by using a Taylor's series. If the expansion is about $u = 0$ Taylor's series becomes McLaurin's series. Thus in the present case

$$y = \frac{u - \left(u - \dfrac{u^2}{2} + \dfrac{u^3}{3} - \dfrac{u^4}{4} + \cdots\right)}{u^2} = \frac{1}{2} - \frac{u}{3} + \frac{u^2}{4} - \cdots \qquad (6\text{–}15)$$

And thus we find that the limit of y as u approaches zero is indeed $\frac{1}{2}$, which gives us the desired limiting-case check of 6–13 and hence of 6–9.

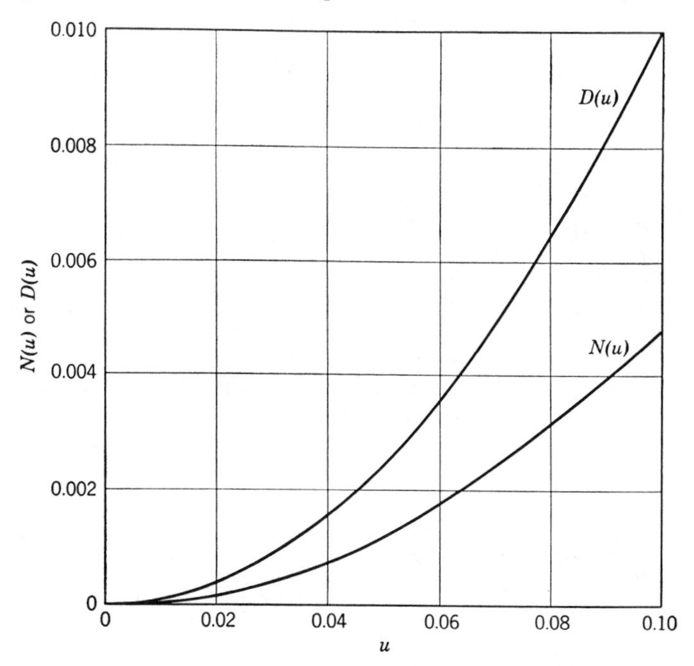

Fig. 6-3. Sketched plot of numerator and denominator of equation 6-14.

Another general method called L'Hospital's rule and treated in most textbooks on advanced calculus is available for dealing with indeterminate forms that often occur in limiting-case checks. L'Hospital's rule states that, if $N(u)$ and $D(u)$ each approaches zero as u approaches a, then

$$\operatorname*{Lim}_{u \to a} \frac{N(u)}{D(u)} = \operatorname*{Lim}_{u \to a} \frac{\dfrac{d}{du} N(u)}{\dfrac{d}{du} D(u)} \qquad (6\text{–}16)$$

Moreover, 6–16 also applies if $N(u)$ and $D(u)$ each approaches infinity as u approaches a, so that the indeterminate form is ∞ / ∞. If the right member of 6–16 also reduces to either of the indeterminate forms 0/0 or ∞/∞ the limit may be found by repeating the process of 6–16 over and over until a limit results which is not indeterminate. However, the repetition may not always yield such a form.

If we apply L'Hospital's rule to 6–14 there results

$$\begin{aligned}\operatorname*{Lim}_{u\to 0} y &= \operatorname*{Lim}_{u\to 0} \frac{\dfrac{d}{du}[u - \ln(1+u)]}{\dfrac{d}{du}u^2} \\ &= \operatorname*{Lim}_{u\to 0} \frac{1-(1+u)^{-1}}{2u} \qquad (6\text{–}17) \\ &= \frac{1-1}{0} = \frac{0}{0}\end{aligned}$$

which is another indeterminate form; hence the process will be repeated. Differentiating 6–17 (before the limit $u = 0$ is substituted) we obtain

$$\begin{aligned}\operatorname*{Lim}_{u\to 0} y &= \operatorname*{Lim}_{u\to 0} \frac{\dfrac{d}{du}[1-(1+u)^{-1}]}{\dfrac{d}{du}(2u)} \\ &= \operatorname*{Lim}_{u\to 0} \frac{(1+u)^{-2}}{2} \qquad (6\text{–}18) \\ &= \frac{1}{2}\end{aligned}$$

which agrees with the limit found from 6–15. Incidentally, 6–17 could also have been evaluated without a second application of L'Hospital's rule merely by clearing fractions, and then substituting $u = 0$.

EXERCISE. Apply 6–16 directly to 6–9 to determine the limits when $(K_E K_T)$ approaches zero and when T approaches zero.

6–6 Symmetry as a Check

Less general in applicability than the methods so far discussed, but sometimes very helpful, is the use of symmetry. Indeed it can be very useful in setting up the equations in the first place, as we have endeavored to show in section 4–9. How symmetry can serve in checking is best shown by some examples.

Consider the resistance between the terminals of the circuit shown in figure 6–4. It is well known to be

$$R_T = R_1 + R_2 + \frac{R_3 R_4}{R_3 + R_4} \tag{6–19}$$

In the circuit R_1 and R_2 are similarly placed, that is, each is in series and therefore we are not surprised that each affects 6–19 in the same way.

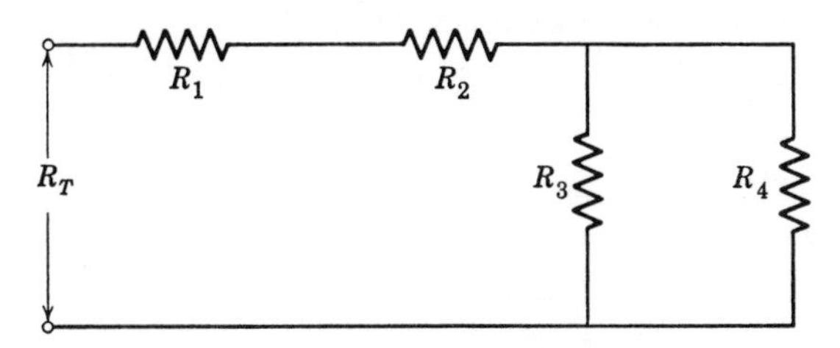

Fig. 6-4. A series-parallel circuit to illustrate checking by symmetry.

Also R_3 and R_4 are similarly placed in the circuit, and of course they are placed similarly in the formula for R_T. By such reasoning we would know at once that the following formula 6–20 could not be correct for the circuit of figure 6–4.

$$R_T = R_1 + R_3 + \frac{R_2 R_4}{R_2 + R_4} \tag{6–20}$$

As with the other methods of checking, however, this one has its limitations; the test of symmetry alone, for example, could not detect the error in

$$R_T = R_3 + R_4 + \frac{R_1 R_2}{R_1 + R_2} \tag{6–21}$$

As another example, consider equation 6–2, the result of the relay-contact problem in section 6–4. In this problem there are two masses M_U and M_L which move as one; accordingly M_U and M_L should occur similarly in equation 6–2 which gives the velocity, and they do. Also, there are two springs having constants K_U and K_L respectively, each acting in the same direction on the combined mass $M_U + M_L$, and these springs respectively have initial compressions C_U and C_L. We expect therefore to find K_U and C_U respectively entering the formula in just the same way as K_L and C_L. Examination of 6–2 shows that this expectation is fulfilled.

As a final illustration of the usefulness of the test for symmetry, consider the following situation. We need a formula for the deflection of a leaf spring, and we copy one from a book:

$$\frac{P}{\delta} = \frac{Ebnh^3 l_1}{6 l_2^2 l_3^2} \tag{6-22}$$

where P = load (lb).
δ = deflection at the load (in.).
E = Young's modulus for the leaf material (lb in.$^{-2}$).
b = width of leaves (in.).
n = number of leaves.
h = thickness of each leaf (in.).
l_1, l_2, and l_3 = lengths (in.) defined in an accompanying diagram, reproduced here as figure 6-5.

Before using the formula we proceed, in good professional style, to check it thoroughly. First we satisfy ourselves that it is correct dimensionally.

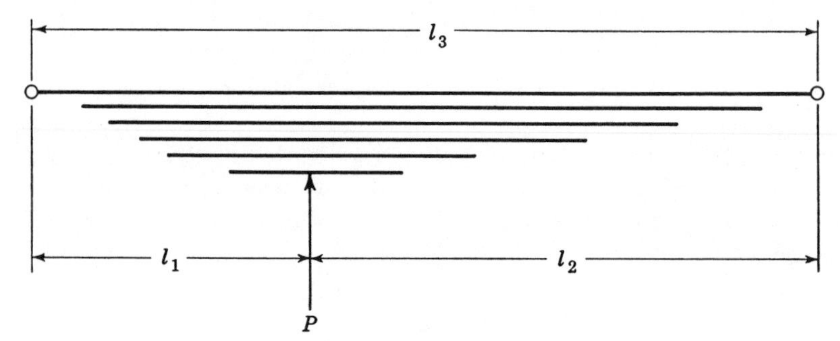

Fig. 6-5. Unsymmetrical leaf spring.

Then our experience tells us that each of the factors E, b, n, and h is reasonably placed in the numerator, for an increase in any one of these would tend to increase the stiffness of the spring, that is, increase the load necessary to produce a given deflection. Turning to the l's, we find reasons for their placement less clear except that having l_3, the total length, in the denominator is reasonable since an increase in it should make the spring less stiff. Then we notice that l_1 and l_2, which surely should enter the formula symmetrically, do not do so, and we conclude that 6-22 is incorrect. Investigating further, we find that in transcribing the formula we inadvertently interchanged the subscripts 1 and 3. Hence, our former conclusion that l_3 should be in the denominator rather than in the numerator where it now is must be reconsidered. We see that a choice of parameters more rational than l_1 and l_2 might have been a fraction f where $f = l_1/l_3$ and $1 - f = l_2/l_3$. Then the formula corrected for the mistake in copying would be

$$\frac{P}{\delta} = \frac{Ebnh^3}{6f^2(1-f)^2 l_3^3} \tag{6-23}$$

and this meets all the tests we have made including that of symmetry, for the factors f and $1 - f$ enter similarly as they should. It must be remarked, however, that we have not shown positively that the formula is correct, nor have we discovered what limitations it may have.

6–7 Check by Solving by an Alternate Method

An attractive possibility for making an overall check of the solution to a problem would be to make a second analysis entirely independent of the first. Unfortunately, however, getting just one solution, let alone two, is about all most of us are able to do, especially if we are working near the limit of our capabilities. Furthermore, in a physical problem restricted by a single set of assumptions, there really is but a single set of principles which govern, so that being able to find wholly independent paths all the way through to the result is not usually possible. Nevertheless, alternative paths through parts of the reasoning often are easy to find and can lead to very helpful partial checks.

Seeking checks through the use of a different method of solution may lead to a shorter and more direct analysis. Also working parts of a problem in different ways is likely to contribute to the learning stage of the analysis. For instance, in the phonograph record-changer problem, section 2–6, the attempt to check equation 2–49 by deriving it differently led to a much more direct path to this equation and on the way turned up an equation, 2–50, which could have been written down at once from a principle of mechanics not used directly in the earlier analysis. Thus not only did we obtain a good check, but we simplified the analysis, and perhaps also learned a useful principle of mechanics.

Another illustration of the value of checking by using an alternate method of solution occurs in the dynamic braking problem, section 6–5, where we can proceed from the differential equation 6–6 to the result, equation 6–9 by two quite independent paths, one of which is indicated in the text, and the other of which is left as an exercise. The latter method is less direct than that given in the text but is perhaps better in that it yields additional information.

Still another illustration also comes from the dynamic braking problem, section 6–5. The solution given there is based on the application of Newton's laws of motion. Suppose we start over again and apply instead the law of conservation of energy, making precisely the same assumptions as before. We can then make the following statement:

The original kinetic energy of the rotating parts must equal the sum of:

1. The kinetic energy remaining at any instant.
2. The work done against the frictional torque up to that instant.

3. The work equivalent of the energy dissipated as heat from the armature circuit resistance.

This may be translated mathematically as

$$\tfrac{1}{2}J\omega_0^2 = \tfrac{1}{2}J\omega^2 + \int_0^\theta T\,d\theta + c\int_0^t i^2R\,dt \tag{6–24}$$

where the symbols are as already defined in section 6–5 with the addition that c is a constant factor to convert energy in watt-seconds or joules to energy in foot-pounds.

To solve 6–24 we can see no direct way of evaluating the last integral; consequently we decide to differentiate with respect to t to remove the integral signs. In doing this we note that the left-hand member and T are constants. Then

$$J\omega\frac{d\omega}{dt} + T\frac{d\theta}{dt} + ci^2R = 0 \tag{6–25}$$

Now we use 6–5 to eliminate i, and observe in doing so that the new solution ceases to be completely independent of the old one. Doing this and multiplying through by dt

$$J\omega\,d\omega + T\,d\theta + \frac{cK_E^2}{R}\,\omega^2\,dt = 0 \tag{6–26}$$

Next we replace $\omega\,dt$ in the last term by its equal $d\theta$, separate variables, and integrate between limits:

$$\int_0^{\theta_s} d\theta = -J\int_{\omega_0}^0 \frac{\omega\,d\omega}{T + \dfrac{cK_E^2}{R}\,\omega} \tag{6–27}$$

Now we observe that, except for one startling difference, this is precisely like 6–8 in the former analysis. The difference is that here we have cK_E^2 where before we had K_EK_T. Is it perhaps possible that $K_T = cK_E$? As may easily be seen, the two are dimensionally the same, as they should be if equal.

To investigate the matter thoroughly we note that both K_T and K_E pertain to the armature as a converter of power from one form to another. Expressed in electrical terms the power converted is the current times the voltage generated; in terms of our symbols this is $iK_E\omega$ measured in watts. The power converted expressed in mechanical terms is the electromagnetic torque produced times the angular velocity, or $K_Ti\omega$ in foot-pounds per second. Equating these two expressions for the power converted after putting in a factor to express watts in foot-pounds per second, evidently the same as c in 6–24, gives

$$K_T i\omega = ciK_E\omega \tag{6–28}$$

from which

$$K_T = cK_E \tag{6–29}$$

Thus we complete our check having shown the equivalence of 6–27 to 6–8; obviously there is no point in pursuing the alternative solution further since at this place it becomes identical with the original one. A by-product of the check is 6–29, which expresses a general truth about d-c motors and generators.

Thus we have seen that sometimes it is possible to check a solution in part by using a different physical principle to start with or by following a different path through the mathematics. When the assumptions underlying the solution remain unchanged, it is usually found, as in the illustrations given, that a new path through a solution will merge somewhere with the original path. If, on the other hand, an alternative solution is based on assumptions sufficiently different, the reasoning all the way through may be different, and the conclusion may be different, too.

6–8 Validity of Assumptions

Essential in the analysis of a problem is checking to see if the assumptions are valid. Sometimes this can be done at the time the assumptions are made; in other cases it may not be possible until the solution is complete.

In the solution of the relay-contact problem, section 6–4, it was assumed specifically that the masses of the springs are negligibly small, and by implication that the moving parts carrying the contacts are rigid. With numerical magnitudes known in a particular case, it would be easy to say at the beginning whether or not these assumptions are valid. Or, if numerical magnitudes are not yet determined, limits might be specified in advance within which the given solution could be expected to hold.

Also, in the solution to the relay-contact problem, it was assumed that conditions were such that the two contacts remain together until the lower stop is reached. Testing the validity of this assumption constitutes a separate problem which could be worked out before or after the determination of the parting velocity. Doing it first could be best since it might turn out that the limits of validity are such as to make the relay impractical; on the other hand, if a particular parting velocity is being designed for, both problems would have to be completed in general terms before numerical magnitudes could be chosen.

EXERCISE. In the relay-contact problem, section 6–4, find the condition for validity of the assumption that the contacts remain together until the lower stop is reached.

In contrast to the above, it is sometimes necessary to carry out a solution at least in part before an original assumption can be checked. For example, in the dynamic-braking problem, section 6–5, it is assumed that the armature current does not become so high as to cause flash-over or other untoward incident. Checking this would require determining the armature current as a function of time, and this involves part of the same reasoning as the main problem.

EXERCISE. In the problem of section 6–5 find the armature current as a function of time.

Sometimes you cannot be sure, until perhaps too late, whether or not your assumptions are valid. In very complex situations, particularly those involving the actions of people, the best you may be able to do is to make several analyses based on different sets of assumptions, and thus try to be prepared for any eventuality.

6–9 Numerical Work

The topic of checks on arithmetic has been reserved for the last because in general the numerical work should not be done until the end of a mathematical analysis. There are three important reasons for carrying the analysis in literal form to the end: first, because only when the work is in literal form can it be checked dimensionally and by limiting cases; second, because by working in literal form a whole class of problems is solved and not just one case; and finally because there is less numerical work and consequently less chance for error when numbers are put in at the end, especially if care is taken to arrange the result for easy calculation.

It goes without saying that the engineer working in good professional style makes full use of the methods for checking arithmetic learned in elementary school. In addition there are three practices which can help enormously in the prevention or detection of errors.

1. PROPER USE OF SIGNIFICANT FIGURES. Engineering calculations are based on data derived from some sort of measurements, and obviously the results can be no more precise than the data. The measurements rarely have an uncertainty of less than 0.1 per cent; frequently the uncertainty is 1 or 2 per cent or even much more. Consequently, except in rare cases, it is a sheer waste of effort to perform calculations to an accuracy of better than 0.1 per cent, that is, three significant figures will nearly always suffice, and often two are enough. Consistent with this is the use of the 10-inch slide rule and the estimation of the third figure (fourth figure when the first is between 1 and 2). Carrying more significant figures than are needed in a given situation simply increases the chance for major error without any compensating benefit. Putting a given amount of effort into

finding and checking two significant figures of a result is immeasurably better than putting the same effort into getting four significant figures and not checking.

2. PLACING THE DECIMAL POINT BY ROUGH CALCULATION. Placing the decimal point correctly is the most important part of any calculation because it fixes the order of magnitude of the result. In a slide-rule calculation it probably is best done by a separate rough calculation which serves not only to place the decimal point but to check the first significant figure of the precise calculation. For example, suppose the following calculation is to be performed:

$$\frac{3.14 \times (141.3)^2 \times 0.0582 \times 10^{-5}}{75.2}$$

First, for the rough calculation set down

$$\frac{3 \times 1.4 \times 1.4 \times 10^4 \times 6 \times 10^{-7}}{8 \times 10}$$

where each factor has been rounded off to one significant figure and powers of 10 have been taken out so as to leave each factor in the range 1 to 10. Then by a rough mental calculation ($1.4 \times 1.4 \approx 2$; $3 \times 6 \div 8 \approx 2+$; and $2 \times 2+ = 4+$) the result is seen to be about 4×10^{-4}. Thus it is known that the result will be something more than 4×10^{-4} but probably not over 5×10^{-4}. Now, performing the operations on the slide rule without regard for decimal point, the significant figures are found to be 486 and thus the result must be 4.86×10^{-4}. Finally, the work is checked by an independent rough calculation based on the original figures. For example, $(141.3)^2 \approx 20{,}000$, $0.0582 \approx \frac{1}{20}$, and therefore the product of these is about 10^3; then $3.14/75.2 \approx 1/25 = 0.04$; therefore the result is approximately $0.04 \times 10^3 \times 10^{-5} = 4 \times 10^{-4}$.

3. PLOTTING GRAPHS TO DETECT ERRORS. If a formula is to be used for a number of calculations, as in the construction of a table, a graph of the result against the variable is an excellent means for detecting errors. If the formula expresses a mathematically continuous function, then a graph of it should be a smooth curve, and a point which does not fall on the curve is a sure indication of error. It may be remarked in passing that making a running plot of data as they are obtained during an experiment is an effective way of catching errors while there is still opportunity to correct them.

For doing arithmetic efficiently cultivate the habit of being careful. A mistake once made may be very hard to find, and simply repeating the same calculation without some variation in its order rarely serves as an effective check. Finally, you must realize that a result wrong because of poor arithmetic may be just as damaging as one due to an error in principle.

6–10 Effectiveness of the Check

In reviewing the checks in a particular case, it is important to consider what has really been checked, and what not, for usually the best of checking will leave some uncertainties and it is well to know what they are. For instance, a common-sense check that says a result should increase with a certain factor can by itself hardly say whether the increase should be linear or as the square or some other way. And again, a series of limiting checks all of which make the same term of a result vanish tells nothing about the correctness of the form of that term except that it should not be present in the limits investigated.

Another thing of which to be wary in checking is your own prejudice. You may work so long and so hard on an analysis that you find it almost impossible to believe that anything could be wrong; perhaps you may come to think the analysis so thoroughly worked out that checks are not necessary. To guard against this, it is desirable to do at least some checking ahead of time, that is, to predict on the basis of careful consideration what is to be expected. The common-sense or experience type of reasoning is likely to be much more reliable when used before your opinions have been colored by the findings of the analysis which you are trying to check. The procedure of the limiting check used ahead of time may be useful in bracketing a result. This may have another advantage if the pressure of time requires you to make a decision before a complete analysis can be finished; such a bracketing even may show that a more precise analysis is not required. Another sort of check in advance which is desirable is to sketch carefully and approximately to scale what you think a calculated curve should look like and to do it before the calculations are started. Obviously this can be a far more effective check than looking wisely at the finished curve and saying "Ah, precisely as I expected!"

C H A P T E R S E V E N

Interpretation of the Mathematics

7–1 Need for Interpretation

The very conciseness which is so valuable in mathematics frequently makes a mathematical result difficult to understand fully. So much can be contained in a brief mathematical formula that its whole significance may not be comprehended without going through a lengthy process of interpretation which may include such steps as calculation of numerical cases, sketching of curves, and expression in words.

Besides the obvious necessity for understanding the results of your own work, it is often necessary to make the meaning clear and convincing to others also if the work is to serve a useful purpose. Conveying the meaning to other persons is a problem in communication, a problem whose solution itself may require full use of professional method. Especially is this likely to be necessary if the communication must be in a written form such as a report, technical paper, or letter. Then you must decide with particular care what it really is you wish to say, and, considering the probable knowledge and points of view of the recipients, the way you can do it best.

But before you can consider trying to make the result of a mathematical analysis clear to other persons, you must make its meaning clear to yourself, and not only must the result be clear but the significance of the steps in the analysis must be clearly understood, too. Some ways of gaining this kind of understanding form the subject of this chapter.

7–2 Use of Numerical Cases

Interpretation in terms of assumed numerical cases is often very helpful even when an analysis has not reached the point where a final quantitative application is to be made.

An awareness of relative magnitudes of the various factors as a mathematical analysis progresses may serve as a guide toward simplification. For instance, in the non-fogging mirror problem, section 2–3, placing numbers in equation 2–19 gives a value, equation 2–20, for temperature drop through the glass that is so small as to suggest that this drop may be neglected and so make possible an improvement in the method for estimating the thermal transient.

Another use of a numerical case is illustrated in the bimetal thermostat problem, section 2–7. There equations 2–75 through 2–78 and figure 2–16 constitute a numerical interpretation of the general mathematical result given by equations 2–73 and 2–74. The numerical values were selected arbitrarily to represent what was believed to be a typical situation. Among the benefits derived from this interpretation were a visualization of the state of stress in the strip and the learning of the significant fact that the stress distribution is independent of the thickness of the strip. Of course, sketches like figure 2–16 might have been made directly from the mathematical relations expressed in literal terms, but there would be so many possibilities that the assumption of a typical numerical case was more useful as it served to focus attention at once on a situation attainable in practice.

It should be clear that, although it is essential to carry mathematical analyses through in literal terms, it is also useful at times to try numerical cases. This serves both to aid understanding of what the mathematics show and to point the way to improvements or simplifications in the analysis that may be possible because of the relative magnitudes of the factors. A valuable by-product of a numerical interpretation is the opportunity it affords for checking by experience, as discussed in section 6–3.

7–3 Sketched Curves as Aids to Understanding

Perhaps the most common device used by engineers in explaining mathematical and other relationships to themselves and others is to sketch curves with the object of visualizing the manner in which the variables affect each other. For this purpose of visual interpretation, the curves are usually not plotted accurately to scale on coordinate paper, but are sketched free hand on plain paper and only roughly to scale. Indeed, the scales sometimes may be distorted deliberately so that some particularly interesting or critical part of the curve may be seen to advantage.

This book contains numerous curves made for interpreting mathematical results. For instance, in the electric accelerometer problem, section 2–2, figure 2–2 was sketched to interpret equation 2–15. In it the engineer showed himself how circuit resistance R affects the response of the accelerometer when a constant acceleration is suddenly applied. With the sketch, he made clear that the steady-state response is unaffected by resistance, and that as the resistance is lowered the response, that is, current versus time, assumes more and more nearly the rectangular form of the acceleration which is applied. It is worth noting that the engineer labeled the sketch carefully even though he intended it only for his own use.

Another typical sketched curve of this sort is in the record-changer problem, section 2–6, where figure 2–13 shows some angular speeds plotted against time. The axes, the separate curves, and significant points on the scales are labeled with the symbols used in the mathematical analysis. This sketch was made during the course of the analysis and brought together some relations already worked out. It then was used in considering the next step in the analysis.

Other examples occur in the problem on the rectifier and resistance network to simulate non-linear resistance, section 4–9, where there is a series of sketches in figures 4–12 through 4–16. The first of these shows the form of one of the branch currents found early in the analysis which must be understood clearly before the analysis can proceed. The next one, figure 4–13, brings together and relates all the branch currents. Figure 4–14 shows the way in which the branch currents in one-half the network combine. Figure 4–15 predicts how the form of the total current will depend on alternative assumptions and is used in deciding which to adopt. Figure 4–16 interprets visually the end result. It should be clearly understood that these sketches are not put in this book to make it easier to follow. Rather they are essential parts of the analysis even though they are no more than rough sketches and not to scale. Each of them contributes to the learning stage of the professional method in that each is used in helping to answer a question such as: "What has been learned so far that is useful in going further?" The sketches of figures 4–14 and 4–15 take part also in planning stages, for each is used in deciding on the next step.

A somewhat different use of rough sketches of curves occurs in finding the starting time of a fan, section 5–2. Here the problem is to solve the differential equation 5–3. Two plans are considered: one a step-by-step process using the equation as it stands, and the other a graphical integration after the variables have been separated. Figures 5–3 and 5–4 respectively interpret the processes that would be involved in these two plans for solution. They were used in gaining understanding of each process so that an intelligent decision between them could be made.

7–4 How to Sketch Curves

In engineering analysis, curves are constructed for various purposes such as conveying information accurately, and, quite differently, visualizing the dependence of one quantity on another. For conveying information accurately values need to be determined with exactness and the corresponding points plotted carefully on suitable coordinate paper. For visualizing quantities, however, usually a more rough-and-ready procedure suffices; free-hand sketching on plain paper is all that is needed. Such sketching of curves is the subject of this section.

The general plan in curve sketching is to separate the mathematical expression to be represented into components whose forms are somewhat familiar, sketch these separately, and then combine them by eye in a final sketch. In handling the components the significant matters are such things as the intercepts with the axes and the values of slopes at special points, or, in the case of a sinusoid, the amplitude, period, and phase angle. In doing the actual sketching it is well to use a soft pencil, and sweeping strokes. If a line intended to be straight bends or wavers a little this fact is not important, nor does it often matter if two lines do not intersect exactly at the spot intended.

Some elementary functions recur so often that it is worth while here to examine their properties from the point of view of sketching. Consider first the exponential function

$$y = A\epsilon^{-mt} \tag{7-1}$$

in the range of values of t starting with zero and increasing to very large values. We may think of t as time, which it often is in such expressions. Clearly $y = A$ when $t = 0$, and as t increases y becomes smaller, as is shown by the solid curve in the upper portion of figure 7-1. Consider now *how* the curve decreases, and compare it with the dashed straight line which starts at the same point A at $t = 0$. Let the points on the t axis, t_1, $2t_1$, $3t_1$, mark off equal intervals of time, during which the straight-line function decreases by equal increments Δ. During these same equal intervals, the exponential curve falls by equal *ratios*. If in time t_1 the curve falls to half its initial value, at $2t_1$ it is half its value at t_1, at $3t_1$ it is half its value at $2t_1$, and so on. The exponential comes down like a geometric progression, the straight line like an arithmetic progression. Since the radioactivity of a substance such as radium is expressed by an exponential function like equation 7-1, it is clear that how long the radioactivity lasts is conveniently expressed in terms of the half-life, that is, the interval t_1 in figure 7-1 during which the function decreases by half. The whole life of the radium or the whole duration of the exponential is infinite, but it may well have decayed to negligible values after 5 to 10 half-lives.

When we have to work with an equation rather than a curve, it is often more convenient to work with a value of t that makes the exponent mt equal to unity, that is, a value of $t = 1/m$. This is called the *time constant* and is the increment of time during which the curve falls to $1/\epsilon$ or $1/2.71828 = 0.368$ of its initial value.

It is clear from the equation that the function $A\epsilon^{-mt}$ is characterized by two quantities, A and m; A gives amplitude, m gives the reciprocal of time required for decay to $1/\epsilon$ of the initial value. Because we are dealing with a function y, equation 7–1, which is decreasing geometrically, its value at the end of an interval of time $1/m$ is $1/\epsilon$ of its value at the beginning of the interval, no matter at what absolute time the interval begins. Thus we

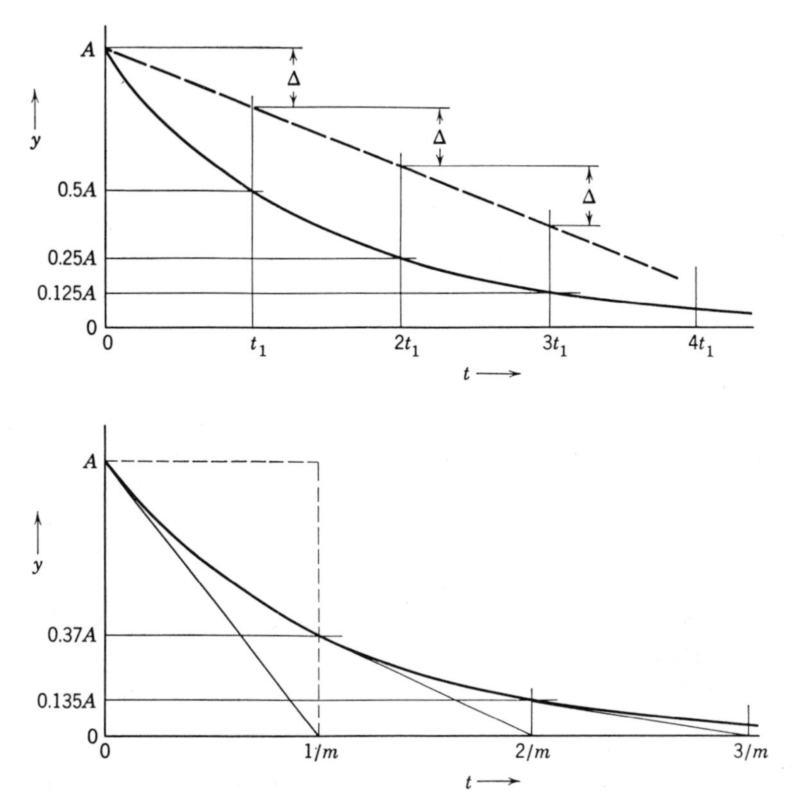

Fig. 7-1. Geometric properties of the function $y = A\epsilon^{-mt}$.

may construct a table of values of y at certain multiples of the time constant.

t	y
0	A
$1/m$	$\epsilon^{-1}A = 0.368A$
$3/m$	$\epsilon^{-3}A = 0.050A$
$5/m$	$\epsilon^{-5}A = 0.007A$

Or, in words, after one time constant the function has fallen to 37 per cent of its initial value (roughly, one-third); after three time constants, to 5 per cent; after five time constants to less than 1 per cent. If we remember these values we can look at an exponential function and tell its duration for all practical purposes; in the lower part of figure 7–1 the curve is shown for three time constants.

The concept of time constant has another important geometric meaning. Consider the line which is tangent to the curve at $t = 0$, $y = A$, and imagine it drawn as in the lower part of figure 7–1 until it intersects the t axis. The intersection occurs at a distance from the origin equal to the time constant $1/m$.

EXERCISE. Verify this statement.

We may therefore say that, if the function decreased linearly, with a slope equal to the actual initial rate of decrease, the function would reach zero in a time equal to the time constant. This relationship holds for any point regarded as the initial one and is shown in figure 7–1 for values of t equal to 0, $1/m$ and $2/m$.

Finally, time constant has still another significant physical meaning. The area under the curve of figure 7–1 out to a point infinitely remote along the y axis is finite (in spite of the infinite base line), and the area under the curve is the product of the initial value A and the time constant. Thus we may say that the area under the exponential curve is equal to the area of a rectangle of height A and base $1/m$, shown by the dotted lines in figure 7–1.

EXERCISE. Verify this statement.

Let us see how to use the properties of the exponential and the idea of time constant to sketch equation 7–1 as in the lower part of figure 7–1. We assume that A and m are known numbers and choose scales recognizing that the maximum height of the curve is A, the base three to five time constants, $3/m$ to $5/m$. We locate values on the time axis 0, $1/m$, $2/m$, $3/m$, etc., and put in the corresponding values of y: A, $0.37A$, $(0.37)^2A$, $(0.37)^3A$, etc. For sketching purposes we interpret $0.37A$ as a little more than one-third of A; $(0.37)^2A$ as a little more than one-third of $0.37A$ or about one-eighth A, and $(0.37)^3A$ a little more than one-third of $(0.37)^2A$ or one-twentieth A. Having located about four points as described above, we sketch lines from each of these points to the point on the t-axis below the next point. These lines determine the slopes of the exponential. Finally, we fair in the curve so that it passes through the points with the slopes called for. Figure 7–1 actually shows the points and construction lines, but after a little experience they need be seen only in the mind's eye.

Suppose we wanted to sketch

$$y = 10\epsilon^{-2t} - 5\epsilon^{-4t}$$

We would choose a convenient scale to show the maximum value of 10 and, say, three time constants of the longer component, which is the first. Then each curve is sketched separately, as in figure 7–2, and the second is subtracted from the first, either by using a divider or by marking intervals

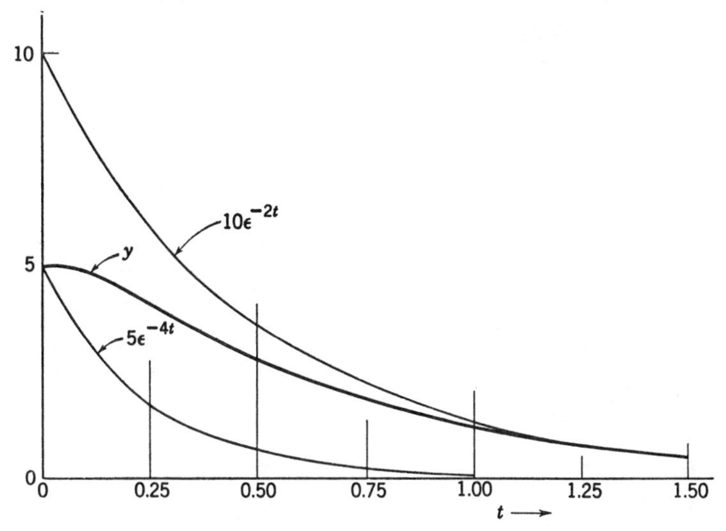

Fig. 7-2. Sketching $y = 10\epsilon^{-2t} - 5\epsilon^{-4t}$.

on a scrap of paper. As a check, or auxiliary bit of information, we see that the slope at $t = 0$ is

$$\left(\frac{dy}{dt}\right)_{t=0} = -20 + 20 = 0$$

Only a few points need to be sketched; it is clear that after the first half of the interval the resultant curve approaches the first of the two components.

Sinusoidal functions occur in such a wide variety of engineering problems that it is important to be able to sketch them easily and with reasonable accuracy. Consider

$$y = B \sin \omega t \tag{7–2}$$

The quantities which characterize this function are the amplitude B and the period or interval of t after which the function begins to repeat itself. The period is the value of t which makes the argument ωt equal 2π; that is, the period is $2\pi/\omega$. For sketching a sine wave, figure 7–3, attention should be focused particularly on the form of the first quarter cycle since the succeeding quarter cycles are merely differently oriented duplications of the first. The first maximum occurs at $\omega t = \pi/2$ radians or 90°. At one-third of this value of ωt, $\pi/6$ radians or 30°, the function reaches precisely half its maximum value, that is, $B/2$. Moreover, the sine wave is very nearly a straight line from the origin to the 30° point; the tangent at the origin passes only about 5 per cent above the function at $\omega t = \pi/6$ and reaches a height equal to the amplitude B in $\omega t = 1$ radian (approximately

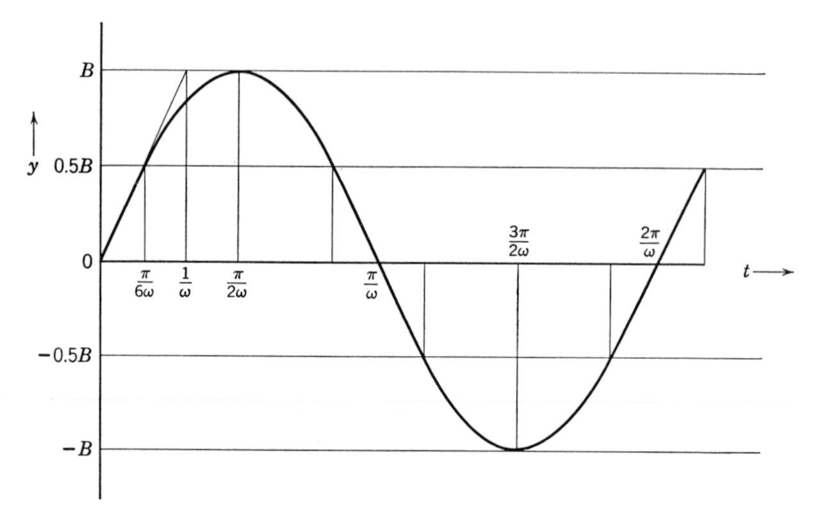

Fig. 7-3. Sketching $y = B \sin \omega t$.

60°). The slope at the maximum is zero, of course. Thus the first quarter of the sine curve is determined well enough for sketching purposes by the point at 90° and the zero slope there; the point at 30° and half amplitude; and the slope through zero passing just above the 30° point.

A cosine wave is merely a sine wave slid along the t-axis to the left by an amount corresponding to 90°. To sketch it, or in general any sine wave with phase angle, locate the zeros, the maxima, and points with appropriate signs 30° each side of the zeros. Then it is easy to sketch in the whole function by paying proper attention to the slopes at the zeros and at the maxima and minima.

As an example of a more complicated function let us sketch

$$y = 3\epsilon^{-1.5t} \cos (5t - \pi/4) \tag{7-3}$$

We observe that this is the product of one function which is $3\epsilon^{-1.5t}$ and another which is $\cos (5t - \pi/4)$. Or if we prefer to think in terms of degrees the second function is $\cos (5t - 45°)$. Our plan is to sketch the two functions separately, then multiply them together by inspection, and sketch the resultant curve.

Which curve shall we sketch first? If we choose the exponential it is more bothersome to put the phase angle into the sinusoidal curve, so we choose to draw the sinusoid first.

What scales are appropriate? Let us suppose we are interested in about three time constants of the exponential or $t = 0$ to $t = 3/1.5 = 2$ sec. In drawing the cosine we can use any convenient vertical scale for amplitude.

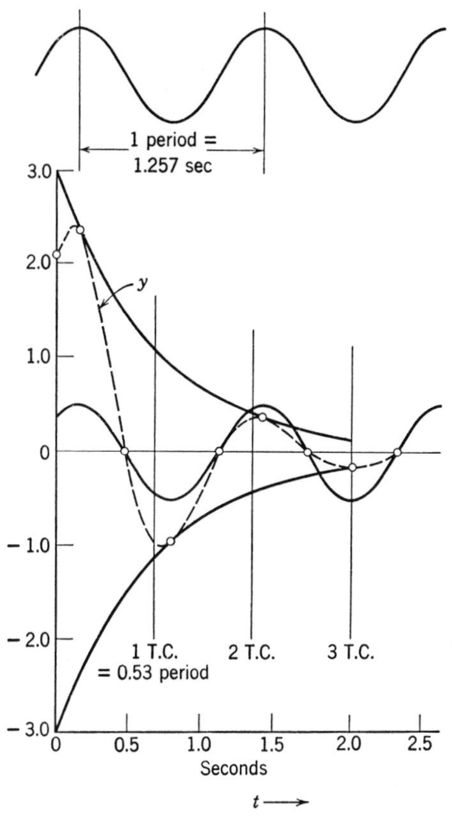

Fig. 7-4. Sketching $y = 3\epsilon^{-1.5t} \cos (5t - \pi/4)$.

The period is $2\pi/5$ or 1.257 sec, so let us draw about two periods. First we draw the sinusoidal function without regard to phase angle, as if it were just cos $5t$. See the upper part of figure 7–4. Two periods are laid off along the time axis; each is divided into four parts; and the maxima, minima, and zeros are located. Then a cosine wave is sketched.

Now the phase angle is put in, not by shifting the curve but by putting the origin in the proper place, shifted to the left from where it would be for a cosine. The shift is 45° and is shown in the lower part of figure 7–4. Now we mark vertical scales up to $y = 3$ and carefully lay off along the

t-axis points at 1.0, 2.0, and 3.0 time constants. One time constant is $1/1.5 = 0.667$ sec or $0.667/1.257 = 0.53$ period of the cosine wave. The exponential is drawn as before, making use of values and slopes. Now we have the two component curves. Multiplication of the two curves is made easy by recognizing that when the cosine is unity the product falls on the exponential, when the cosine is zero the product is zero, when the cosine is minus unity the product has the value of the exponential but with negative sign. It is helpful to draw the reflection of the exponential in the t-axis and proceed to locate the circled points of the product curve, which is then drawn with a dashed line. At our desks we would use colored pencils.

It is to be observed that the product curve always falls within the two envelope curves and has maxima and minima before the points of tangency with the envelopes. The value of y at $t = 0$ is probably most easily found by calculation as

$$y = 3 \cos\left(-\frac{\pi}{4}\right) = 3 \times 0.707 = 2.1$$

The slope at $t = 0$ should be checked, at least for sign,

$$\left(\frac{dy}{dt}\right)_{t=0} = 3\left[-1.5 \cos\left(-\frac{\pi}{4}\right) - 5 \sin\left(-\frac{\pi}{4}\right)\right]$$

$$= 3(-1.5 \times 0.707 + 5 \times 0.707) = +7.4$$

and this agrees with our sketch.

Finally, we might want to place units on the time scale that are round numbers such as 0.5, 1.0, 1.5 seconds. We find the length of this interval by proportion, sometimes by eye, sometimes by the use of a scale.

Curves drawn in this way are primarily sketches for the purpose of visualizing functions, yet in the authors' experiences they are usually surprisingly accurate, often with but a few per cent of error, when only ordinary care is used. It is helpful to use paper that is lined in one direction, but graph paper is not desirable because at least one change of scale may need to be made and the commonly encountered numbers are not simple multiples of one another.

EXERCISE. If no more points are to be used than are shown in figure 7–4, need the cosine wave itself be sketched? What operations need to be done with relative care?

Although the discussion in this section has been directed toward making rough sketches of mathematical functions, the kinds of procedures described are recommended as preliminary also to the construction of precise plots. These procedures serve to locate the critical points accurately and to give an early indication of the general form of the curve, which is very helpful in showing where more points need to be calculated.

7–5 Curves in Dimensionless-Ratio Form

When the effects on a result of a number of separate factors is to be studied, a curve in dimensionless-ratio form may be very helpful. What is meant will be shown first by an example taken from the problem about the electric accelerometer, section 2–2, where figure 2–2 is an interpretation of equation 2–15 intended to show the effect of different values of resistance

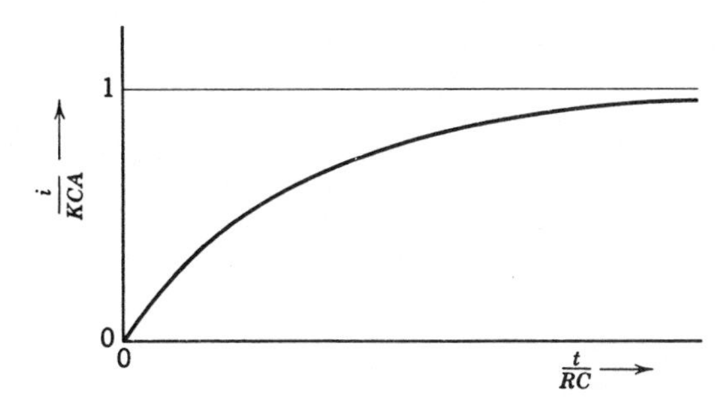

Fig. 7-5. Dimensionless plot of equation 2-15.

R. Another way of showing the effect of R as well as of the other factors is the following.

First, equation 2–15 is put in dimensionless form by dividing through by the factor KCA; thus

$$\frac{i}{KCA} = 1 - \epsilon^{-t/RC} \tag{7–4}$$

Then by the methods of section 7–4 the quantity i/KCA is sketched as a function of the quantity t/RC, as in figure 7–5. Now it will be observed that the plot figure 2–2, which is a family of curves with parameter R, has been reduced to the single curve in figure 7–5. Other families of curves of i versus t with parameters C, K, and A are also contained in this single curve.

Figure 7–5 may be described as a plot relating the dimensionless variables i/KCA and t/RC. These variables are, respectively, the ratio of the instantaneous current to the final steady-state current ($= KCA$) and the ratio of the time to the time constant of the circuit ($= RC$). From another point of view, current has been plotted not to a scale laid off in amperes but to one where the final steady current is taken as the unit. Similarly, time is measured in time constants of the circuit rather than in seconds.

EXERCISE. Sketch and interpret a dimensionless ratio plot of equation 2-25 in section 2-3.

A tremendous advantage of expression in terms of dimensionless ratios comes when a quantitative plot is to be made, because then the use of dimensionless variables usually both reduces and simplifies the computations. For instance, to plot figure 7-5 precisely, all that need be done is to read values of ϵ^{-x} from a table or log-log slide rule, subtract them from 1, and plot the result against x.

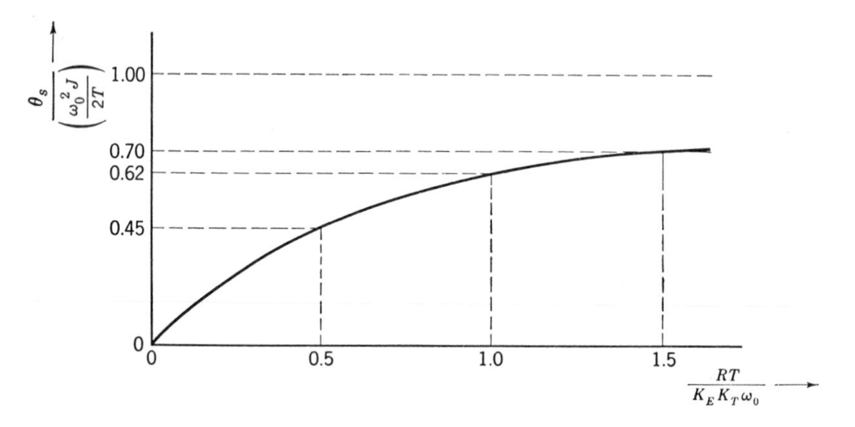

Fig. 7-6. Dimensionless plot of equation 7-5.

As another example of interpretation using dimensionless-ratio plots, consider the result of the problem on the dynamic braking of a motor in section 6-5. The problem there is to find how the angle turned during braking is affected by the magnitude of the braking resistance. The mathematical result is equation 6-9, which gives the angle θ_s as a function of the resistance R and other constants of the system. For a general case, this might be interpreted by plotting θ_s as a function of R for each of a number of assumed values for the other constants, and thus families of curves would be obtained.

To interpret the result as a single curve, equation 6-9 may be changed to the form

$$\frac{\theta_s}{\dfrac{\omega_0{}^2 J}{2T}} = 2\,\frac{RT}{K_E K_T \omega_0}\left[1 - \frac{RT}{K_E K_T \omega_0}\ln\left(1 + \frac{K_E K_T \omega_0}{RT}\right)\right] \qquad (7\text{-}5)$$

by dividing both sides by the factor $\omega_0{}^2 J/2T$ which, as shown in section 6-5, is the angle that would be turned if there were only the constant friction torque T and no electromagnetic torque. Now a plot of 7-5, sketched in figure 7–6, shows the dimensionless variable $\theta_s/(\omega_0{}^2 J/2T)$ as a function of the dimensionless variable $RT/K_E K_T \omega_0$, which is proportional to R.

Figure 7–6 may be checked to see that it is correct at the limiting values zero and infinity of $RT/K_E K_T \omega_0$ by comparison with the limiting case checks made in section 6–5; indeed, the construction of such dimensionless plots is done best in conjunction with limiting-case and dimensional checking.

The foregoing reduction of the mathematical result to the dimensionless plot, figure 7–6, is not unique; that is, there is not just one dimensionless plot possible for a given equation. Clearly the same information might be plotted against abscissas which are the reciprocals of the ones shown. The resulting curve would be quite different and for some purposes more useful. But curves with even more profound differences are possible, too.

EXERCISE. Interpret equation 6–9 by a sketched plot in terms of the dimensionless variables $\theta_s \Big/ \dfrac{J R \omega_0}{K_E K_T}$ and $\dfrac{K_E K_T \omega_0}{RT}$.

7–6 Verbalization as an Aid to Physical Interpretation

The dimensionless variable $\theta_s \Big/ \dfrac{\omega_0{}^2 J}{2T}$ in equation 7–5 has a simple physical significance. As shown in the derivation of 6–12, it is the ratio of the angle turned to the angle that would be turned in the absence of electromagnetic torque; in fact 7–5 was formed from 6–9 deliberately in such a way that there would be this physical significance. Now we may ask whether the other variable in 7–5, $\dfrac{RT}{K_E K_T \omega_0}$, has any simple physical interpretation; and the answer is that it does, as we shall now show.

Suppose that we arrange this variable in the equivalent form

$$\frac{T}{\dfrac{K_E \omega_0}{R} K_T} \tag{7–6}$$

and try to interpret the denominator physically. K_E is the voltage generated per unit speed, and ω_0 is the initial speed; therefore $K_E \omega_0$ is the voltage generated initially in the armature circuit. R is the resistance of the armature circuit; consequently $K_E \omega_0/R$ is the initial value of armature current. K_T is the torque developed per unit of armature current, and therefore $[K_E \omega_0/R] K_T$ is the initial value of torque developed electromagnetically. The numerator T is the friction torque; therefore 7–6 is the ratio of the friction torque to the initial value of the electromagnetic torque. Finally, we may strip the last vestiges of the mathematics from figure 7–6 and resketch it as figure 7–7, which shows a physical rather than a mathematical relationship, although, of course, the curve is of precisely the same shape as the one in figure 7–6 and both come from the same mathematical analysis.

The process above by which the physical significance of a mathematical result is ascertained through verbalization depends on careful attention to the precise definitions of symbols and on continued awareness of the significance of groups of factors already established, particularly in connection with checks. The process may be carried out mentally if the

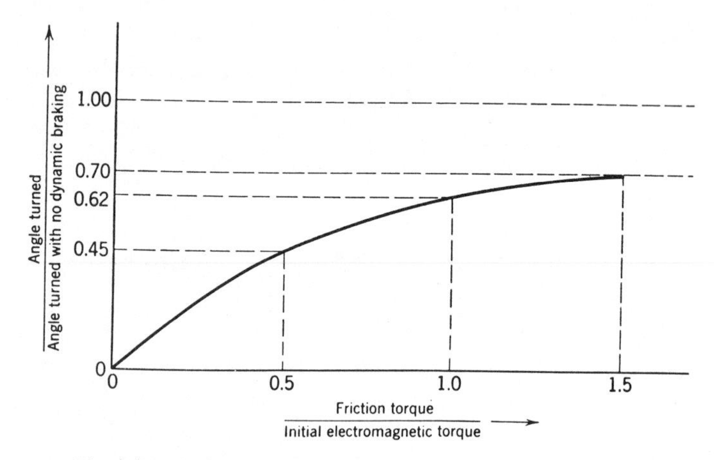

Fig. 7-7. A physical relationship deduced from equation 7-5.

situation is simple, but is best written out carefully in precise English. The reason for careful writing here is exactly the same as that for making a careful verbal statement of principle before constructing a mathematical equation, as discussed in section 4–2. In each situation the need is for particularly clear thinking, and precise writing is a powerful means for clarifying thought.

Exercise. Referring to the derivation of 6–10, make a physical interpretation of the variables specified in the exercise at the end of section 7–5.

There are many other examples of verbalization in this book in connection with interpreting mathematical expressions. For instance, in the roller bearing problem, section 2–4, the engineer made a careful verbal interpretation of equation 2–28 to help himself see why this equation is so simple when his derivation of it was so labored. From this verbal interpretation, it became clear to him that his analysis could have been more direct.

An example of verbal interpretation of a somewhat different sort occurs in section 2–2, the problem on the electric accelerometer. There equation 2–5 is the end of a mathematical analysis. After some checks, the engineer makes a statement which interprets this equation and includes a summary of the assumptions on which it is based. The engineer then goes on to a more general verbal interpretation of what he has learned: how to differentiate by electrical means. These interpretations he makes as part of his own learning process, the last stage of the professional method. Such interpretations may also be useful later when a report of the work must be written so that other people, too, can easily understand what has been accomplished.

7–7 An Illustration: Relay Coil for a Flame Detector

Interpretation of mathematical expressions plays an especially prominent part in the analysis of the following problem, which is presented to illustrate some of the points made so far in this chapter.

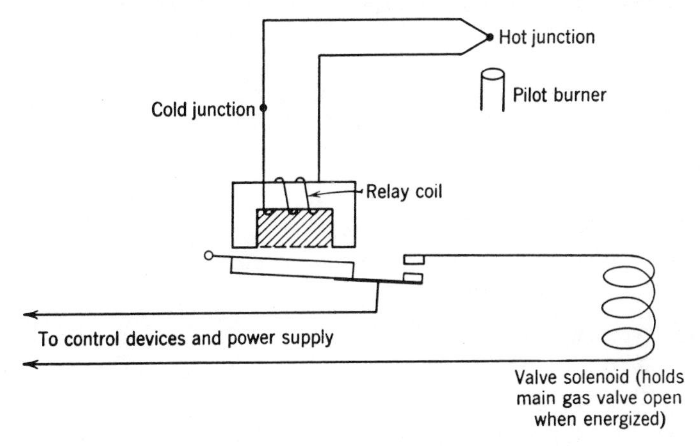

Fig. 7-8. Arrangement for assuring that main gas valve can be opened only when pilot flame is present.

A safety device for a gas-fired domestic hot-water heater is being designed. The arrangement is shown in figure 7–8. It consists of a thermocouple placed close to the flame of the gas pilot and connected to the coil of a small electromagnetic relay whose contacts control a solenoid-operated valve in the gas pipe to the main burner. As long as the pilot is lit, the electromotive force developed by the thermocouple is to be enough to energize the relay, which in turn maintains the circuit to the valve solenoid. If the pilot light is extinguished, the small relay drops out, opening the circuit to the valve solenoid and thus shutting off the main gas supply.

The design has progressed to the point where the thermal electromotive force to be expected has been determined and a tentative magnetic design for the relay has been completed. The magnetic design includes a determination of the ampere-turns that the coil must provide to keep the relay closed and the size of the space available for the coil, the shaded area shown in figure 7–8, which is called the window.

The problem which remains is to determine the size of coil, that is, number of turns and wire size, necessary to match the thermocouple output with the magnetic requirements of the relay.

With the known voltage of the thermocouple applied to the circuit comprising the relay coil and its leads, the current must be such that it produces the desired number of ampere-turns. But the current which flows depends upon the resistance of the circuit, and this, in turn, is related to the number of turns in the coil. Application of Ohm's law to the circuit appears to be a promising plan.

The current is given by the thermocouple voltage, assumed independent of current, divided by the resistance of the circuit; in symbols:

$$I = \frac{E}{R_C + R_E} \tag{7–7}$$

where I = current (amps).
E = thermocouple voltage (volts).
R_C = resistance of the coil (ohms).
R_E = resistance external to the coil composed of the thermocouple and leads (ohms).

If we let N be the number of turns in the coil, and F be the necessary magnetomotive force in ampere-turns which is to be produced by the coil, then

$$F = NI \tag{7–8}$$

In equation 7–8 we can substitute for I the value given by 7–7; thus

$$F = \frac{NE}{R_C + R_E} \tag{7–9}$$

At this point, we envisage a coil of a considerable number of turns, and, since the thermocouple and coil are to be located near each other, it appears reasonable to neglect the external resistance R_E in comparison with the coil resistance. Making this approximation, we obtain from 7–9 the relation:

$$\frac{R_C}{N} = \frac{E}{F} \tag{7–10}$$

Now to determine the design of the coil, that is, R_C and N separately, it appears that we must have additional independent relationships involving R_C or N or both, and we might get them by considering the limitations of space into which the coil must fit.

But before seeking another relationship let us pause to interpret 7–10. The left-hand member is the resistance of the coil per turn. Thus since E and F are each constant, equation 7–10 says that the resistance per turn must have a magnitude determined by the given values of E and F. The implication of this is perhaps somewhat startling; the coil can have any number of turns, for instance, one turn, and yet satisfy the voltage and magnetomotive-force relations. Checking physically, it is clear that, in the absence of external resistance, if one turn produces the desired number of ampere-turns, then ten turns of the same wire (same resistance per turn) will draw one-tenth the current, but still will produce the desired number of ampere-turns.

In further interpretation of equation 7–10, we observe that to design the coil we would ascertain from the design of the magnetic structure what mean length of turn would be necessary to surround the core. With this length of wire known, we could use a copper-wire resistance table to find the wire size to give a resistance equal to the known E/F. Then we could use any convenient number of turns of this wire, but using a large enough number to insure that the total resistance of the coil is sufficiently high to justify the neglect of the external resistance R_E. The resistance of the coil may be made larger by increasing the number of turns. But the space required by the coil is directly proportional to the number of turns, and, since the magnetic circuit is specified, there is a limit to the number of turns that may be employed. Indeed, there is no assurance from 7–10 alone that there will be room for even one turn of the cross section required! Interpreting 7–10 has yielded much information and insight into our problem, but has pointed even more strongly to the need for considering space limitations.

In considering a way to do this we reason that, to get the greatest magnetomotive force, it will be desirable to make the coil resistance as low as possible or to provide as many turns as possible, and so we assume a coil which will completely fill the space available. To relate the resistance of such a coil to the dimensions of the space and the number of turns, we express, first, the total length L (centimeters) of wire as

$$L = Nm \tag{7–11}$$

where m is the mean length (cm) of the turns and can be found from the dimensions of the core which are presumed to be known. Next, the cross-sectional area a of the wire (cm^2) is

$$a = As/N \tag{7–12}$$

where A is the known area of the window in (cm^2) and s is a space factor, less than unity and dimensionless, which allows for space to be occupied by insulation and for the wire being round whereas the window is rectangular. Then the resistance of a coil which fills the space is

$$R_C = \rho \frac{L}{a} = \frac{\rho N m}{As/N} = \frac{\rho m}{sA} N^2 \tag{7-13}$$

where ρ is the resistivity of the wire material, copper, (ohm-centimeters). Now, substituting 7-13 into 7-10,

$$\frac{\rho m}{sA} N = \frac{E}{F}$$

and, solving for N,

$$N = \frac{E}{F} \frac{sA}{\rho m} \tag{7-14}$$

which is the desired solution for N in terms of quantities which either are given or are matters of design choice. The resistance of the coil may be found from 7-14 and 7-10 as

$$R_C = \left(\frac{E}{F}\right)^2 \frac{sA}{\rho m} \tag{7-15}$$

It must be remembered that 7-14 and 7-15 are valid only if the external resistance may be neglected in comparison with R_C given by 7-15. Moreover, N must be greater than unity; indeed, only integral values of N would lend themselves to easy realization in practice.

If it turns out that R_E cannot be neglected in comparison with R_C, we must return to equation 7-9 and make an analysis without neglecting R_E. Substituting 7-13 into 7-9,

$$F = \frac{NEsA}{\rho m N^2 + sAR_E} \tag{7-16}$$

from which we can get N in terms of known factors. Clearing of fractions and dividing through by the coefficient of N^2,

$$N^2 - \frac{EsA}{F\rho m} N = -\frac{sAR_E}{\rho m} \tag{7-17}$$

and, solving the quadratic equation in N,

$$N = \frac{EsA}{2F\rho m} \pm \sqrt{\left(\frac{EsA}{2F\rho m}\right)^2 - \frac{sAR_E}{\rho m}} \tag{7-18}$$

This is the result we seek; it gives the number of turns N necessary to produce the required number of ampere-turns F. With N determined the cross-sectional area of the wire can be found from 7–12.

Before attempting to discover the full meaning of 7–18 we must check it. In checking dimensionally we treat N as dimensionless and consistently with this measure F in amperes and then find that 7–18 is correct in this respect. In the limiting case when $R_E = 0$, equation 7–18 should give a result consistent with the earlier solution, 7–14. Putting $R_E = 0$ in 7–18 gives

$$N = \frac{EsA}{F\rho m}$$

agreeing with 7–14, or zero, depending whether the plus or the minus sign in 7–18 is used. At this point in the analysis, we do not know why 7–18 reduces also to $N = 0$ in the limiting case where $R_E = 0$. In fact, we do not understand the significance of the double sign in 7–18, but we believe that there *is* a significance, and, until we have discovered what it is, we have no right to discard either sign.

Sometimes an extra zero solution is introduced by algebraic manipulations; hence we decide to investigate whether this may have happened in this case. First we check to see whether $N = 0$ satisfies one of our original equations, 7–16, after R_E has been set equal to zero. We find that it does not; it leaves the left member as F but makes the right member infinite after division of numerator and denominator by N. Yet on examination we find that $N = 0$, $R_E = 0$ does satisfy the next equation, 7–17, and thus finally locate the source of the spurious root in the operation of clearing fractions when $R_E = 0$. In that case 7–16 gives

$$F = \frac{NEsA}{\rho m N^2} = \frac{EsA}{\rho m N}$$

which is not satisfied by $N = 0$. But if we multiply both sides by N^2 without any cancellation we get

$$N^2 F = \frac{NEsA}{\rho m}$$

which is satisfied by $N = 0$. The multiplication of both sides by a quantity which could assume a zero value gives us a spurious solution.

Exercise. For the case $R_E \neq 0$ show that both values of N found from 7–18, with plus sign and minus sign, satisfy equation 7–16.

Proceeding to see what we can learn of the physical meaning of equation 7–18, including the significance of the double sign, we decide to apply the technique of section 7–5, that is, put 7–18 in dimensionless-ratio form. In other words, we seek a number of turns by which to divide both sides of the equation so that the left member will be the ratio of the number of turns N to some other number of turns. From 7–14, we know that the group of factors $(EsA)/(F\rho m)$ is the number of turns that would be calculated assuming $R_E = 0$. Calling this number N_0 and dividing 7–18 through by it, we get

$$\frac{N}{N_0} = \frac{1}{2} \pm \sqrt{\frac{1}{4} - \frac{sA}{N_0^2 \rho m} R_E} \tag{7–19}$$

Now it is clear that, if 7–19 is to be dimensionally homogeneous, the coefficient of R_E must have the dimensions of reciprocal resistance, and we ask ourselves if this coefficient has any simple physical significance. We see at once that it does by comparing it with equation 7–13, which gives the coil resistance. Thus, the coefficient of R_E in 7–19 is the reciprocal of the coil resistance that would be calculated using a number of turns N_0, that is, by ignoring the effect of the external resistance R_E. Designating this value of coil resistance by R_0, 7–19 becomes

$$\frac{N}{N_0} = \frac{1}{2} \pm \sqrt{\frac{1}{4} - \frac{R_E}{R_0}} \tag{7–20}$$

The solid curve in figure 7–9 is a sketch of 7–20; it is a parabola, the upper branch corresponding to the plus sign, the lower to the minus sign.

For a value of R_E/R_0 between zero and 0.25 let us see how there can be two satisfactory values of N, and the significance of values between them. Suppose the numerical data were such that we would calculate N_0 to be 100 turns. This means that with the 100-turn coil directly connected to the source with leads of zero resistance, that is, $R_E = 0$, we would get the required ampere-turns F. Furthermore we could calculate the resistance R_0 of such a coil. Now imagine that we used a different coil, this time with only 10 turns but still filling the window of the magnetic core. By 7–13 the resistance of the new coil would be $0.01R_0$ (length of wire 0.1, cross-sectional area 10.0 times the original wire). With R_E still zero the new coil would draw 100 times as much current when connected to the source. Since the turns have been reduced to $0.1N_0$ the number of ampere-turns would be 10.0 times the original value, that is, the new coil would give

$$F_1 = 10F$$

Now we can bring the ampere-turns from F_1 down to the original value F by adding external resistance until the total resistance is 10.0 times that of the new coil alone. That is, we can add external resistance

$$R_E = 10(0.01R_0) - 0.01R_0 = 0.09R_0$$

To summarize, if $N/N_0 = 0.1$ with $R_E/R_0 = 0.09$, a point on the lower branch of the parabola, we still get the required ampere-turns F. With less external resistance we would get more than the required ampere-turns.

Suppose, instead, that the new coil had $0.9N_0$ turns, which would make its resistance $0.81R_0$ ohm. Without external resistance this coil would

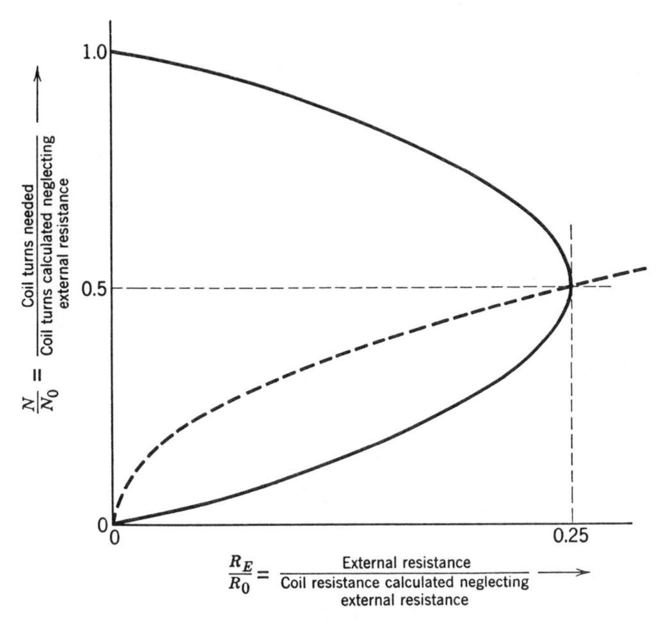

Fig. 7-9. An interpretation of the mathematical results.

draw 1/0.81 times as much current, and since the turns have been reduced to $0.9N_0$ the ampere-turns produced would be

$$F_2 = \frac{0.9}{0.81} F = \frac{F}{0.9}$$

Now the ampere-turns could be brought down to the desired value F by adding external resistance until the total resistance is 1/0.9 times the coil resistance; that is

$$R_E = \frac{0.81}{0.9} R_0 - 0.81R_0 = 0.09R_0$$

With this value of external resistance we get the desired ampere-turns F; with less resistance, more ampere-turns.

Thus we find that we can have N/N_0 either 0.1 or 0.9, corresponding to the same R_E/R_0 of 0.09. We have deduced two values, which agree with 7–20 as they should, and we have evidence that, inside the area bounded by the curve of figure 7–9, the points correspond to a design that gives more than the required ampere-turns; within the area bounded by the parabola the ampere-turns must have a maximum with respect to N for any fixed value of R_E.

Indeed it now appears that, rather than design for a number of turns to give just the necessary ampere-turns, it would be better to design for the maximum possible ampere-turns in order to obtain a factor of safety, especially as in this case the maximum apparently can be achieved at no extra cost. To find the maximum F we differentiate 7–16 with respect to N:

$$\frac{dF}{dN} = \frac{(\rho m N^2 + sAR_E)EsA - NEsA(2N\rho m)}{(\rho m N^2 + sAR_E)^2} \tag{7–21}$$

Then setting this equal to zero we find

$$0 = sAR_E - N^2\rho m \tag{7–22}$$

from which the condition for maximum F is

$$N = \sqrt{\frac{sAR_E}{\rho m}} \tag{7–23}$$

where we have discarded the negative root since a negative number of turns has no physical significance here. The value of the maximum, F_m, we find by substituting the value of N given by 7–23 back into 7–16.

$$F_m = \frac{EsA\sqrt{\dfrac{sAR_E}{\rho m}}}{sAR_E + sAR_E} = \frac{E}{2R_E}\sqrt{\frac{sAR_E}{\rho m}} \tag{7–24}$$

$$= \frac{EN}{2R_E} \tag{7–25}$$

In comparing 7–25 with 7–9 or by combining 7–13 and 7–23 we notice that maximum ampere-turns occur when the turns are chosen such that the coil resistance R_C equals the external resistance R_E. As a physical check we realize after a little thought that this is consistent with the fact that maximum transfer of power from a battery of fixed voltage to a resistance load occurs when the load resistance is equal to the internal resistance of the battery.

We can express these results pertaining to F_m in terms of the parameters N_0 and R_0 in the following way. From 7–13 and 7–15 we obtain, respectively,

$$R_0 = \frac{\rho m}{sA} N_0^2 \tag{7-26}$$

and

$$R_0 = \left(\frac{E}{F}\right)^2 \frac{sA}{\rho m} \tag{7-27}$$

where, as before, R_0 is the resistance and N_0 the number of turns of a coil calculated to produce F ampere-turns on the assumption that $R_E = 0$. Now solving 7–26 for the factor $sA/\rho m$ and substituting this in 7–23 gives

$$\frac{N}{N_0} = \sqrt{\frac{R_E}{R_0}} \tag{7-28}$$

as the condition for maximum ampere-turns F_m. This condition is sketched as the dashed curve in figure 7–9.

To express F_m in these terms we solve 7–27 for the factor $sA/\rho m$, substitute the result in 7–24, and so obtain

$$\frac{F_m}{F} = \frac{1}{2\sqrt{\dfrac{R_E}{R_0}}} \tag{7-29}$$

Summarizing, we can say that to design a coil to produce at least the specified number of ampere-turns F we may choose N such that N/N_0 falls anywhere within the area bounded by the solid curve in figure 7–9. To obtain the greatest possible number of ampere-turns, N must be chosen so that N/N_0 falls on the dashed curve. The solid curve that describes the limit outside of which the specified F cannot be obtained and the curve for maximum F intersect at $R_E/R_0 = 0.25$, $N/N_0 = 0.5$. At this intersection we see from 7–29 that $F_m = F$ and at greater values of R_E/R_0 the maximum value of ampere-turns F_m is less than the value F which is desired. Thus, if $R_E > 0.25R_0$ some modification of the conditions must be made if a workable design is to be had.

7–8 Interpretation of the Effects of Damping in a Mechanical System

As the final illustration of the methods discussed in this chapter we shall make an interpretation of the mathematical expressions obtained in sections 5–8 and 5–13 for certain behaviors of the mass, spring, and dashpot system shown in figures 5–12 and 5–19.

Consider first the vibration of the system in figure 5–12 following release with zero initial velocity from an initial displacement x_0. Mathematical expressions for the resulting motion are obtained by putting $v_0 = 0$ in equations 5–83, 5–85, and 5–95, respectively, depending on whether the damping is less than, greater than, or equal to critical damping.

For less than critical damping, $0 \leq D < 2\sqrt{KM}$, the expression we wish to interpret, obtained from 5–83 by putting $v_0 = 0$, is

$$x = x_0\epsilon^{-Dt/2M}\left[\cos\sqrt{\frac{K}{M} - \left(\frac{D}{2M}\right)^2}\,t + \frac{\frac{D}{2M}}{\sqrt{\frac{K}{M} - \left(\frac{D}{2M}\right)^2}}\sin\sqrt{\frac{K}{M} - \left(\frac{D}{2M}\right)^2}\,t\right] \tag{7–30}$$

To interpret 7–30 fully we might plot x as a function of t for each of a series of values of the separate factors x_0, D, M, and K, thus obtaining a great multiplicity of curves. Closer scrutiny shows, however, that the task would be simplified if we recognized that K occurs only in the combination K/M and D only in $D/2M$, thus reducing the total number of possible variables by one. But the work in section 5–8, particularly figure 5–13, suggests that still further simplification might be achieved by expressing the damping actually present as a fraction of critical damping for the system, that is, the value of D which causes the radicals in 7–30 to be zero. Expressing the damping as a fraction of critical damping is the plan we propose to try.

Thus let

$$D = \delta(2\sqrt{KM}) \tag{7–31}$$

where δ is the ratio of the actual damping present to the critical value $2\sqrt{KM}$, and in the present case can have any value in the range between zero and one.

Exercise. With regard to the units of D, K, and M given in section 5–8, check to see that δ as defined by 7–31 is dimensionless.

In terms of δ the factor $D/2M$ which occurs in 7–30 may be expressed as

$$\frac{D}{2M} = \frac{\delta(2\sqrt{KM})}{2M} = \delta\sqrt{\frac{K}{M}} \tag{7–32}$$

Making this substitution in 7–30 we obtain

$$x = x_0\epsilon^{-\delta\sqrt{K/M}t}\left[\cos\sqrt{1-\delta^2}\sqrt{\frac{K}{M}}\,t + \frac{\delta}{\sqrt{1-\delta^2}}\sin\sqrt{1-\delta^2}\sqrt{\frac{K}{M}}\,t\right] \tag{7–33}$$

Here we notice that K, M, and t occur only in the combination $\sqrt{K/M}\, t$, and this is dimensionless. The quantity $\sqrt{K/M}$ is the *undamped natural angular frequency* already met in section 5–7. We shall represent it by β_0, and in these terms we see that time in 7–33 is represented by the dimensionless variable $\beta_0 t$ which may be said to be measured in undamped radians. Furthermore, displacement x may be incorporated in a dimensionless variable x/x_0 simply by dividing 7–33 through by x_0. Thus the result of our work is

$$\frac{x}{x_0} = \epsilon^{-\delta(\beta_0 t)} \left[\cos \sqrt{1 - \delta^2}\, (\beta_0 t) + \frac{\delta}{\sqrt{1 - \delta^2}} \sin \sqrt{1 - \delta^2}\, (\beta_0 t) \right] \quad (7\text{–}34)$$

The foregoing choice of dimensionless quantities is not the only one possible. Our selection was aimed toward the objective of seeing how displacement x as a function of time t is influenced by the damping D. For this purpose having each of the factors x, t, and D appear in only one dimensionless quantity and to the first power is clearly desirable, and in this case possible. Another choice of dimensionless quantities would have been

$$\sqrt{\frac{K}{M} - \left(\frac{D}{2M}\right)^2}\, t \qquad \text{instead of} \qquad \sqrt{\frac{K}{M}}\, t$$

and

$$\frac{\dfrac{D}{2M}}{\sqrt{\dfrac{K}{M} - \left(\dfrac{D}{2M}\right)^2}} \qquad \text{instead of} \qquad \frac{D}{2\sqrt{KM}}$$

but with this selection it would be difficult to see the effect of D since it occurs in two of the quantities instead of in only one, and not linearly.

For purposes of sketching or plotting it is convenient to combine the trigonometric terms in 7–34 to give the form

$$\frac{x}{x_0} = \frac{1}{\sqrt{1 - \delta^2}}\, \epsilon^{-\delta(\beta_0 t)} \cos \left[\sqrt{1 - \delta^2}\, \beta_0 t - \sin^{-1} \delta\right] \quad (7\text{–}35)$$

Exercise. Verify the derivation of 7–35 from 7–34 with the aid of a right triangle with hypotenuse 1 and sides δ and $\sqrt{1 - \delta^2}$, and the formula for the cosine of the difference of two angles.

Result 7–34 or 7–35 is interpreted in figure 7–10 as the curves for which δ, the ratio of the actual damping to critical damping, has values less than one. These curves show x/x_0, the ratio of displacement at any time to the initial displacement, as functions of $\beta_0 t$, the angle or argument of the cosine for the case where the damping is zero. This family of curves is for $v_0 = 0$, that is, zero initial velocity.

The family of curves in figure 7-10 may be completed for values of damping greater than critical and for the special case of critical damping by

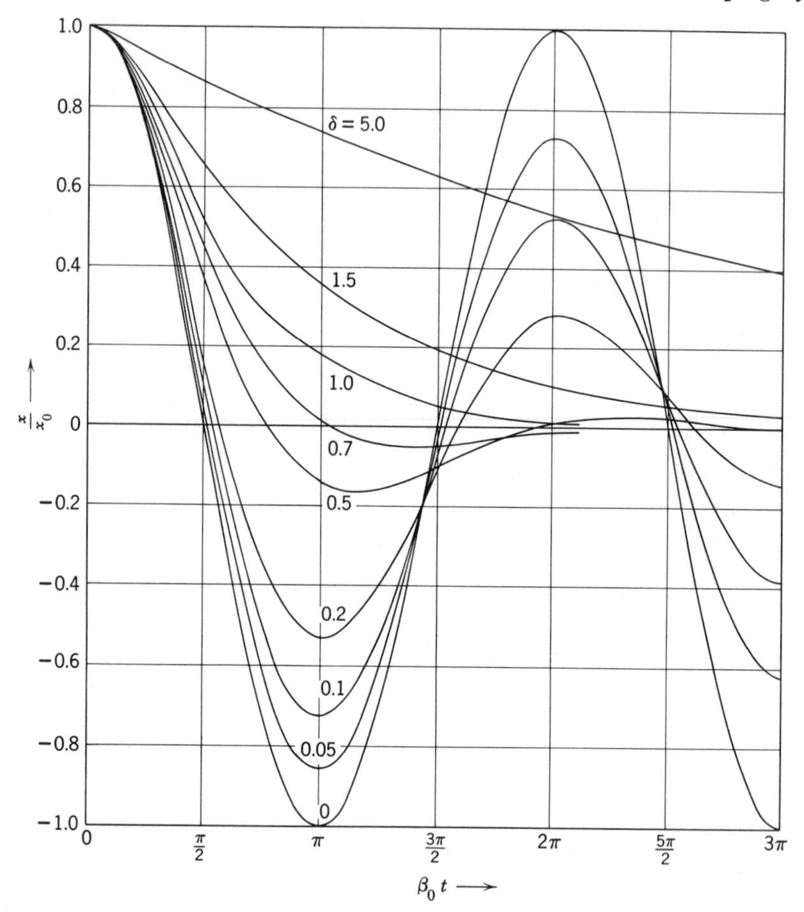

Fig. 7-10. Free vibration of the system in figure 5-12 following an initial displacement.

reasoning parallel to that just developed and in terms of exactly the same dimensionless variables. In these terms we obtain the following, respectively, from 5-85 and 5-95 for the case of zero initial velocity, $v_0 = 0$. For $\delta > 1$,

$$\frac{x}{x_0} = \frac{(\delta + \sqrt{\delta^2 - 1})\,\epsilon^{-(\delta - \sqrt{\delta^2-1})\beta_0 t} - (\delta - \sqrt{\delta^2 - 1})\,\epsilon^{-(\delta + \sqrt{\delta^2-1})\beta_0 t}}{2\sqrt{\delta^2 - 1}} \tag{7-36}$$

and, for $\delta = 1$

$$\frac{x}{x_0} = \epsilon^{-\beta_0 t}[1 + \beta_0 t] \tag{7-37}$$

EXERCISE. Verify 7-36 and 7-37 by introducing the dimensionless variables in 5-85 and 5-95 for the case where $v_0 = 0$.

EXERCISE. Put 5-86 in dimensionless form for the case $v_0 = 0$ and so obtain an expression equivalent to 7-36 but in terms of hyperbolic functions.

In checking the curves in figure 7-10 for physical reasonableness we see that all start as they should from the same value of x/x_0, that is, the same initial displacement, and that all start horizontally, corresponding to the fact that the initial velocity v_0 has been taken equal to zero. It is especially interesting to note the smooth transition in the form of the curves throughout the range of damping in spite of the apparently violent change in the mathematical form as critical damping is passed. The curve for critical damping, $\delta = 1$, simply separates the family of curves into two groups: those which pass into negative values of x/x_0, and those which remain positive throughout. As the damping is increased from zero toward the critical value the chief effect is to increase the rate at which the amplitude of the oscillations decays. Another effect, which becomes very pronounced as δ gets close to 1, is to lengthen the period of the oscillation.

If we are interested in the behavior of the system in figure 5-12 after the imparting of an initial velocity instead of an initial displacement we can again begin with equations 5-83, 5-85, and 5-95, this time setting $x_0 = 0$ but retaining v_0. The equation which applies for less than critical damping, 5-83, then takes the form

$$x = \epsilon^{-Dt/2M} \frac{v_0}{\sqrt{\frac{K}{M} - \left(\frac{D}{2M}\right)^2}} \sin \sqrt{\frac{K}{M} - \left(\frac{D}{2M}\right)^2}\, t \tag{7-38}$$

As before, we introduce the ratio of damping to critical damping, $\delta = D/2\sqrt{KM}$, and the undamped natural frequency, $\beta_0 = \sqrt{K/M}$; and 7-38 becomes

$$x = \epsilon^{-\delta\beta_0 t} \frac{v_0}{\beta_0\sqrt{1 - \delta^2}} \sin \sqrt{1 - \delta^2}\, \beta_0 t \tag{7-39}$$

We may make this equation dimensionless by dividing through by the factor v_0/β_0, which has the dimensions of length. Thus

$$\frac{x}{v_0/\beta_0} = \frac{\epsilon^{-\delta\beta_0 t}}{\sqrt{1 - \delta^2}} \sin \sqrt{1 - \delta^2}\, \beta_0 t \tag{7-40}$$

What is the significance of the length v_0/β_0 which appears in the denominator of the left side of 7–40? We notice that, if $\delta = 0$, the case of no damping, the right side of 7–40 reduces simply to $\sin \beta_0 t$. From this we see that the length v_0/β_0 is the maximum value of the displacement x in the case of no damping.

EXERCISE. Check this conclusion about v_0/β_0 independently by applying the condition that in the case of no damping the strain energy in the spring at maximum deflection must equal the kinetic energy initially imparted to the system to set it in motion.

Thus we find that the left side of 7–40 is the ratio of the displacement at any instant to the maximum displacement that would occur in the case of zero damping.

In similar fashion we can derive dimensionless expressions for the motion following an initial velocity for the cases of $\delta > 1$ and $\delta = 1$ by starting with 5–85 and 5–95. The results are, respectively,

$$\frac{x}{v_0/\beta_0} = \frac{\epsilon^{-(\delta-\sqrt{\delta^2-1})\beta_0 t} - \epsilon^{-(\delta+\sqrt{\delta^2-1})\beta_0 t}}{2\sqrt{\delta^2-1}} \tag{7–41}$$

$$\frac{x}{v_0/\beta_0} = \epsilon^{-\beta_0 t}\beta_0 t \tag{7–42}$$

These results, 7–40, 7–41, and 7–42, are plotted for a series of values of δ in figure 7–11.

What have we accomplished by our interpretation of the mathematical expressions for the free vibration of the mass, spring, and dashpot system? Comparing the original equation, 7–30, with the one derived from it in dimensionless form, 7–34, we see that the quantities involved have been reduced from x, x_0, D, M, K, and t, a total of six, to the three dimensionless quantities x/x_0, δ, and $\beta_0 t$. This made possible the representation of the result by a single family of curves, figure 7–10, from which x as a function of t can be visualized for all possible values of x_0, D, M, and K. Had we not reduced 7–30 to the dimensionless form 7–34, but had attempted to represent x as a function of t directly, a separate set of curves each somewhat like figure 7–10 would have been necessary to show what happens for each of the infinitude of combinations of values of x_0, D, M, and K. Statements exactly parallel to the above can, of course, be made with respect to the case represented by figure 7–11.

Now let us extend the application of these ideas to the interpretation of the mathematical result found in section 5–13 for the steady-state vibration of the same mass, spring, and dashpot system under the action of a force varying sinusoidally with time, and given by $F \cos \omega t$. The physical set-up is shown in figure 5–19, and the mathematical result is expressed by

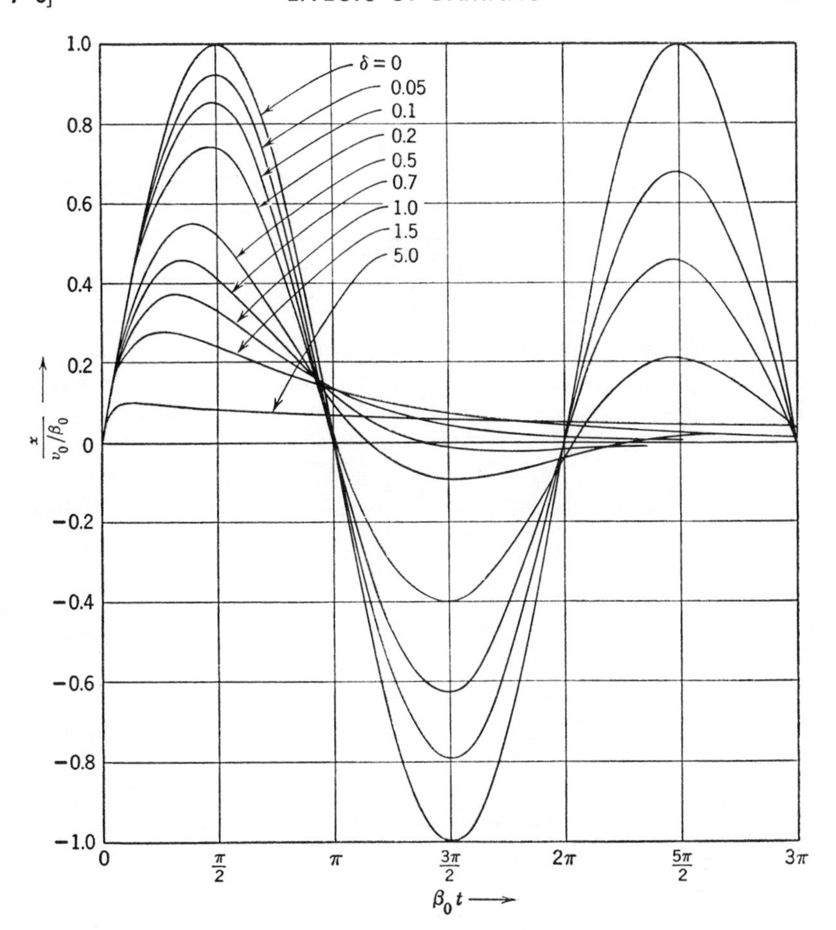

Fig. 7-11. Free vibration of the system in figure 5-12 following an initial velocity.

equations 5–159 and 5–156. These equations, repeated here for convenience, are

$$x = \frac{F}{\sqrt{(K - \omega^2 M)^2 + \omega^2 D^2}} \cos(\omega t - \phi) \tag{7-43}$$

where

$$\phi = \tan^{-1} \frac{\omega D}{K - \omega^2 M} \tag{7-44}$$

The problem we set outselves here is to interpret how the forced vibration thus expressed mathematically depends on the amplitude F and the angular frequency ω of the applied force and on the constants of the system K, M, and D.

First we observe that the motion as expressed by 7–43 is sinusoidal and has the same angular frequency ω as the applied force but lags behind the force by a phase angle ϕ. To describe the effects of F, ω, K, M, and D on this motion it will thus suffice if we describe the effects on the amplitude, given by

$$X = \frac{F}{\sqrt{(K - \omega^2 M)^2 + (\omega D)^2}} \tag{7-45}$$

and on the phase angle ϕ, given by 7–44.

Let us see what happens if we replace D in 7–45 by its expression in terms of the damping ratio δ, which worked well in the case of the free vibrations and which is given by 7–31. Making this substitution, 7–45 becomes

$$X = \frac{F}{\sqrt{(K - \omega^2 M)^2 + 4\delta^2\omega^2 KM}} \tag{7-46}$$

If we factor K out of the denominator we can make the undamped angular frequency $\sqrt{K/M} = \beta_0$ appear; thus

$$\begin{aligned} X &= \frac{F}{K\sqrt{\left(1 - \omega^2 \dfrac{M}{K}\right)^2 + 4\delta^2\omega^2 \dfrac{M}{K}}} \\ &= \frac{F}{K\sqrt{\left[1 - \left(\dfrac{\omega}{\beta_0}\right)^2\right]^2 + 4\delta^2\left(\dfrac{\omega}{\beta_0}\right)^2}} \end{aligned} \tag{7-47}$$

The final step in making the expression dimensionless is evidently to divide through by F/K with the result

$$\frac{X}{F/K} = \frac{1}{\sqrt{\left[1 - \left(\dfrac{\omega}{\beta_0}\right)^2\right]^2 + 4\delta^2\left(\dfrac{\omega}{\beta_0}\right)^2}} \tag{7-48}$$

The left-hand member may be interpreted as follows: The factor F/K is the amount by which the spring would be deflected by the steady application of a constant force of magnitude F. This is the same as saying that F/K is the steady-state motion which results when the frequency of the applied force is zero. Thus the left side of 7–48 is the ratio of the amplitude of the steady-state vibration to the steady deflection that would be produced by a constant force of magnitude equal to the amplitude of the impressed alternating force. This may be checked by putting $\omega = 0$ in 7–48. Thus we have replaced relation 7–45 involving the six quantities X, F, K, ω, M, and D by a relation, 7–48, among the three dimensionless quantities $X/(F/K)$, ω/β_0, and δ.

In similar fashion equation 7–44 for the phase angle may be put in dimensionless terms with the result

$$\phi = \tan^{-1} \frac{2\delta\left(\dfrac{\omega}{\beta_0}\right)}{1 - \left(\dfrac{\omega}{\beta_0}\right)^2} \tag{7–49}$$

Finally, we may represent 7–48 and 7–49 by the families of curves in figure 7–12 which show the amplitude and phase angle as functions of the applied frequency with damping as parameter, all in dimensionless terms.

All the curves for dimensionless amplitude $X/(F/K)$ start at unity regardless of the amount of damping, corresponding to the fact that at this point all represent merely the steady deflection of the spring under the action of a constant force. At the other end of the frequency scale, as ω/β_0 approaches infinity, the amplitude approaches zero, corresponding to the familiar fact that it is difficult to cause a massive system, one having low natural frequency, to vibrate at high frequency. In the mid-portion of the frequency range we have the phenomenon of resonance. With damping of zero, the amplitude of the steady-state vibration becomes infinite when the forcing frequency equals the undamped natural frequency. Thus the quantity $\beta_0 = \sqrt{K/M}$, in addition to being the undamped natural frequency, has another important physical significance: it is the resonant frequency at zero damping. As the damping is increased the frequency corresponding to maximum amplitude shifts gradually into the frequency range $\omega/\beta_0 < 1$. The phenomenon of resonance disappears altogether when the damping ratio δ exceeds $1/\sqrt{2} = 0.707$, that is, with δ greater than this value, the amplitude is always less than its value at zero forcing frequency.

The curves in figure 7–12, showing the phase angle ϕ by which the displacement lags behind the force, reveal still another significance of β_0. When the forcing frequency equals the undamped natural frequency of the system the displacement lags behind the force by $\pi/2$ radians ($= 90°$) no matter what value of damping is present. When $\omega/\beta_0 < 1$ the angle ϕ is always less than $\pi/2$; when $\omega/\beta_0 > 1$, ϕ is greater than $\pi/2$ approaching π radians ($= 180°$) at large values of the forcing frequency. In the ideal case of no damping, $\delta = 0$, the shift of ϕ from 0 to π is discontinuous as the forcing frequency ω passes through the value β_0.

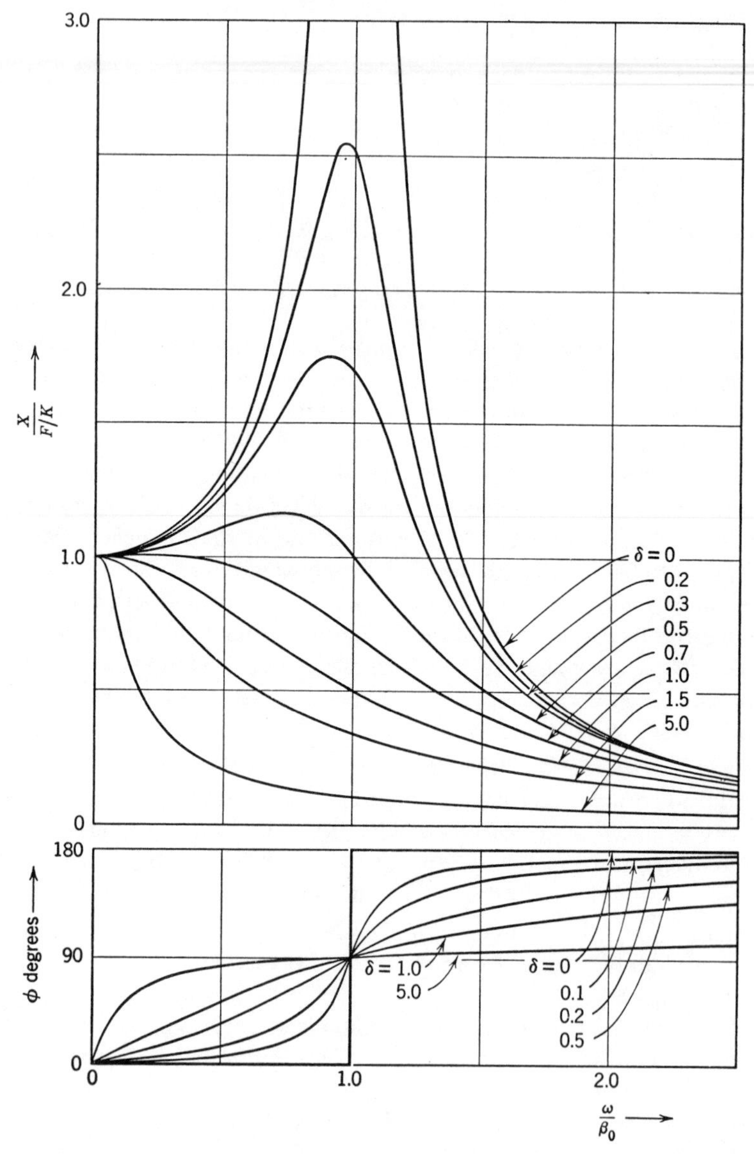

Fig. 7-12. Amplitude and phase angle for the steady-state forced vibration of the system in figure 5-19.

At the beginning of this section we studied the behavior of the system shown in figure 5–12 as it returns to equilibrium after an initial disturbance. This behavior is illustrated by the curves in figures 7–10 and 7–11 and is called *free vibration*, that is, vibration which results when the system is free of any driving force. Later in the section we studied the behavior of the same system under the action of a sinusoidal driving force. We limited our attention to the steady-state motion and obtained the curves in figure 7–12. This behavior is called the *forced vibration* of the system. We found in both the free and the forced vibration that the results of the mathematical analysis were simplified by the introduction of the damping ratio $\delta = \dfrac{D}{2\sqrt{KM}}$ and the undamped natural frequency $\beta_0 = \sqrt{\dfrac{K}{M}}$. These two parameters, δ and β_0, replace the three quantities M, K, and D in terms of which the original equations were expressed.

The parameters δ and β_0 are also advantageous in the experimental investigation of practical systems such as that which is idealized by figures 5–12 and 5–19. Although it may be relatively easy in an actual system to determine M and K by direct measurement, the direct determination of D is likely to be less simple since the damping may be produced by various hard-to-isolate frictional effects rather than by a dashpot. In such a case, however, the parameters δ and β_0 may be quite easy to determine by experiment. One way is to measure the free vibration of the system and then calculate δ and β_0 from observations of the period of the oscillations and the rate at which they decay. Another way is to drive the system at various frequencies and from observations of the steady-state amplitude calculate δ and β_0. The latter method is often used in the study of dynamic systems such as servomechanisms. Regardless of how β_0 and δ are determined, however, the values may be used to calculate the free or forced vibrations of the system in figures 5–12 and 5–19 under any conditions.

EXERCISE. Assuming the damping to be small ($\delta < 0.2$) describe how to find δ and β_0 from a curve which shows steady-state amplitude as a function of frequency.

7–9 Extension of the Interpretation: Dimensional Analysis and Analogs

In section 7–8 we found that, by rearranging the mathematical expressions for the behavior of the damped mass-and-spring system so that the variables and parameters appear only in dimensionless groups, the number of quantities involved is significantly reduced. A result of this is, for example, that any one of the curves in figure 7–10 can pertain to a wide range in sizes of the spring, mass, and dashpot, the only requirement being that for anyone of the curves the combination of quantities $\delta = D/2\sqrt{MK}$ must be the same. That is, a single curve is applicable equally to all systems like figure 5–12 composed of masses and springs utterly different from one system to the next provided only that in each case the damping is adjusted to make δ the same for all.

The above statement has a very important implication. Suppose we were unable to solve the differential equation from which figure 7–10 is derived and were forced to attack the problem experimentally. We could proceed by measurements on an arrangement composed of a single mass and spring of convenient size and an adjustable dashpot. Then if we plotted the data in terms of the dimensionless quantities x/x_0, $\beta_0 t$, and δ we should be able to obtain experimentally the entire family of curves in figure 7–10. But, as we have seen in section 7–8, the curves pertain to all systems like figure 5–12, not merely to systems having particular values of mass-and-spring constant such as might be employed in the experiments. This illustrates how the behavior of one system, called the prototype, may be predicted from an experimental investigation of another, called the model. The model must have certain physical similarities to the prototype, for instance, in the present illustration δ must be the same for both, but it may differ greatly in other ways, for example, in size. The procedures for deciding how to arrange the variables of a problem systematically in proper dimensionless groups so that model experiments will be meaningful are dealt with in dimensional analysis.* Where physical understanding or knowledge of mathematics is insufficient to permit a precise mathematical formulation, dimensional analysis can often serve as a powerful weapon for attacking a problem. To use dimensional analysis the variables of the problem must be known, and thus, as in all engineering analysis, there is no substitute for the most thorough physical understanding possible under the circumstances.

In addition to applying to all mechanical systems composed of a mass, spring, and dashpot arranged as in figures 5–12 and 5–19, the curves in figures 7–10, 7–11, and 7–12 have still greater generality. To see this we shall analyze a new problem.

Consider an electric circuit, figure 7–13, composed of inductance L, resistance R, capacitance C, and a switch, all connected in series. Initially the switch is open and the condenser holds a charge q_0. The switch is closed suddenly and we wish to find the charge q on the condenser as a function of time t thereafter.

By Kirchhoff's voltage law the sum of the voltage drops around the circuit at any instant must equal zero. After the switch is closed the voltage

* For a fundamental treatment see Langhaar, *Dimensional Analysis and Theory of Models*, Wiley, New York, 1951.

drop across it may be assumed to be zero. The current in the circuit at any instant is equal to the time rate of change of charge on the condenser, dq/dt, so we can express Kirchhoff's voltage law by the equation

$$L\frac{d^2q}{dt^2} + R\frac{dq}{dt} + \frac{q}{C} = 0 \tag{7-50}$$

To find q as a function of t, this equation must be solved subject to the initial conditions that the charge q is q_0 and the current dq/dt is zero.

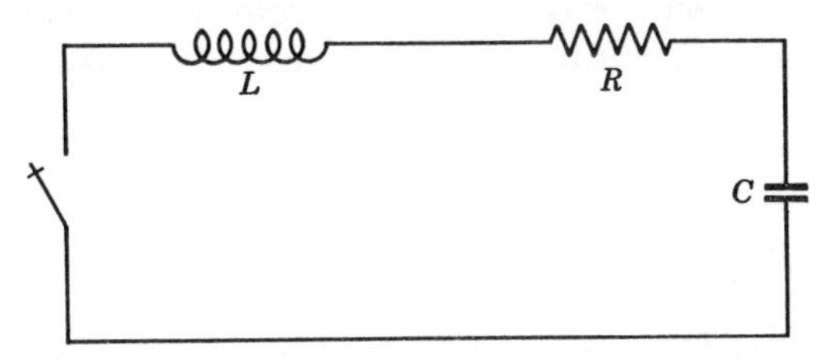

Fig. 7-13. A circuit analogous to the mechanical system in figure 5-12.

Equation 7–50 is a linear differential equation with constant coefficients and can be solved easily by the methods discussed in section 5–9. Rather than apply those methods directly in this case, however, we decide on another course of action. We recall from section 5–8, the study of the damped mechanical system depicted in figure 5–12, that we developed a differential equation exactly similar in form to 7–50. That equation is 5–67 and is repeated here for convenience. It is

$$M\frac{d^2x}{dt^2} + D\frac{dx}{dt} + Kx = 0 \tag{7-51}$$

and we solved it subject to the initial conditions $x = x_0$ and $dx/dt = v_0$.

Upon comparing the differential equations 7—50 and 7–51 and the initial conditions specified for each we see that if we consider the special case of $v_0 = 0$ for 7–51 the two equations and the respective initial conditions are exactly equivalent in mathematical form. That is, equation 7–50 and its initial conditions can be converted into 7–51 and its initial conditions by replacing the symbols in the first with those from the second as follows:

Replace L with M

Replace R with D

Replace $1/C$ with K

Replace q with x

Replace q_0 with x_0

Evidently t occurs similarly in both equations and so can be left unchanged. Since the two equations 7–50 and 7–51 and the respective initial conditions are thus of identical mathematical form, the solution to 7–50 can be obtained from the solution to 7–51 simply by a substitution of symbols opposite to that in the foregoing tabulation.

The solution to 7–51 for the case we want, that is, $v_0 = 0$, is given in dimensionless form by equations 7–35, 7–36, and 7–37, which are represented by the curves plotted in figure 7–10. To utilize this result as a solution to 7–50 we must make a replacement of symbols in the definitions of the dimensionless quantities opposite to that used in converting 7–50 to 7–51; thus:

Replace $\dfrac{x}{x_0}$ by $\dfrac{q}{q_0}$

Replace $\beta_0 t = t\sqrt{\dfrac{K}{M}}$ by $\beta_0 t = \dfrac{t}{\sqrt{LC}}$

Replace $\delta = \dfrac{D}{2\sqrt{KM}}$ by $\delta = \dfrac{R}{2\sqrt{\dfrac{L}{C}}}$

With these simple changes in the definitions of the quantities plotted we see that the curves of figure 7–10, and, of course, the equations from which the curves come, represent the solution to the electric-circuit problem we set for ourselves.

It is not hard to extend the above reasoning to show that the families of dimensionless curves in figures 7–11 and 7–12 also may represent the solutions to electric-circuit problems as well as the solutions to the mechanical problems for which they were derived. Mechanical systems and electric circuits which bear this kind of relation to one another are said to be analogous. The circuit in figure 7–13 is an electrical analog of the mechanical system in figure 5–12, and conversely the mechanical system is a mechanical analog of the circuit.

A mechanical system may also be the analog of another mechanical system; for instance, the system in figure 5–12 may have a rotational analog comprising a moment of inertia, a torsional spring, and a damping device which produces a torque.

EXERCISE. Draw a diagram of a rotational system analogous to figure 5–12, set up its differential equation, and show how to use figures 7–10 and 7–11 as the solution.

EXERCISE. See whether you can construct another electrical analog to the systems of figures 5–12 and 5–19 using a circuit consisting of inductance, capacitance, and resistance, all in parallel.

Analogs are valuable aids to thought in unfamiliar fields; for instance, it is easier to understand and interpret electric-circuit phenomena if you are already familiar with the behavior of analogous mechanical systems; and the reverse is equally true.

But analogs have an even more practical value. They are frequently used in analog computers to solve problems experimentally which are too difficult or too time consuming for completely analytical methods. Electric circuits are used most commonly to construct the analogs because of their flexibility and the ease and accuracy with which electrical measurements and records can be made. The situation represented by the analogous circuit may be mechanical, thermal, electrical, a combination of these, or indeed almost anything capable of mathematical formulation.

By this further interpretation of the results of the problem in section 7–8 we have had brief views of two powerful engineering tools: dimensional analysis and analog computers. Although the nature of these tools implies experimental work, they must not be regarded as substitutes for analysis; rather they extend the range of engineering analysis and can be used to their fullest only by those who have mastered the kinds of thinking treated in this book.

7–10 Learning from Mathematical Results: Summary

To be most useful for engineering purposes, the full physical significance of a mathematical result must be thoroughly understood. To gain the necessary understanding sketched plots, especially plots in dimensionless-ratio form, are usually helpful and often lead to simplification. Finding suitable dimensionless ratios is likely to be a by-product of dimensional and limiting-case checking. It is frequently possible to attach simple physical meanings to the dimensionless ratios, and, although ingenuity helps, meanings may often be found by a systematic process of verbalization in which the meanings of groups of factors are deduced from the definitions of the separate symbols.

The above remarks apply not only to the final results of a mathematical analysis but to intermediate results as well. In fact, as a mathematical analysis is developed you should be alert to the possibility that at any stage interpretation in physical terms may provide opportunity for checking the work up to that place and may point the way to important simplification of the analysis which lies ahead.

Not until you yourself thoroughly comprehend the meaning of what you have done are you ready to tackle the problem of interpreting your results for other people. In solving this exceedingly important problem of communication of ideas, the devices you use to help your own understanding may be very useful. On the other hand, they may not be; it depends entirely on what the points of view of the individuals comprising your audience are likely to be.

Problems

1. Accelerometer with Two Liquids

It is proposed to increase the sensitivity of the liquid-level accelerometer discussed in section 1–2 by replacing the air in the upper part with a second liquid having a density less than that of the liquid in the lower part. The two liquids would not mix, for example, kerosene and water, and they would have different colors. It is believed that the sensitivity would be particularly high if the densities of the two liquids were nearly equal. In this modified design the upper horizontal tube would have no constriction but would be a smooth cylindrical glass tube like the bottom one. Constrictions if necessary to provide damping could be put at the tops of the vertical tubes.

Study this proposal and make a recommendation.

2. Accelerometer with Dissimilar Vertical Tubes

A modification of the liquid-level accelerometer of section 1–2 has been produced commercially in which one of the vertical tubes has a diameter about ten times that of the other. Is there any advantage in this design?

3. Modified Electric Accelerometer

In the search for a way to measure acceleration of an automobile by electrical means, it appears that the solution given in section 2–2 may not be acceptable because of the large capacitor required. It is proposed to substitute for the capacitor a small d-c motor with permanent magnet field, and it is claimed that the indication of the ammeter will be proportional to acceleration. The motor shaft would carry a small flywheel but otherwise would be unloaded. Investigate this proposal.

4. Dielectric Voltmeter

It is suggested that an electrostatic voltmeter might be made by utilizing the force tending to draw a plate of dielectric into the space between the parallel plates of a condenser. The force on the dielectric plate would be balanced by means of a spring so that the position of the plate would be a measure of the voltage. An advantage would be that no electrical connections would have to be made to the moving element. As a preliminary to the serious consideration of the suggestion, find how the force on the dielectric depends on significant quantities, assuming for simplicity the geometry shown in figure P-4.

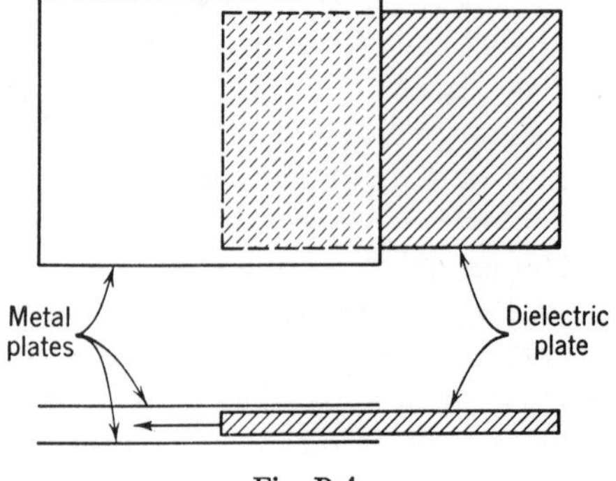

Fig. P-4.

5. Closing Contacts

The moving contact of a certain relay is carried on a massive element which rotates freely about a vertical axis. The other contact, which is very light, is mounted on a spring. Figure P–5 shows the contacts in the open position. If a closing torque which remains constant throughout the duration of motion is suddenly applied to the moving element, will the contacts separate once they have closed?

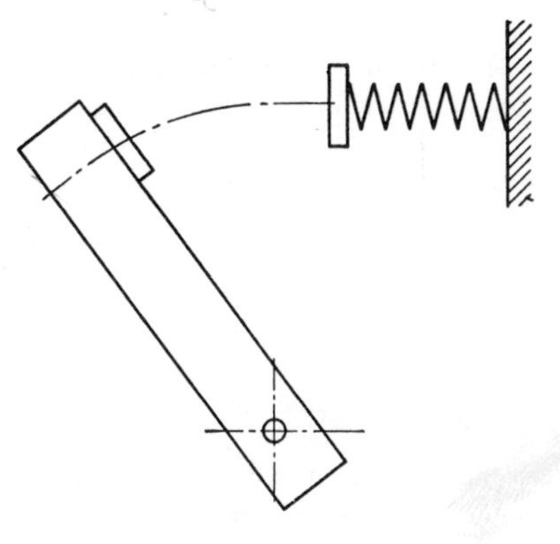

Fig. P-5.

6. Scheme for Obtaining a Speed-Torque Curve

The following method is proposed for measuring the speed-torque curve of a motor: A very small d-c generator with permanent magnet field (electric tachometer) is coupled to the motor shaft, thus giving a voltage proportional to speed. This voltage, together with a timing wave, is recorded on an oscillogram as the motor speeds up after being connected to the line. The motor is not coupled to a load during this test.

Data measured from an oscillogram taken in this way are as follows:

Time, seconds	*Tachometer Voltage, volts*
0.0	0
0.0122	5
0.0241	10
0.0357	15
0.0469	20
0.0577	25
0.0682	30
0.0783	35
0.0880	40
0.0973	45
0.1061	50
0.1143	55
0.1221	60
0.1295	65
0.1365	70
0.1432	75
0.1497	80
0.1563	85
0.1634	90
0.1721	95
0.1783	97.5

The voltage of the tachometer is 50 when the speed is 900 rpm. The WR^2 of the motor is 0.5 lb ft.^{2} Obtain the speed-torque curve from these data.

7. Speed-Torque Curve on Oscilloscope

Referring to problem 6, can you devise a scheme by which a cathode-ray oscilloscope may be made to trace the speed-torque curve of the motor directly?

8. Falling Elevator

Derive a formula for the acceleration of a passenger elevator which would result if both the brakes and power supply failed, allowing the system to move freely with negligible friction. The counterweight overbalances the car when the car is empty, but is overbalanced when the car is fully loaded. The driving motor is geared to the drum over which the supporting cables are wrapped.

9. Motor on Intermittent Load

The life of electrical insulation depends critically on the temperature at which it operates. For a particular insulation the life expectancy is 20 years when operated continuously at 105°C. The life expectancy decreases by a factor of 2 for every 8°C increase above 105°C and increases by a factor of 2 for every 8°C decrease below 105°C.

A motor employing this insulation on its windings is being considered for an application where the life expectancy desired is 20 years. The motor is to operate cyclically at two different loads, one existing for 10 per cent of the time and the other for 90 per cent. The temperature rise above ambient at the hottest spot in the insulation for the short-time load is expected to be double that for the long-time load. The ambient temperature is 40°C. The thermal-transient times are believed to be short compared to the load-cycle time.

How high a value for the upper temperature of the hottest spot is permissible?

10. Supersynchronous Motor

Figure P-10 shows a special form of synchronous motor designed to overcome the handicap of low starting torque inherent in ordinary synchronous motors. The outer

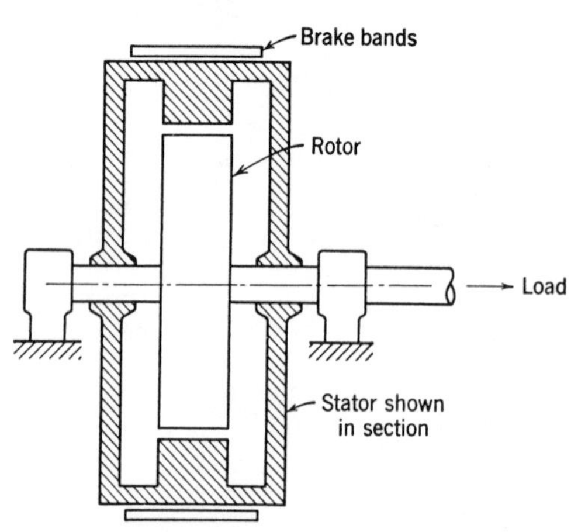

Fig. P-10.

element of the motor, the "stator," is supported in bearings so that it, as well as the rotor, can turn. Brake bands are provided around the stator so that it can be stopped and held stationary if desired. Electrical connections are made through slip rings.

The procedure in starting a heavy load is as follows: The brake is released and the motor connected to the line. The rotor remains at rest while the stator comes up to synchronous speed. When the motor is in synchronism and ready to develop full-load torque the brake is applied, and the stator is slowly brought to rest while the rotor and load gradually pick up speed. The speed of the rotor with respect to the stator remains equal to synchronous speed.

In terms of rated motor torque and other significant factors, determine the brake torque, assumed constant, required to cause a heavy inertia load such as a flywheel to reach synchronous speed as quickly as possible without exceeding the rated torque of the motor. Also determine the time required and the amount of energy which must be dissipated in the brake. The inertia of the stator is not negligible although it is small compared to the inertia of rotor and load.

11. Shaking Conveyor

A special type of conveyor is proposed for moving cardboard cartons of soap away from the packing machine. The conveyor is to consist of a long flat steel plate supported in a horizontal plane and arranged to be shaken rapidly back and forth lengthwise with a short reciprocating motion such that the cartons will progress along the conveyor in the desired direction.

Devise, if possible, a suitable motion for the conveyor. Do not design the mechanism to produce the motion.

12. Ping-Pong Ball

It is desired to cast centrifugally table-tennis balls, using a new type of casting resin. The mold, containing a spherical cavity, is to be turned sufficiently fast that centrifugal force will cause the resin to spread evenly over the surface. It was suggested that rotation about a single axis would not produce a uniform spreading, and hence it might be necessary to mount the mold in gimbals and turn it about two axes simultaneously. Discuss the feasibility of this method of casting, using one and two axes of rotation.

13. Arc Energy in Switching

A series circuit consisting of a resistor, a solenoid, a switch, and a battery is interrupted by opening the switch. ($L = 1.0$ henry, $R = 20$ ohms, $E = 10$ volts.) Assuming that the arc which forms as the switch opens has a constant drop of 20 volts independent of current, find how the current decays as a function of time. Determine the energy dissipated in the arc, and give an energy balance for the decay period.

14. Charging a Condenser

It is claimed by one student that when a condenser is charged through a resistance the amount of energy dissipated in the resistance is always exactly equal to that stored in the condenser. Another student maintains that this is true only when the charging is from a source of constant potential. Establish which one is correct.

15. Starting an Unloaded Motor

The field circuit of an unloaded d-c shunt motor is closed, and after the field current has reached its steady value the armature circuit is closed through a constant resistance. An engineer says that under these conditions there is a simple relation between the heat developed in the armature circuit resistance and the rotational kinetic energy. Is he right?

16. MATCHING A MOTOR AND LOAD

A mixing machine which must be started when full is to be driven by an electric motor which is not ideally suited to the job but which happens to be available. The motor will drive the mixer through a chain drive, and it is desired to determine a suitable speed ratio.

The mixer load is such that the torque required decreases as the speed increases, as shown by the torque-speed characteristic tabulated below.

The motor to be used is a single-phase a-c motor rated 1½ hp, which starts as a repulsion motor and runs as an induction motor. The starting and running-torque characteristics are given below. Transfer between the characteristics is effected automatically by means of a centrifugal switch mounted on the motor rotor, which may operate anywhere within the range 1000 to 1200 rpm.

The power input to the mixing machine under steady running conditions should be as great as possible within the limitations imposed by the available motor. How much will this power be and what speed ratio do you recommend?

The type of chain drive to be used has a high efficiency, and its losses may be neglected at least as a first approximation.

MIXING MACHINE CHARACTERISTIC

Speed, rpm	*Torque, lb ft*
0	36
100	30
200	25
300	21
400	18
500	15
600	12
700	10
800	9
900	8.5
1000	8

MOTOR CHARACTERISTICS
Repulsion Start—Induction Run
Rated 1.5 hp at 1750 rpm

Speed, rpm	*Starting-Connection Torque, lb ft*	*Running-Connection Torque, lb ft*
0	13.0	
200	14.2	
400	13.6	
600	11.3	2.8
800	9.2	4.0
1000	7.2	5.4
1200	6.0	7.2
1400	5.2	9.4
1600		11.2
1700		9.0
1750		4.5
1800		0

17. Devising an Equivalent Circuit

It is desired to find an equivalent electric circuit for a device under test. When an electromotive force of 90 volts is applied to the device the current rises abruptly to 0.03 ampere, from which it falls exponentially to 0.01 ampere, the time constant being 0.02 second. Refer to figure P–17. When the battery after having been connected for

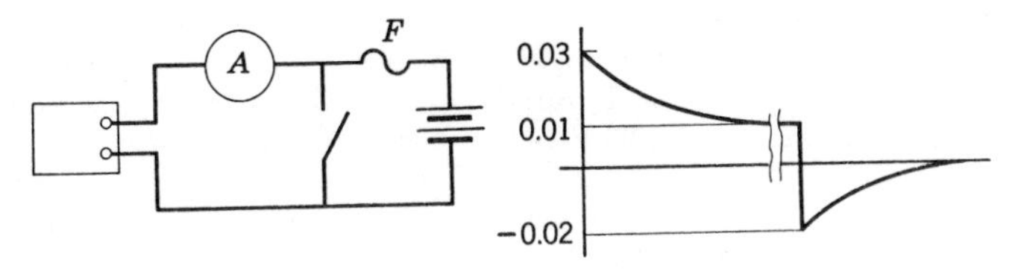

Fig. P-17.

several minutes is removed from the circuit by short circuiting and allowing a fuse F to blow, the current suddenly reverses to the value of 0.02 ampere and then decays to zero exponentially, with the same time constant as before.

Give an equivalent circuit, specifying values for the parameters. Is your circuit the only one possible?

18. Temperature-Change Thermostat

Figure P–18 shows a temperature-change thermostat for use in domestic-heater safety controls.* It comprises a relatively high expansion U-shaped element of stainless steel

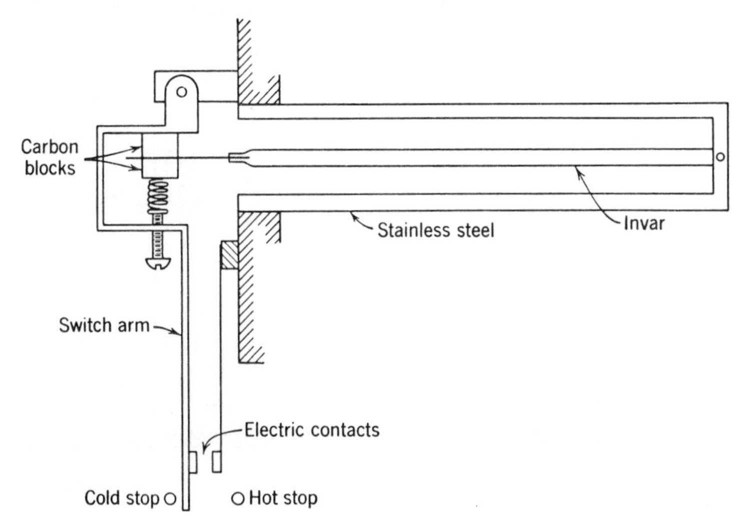

Fig. P-18.

secured at the open end and a relatively low expansion element of Invar secured to the bottom of the U-shaped element. Upon application of heat, the legs of the U lengthen,

* Described in "Flame Detectors for Domestic Fuel Burner Safety Devices," by J. A. Deubel, *AIEE Transactions*, Vol. 69, 1950, p. 220.

pulling the Invar piece with them and providing a resultant motion of about 0.001 inch per 20°F temperature change. This motion is coupled to the electric switch through a "slip-friction" transmission and a multiplying lever of approximately 20 to 1 ratio. The "slip-friction" transmission consists of two carbon blocks secured to the switch arm and arranged to have sandwiched between them a flexible blade fastened to the Invar element. When there is a differential expansion the Invar element moves and the switch arm follows because of the force transmitted through the sandwich until the arm engages either of the two limit stops. Further motion of the Invar element results in slipping the flexible blade between the carbon blocks. The switch is arranged to close on a rise in temperature of about 25°F and to open on a fall of temperature of about 75°F. Total travel between stops is approximately 1/10 inch.

The contacts are in the circuit which controls the fuel valve, so that if the temperature falls because the flame goes out or does not rise because of ignition failure the fuel supply will be cut off. To start the heater it is, of course, necessary to momentarily close a circuit around the contacts so as to open the fuel valve initially.

How can calibration be accomplished; that is, what must be altered and by what amount in order that the switch will open or close with temperature changes different from those given above?

19. Shear Load on Pins

The following situation occurs in the design of a line of equipment for mass production A stiff rectangular plate carried by four pins supports an off-center load W, as shown in figure P–19. The plate slips easily between the vertical members a, but the clearance is not enough to permit appreciable tipping of the plate out of the vertical plane. The four holes in the plates and their mates in the supports are not located precisely, but they are enough larger than the pins that the pins can be slipped in easily during assembly before W is applied.

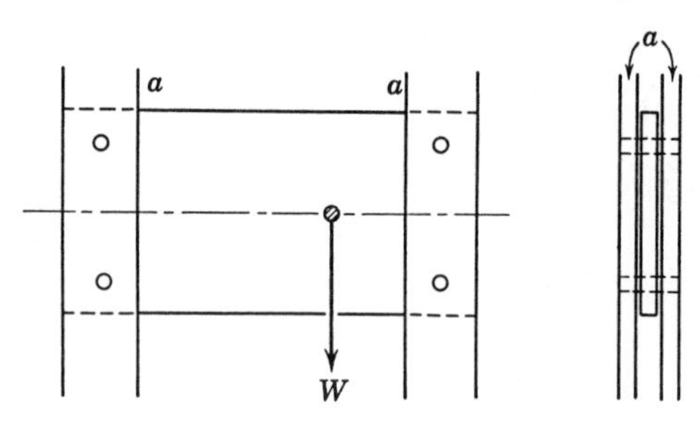

Fig. P-19.

We wish to know the maximum shearing load in terms of W that any one of the pins may have to support.

20. Fluid Drive

A fluid coupling is being considered for connecting a motor to a load, as shown in figure P–20. The motor shaft 1 is fastened to the coupling housing 3 and vanes 5 which when turned tend to carry with them the oil in space 4. The oil in motion exerts a torque on vanes fixed to the end of load shaft 2. For transmission of torque from shaft 1 to shaft 2 there must be a difference in speeds.

It is desired to compare the steady operating speeds of motor and load when the fluid coupling is employed with what the speed would be with a solid coupling. For simplicity the various torques may be assumed to be given by the following linear approximations:

$$\text{Torque required by load} = 0.02N_2 \text{ lb ft}$$

$$\text{Torque supplied by motor} = 25 - 0.01N_1 \text{ lb ft}$$

$$\text{Coupling torque} = 0.2(N_1 - N_2) \text{ lb ft}$$

In these expressions N_1 and N_2 represent the speeds of shafts 1 and 2 in rpm.

At steady (equilibrium) speed, how many horsepower will be transmitted to the load?

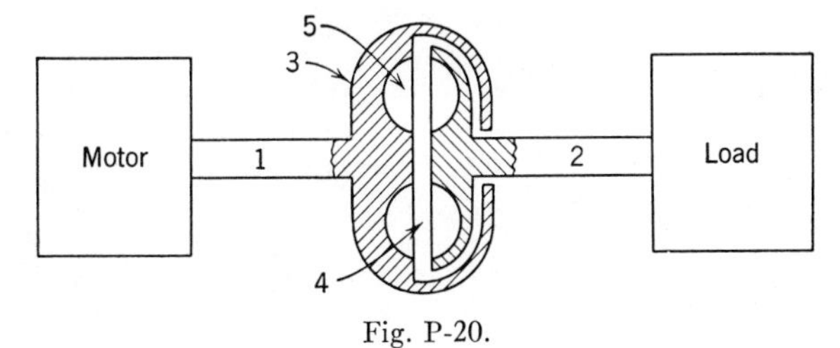

Fig. P-20.

21. Hydraulic Power Transmission

A hydraulic power transmission system is to consist of a positive displacement rotary pump P and two identical hydraulic motors $M1$ and $M2$, as shown in figure P–21. To

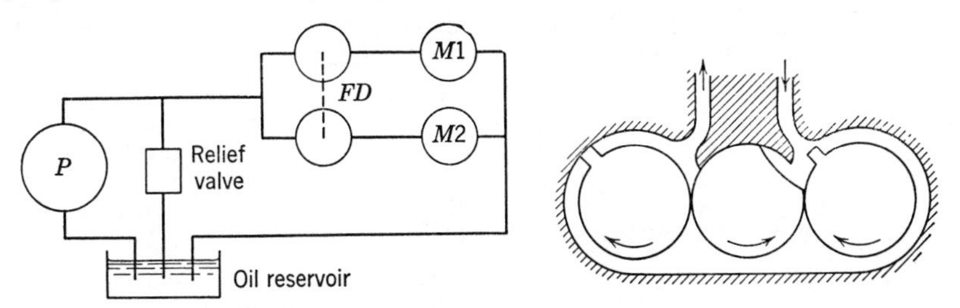

Fig. P-21.

give the desired motor characteristics a flow divider FD will be provided, consisting of two units each exactly like one of the motors and coupled together by a common shaft but otherwise free to turn.

Each motor contains three rotating elements geared together in one-to-one ratio, the output being taken from the central shaft. The clearances between the cylindrical surfaces of the rotors and between the case and the projections on the outer rotors are so small that leakage is negligible. The central rotor is notched so as to allow the projections on the outer rotors to pass it. The design is such that at least one of the projections is always mating with the case, thus providing a barrier continuously between the high- and low-pressure sides.

The pump is similar to the motors in design, is to be driven at constant speed, and will develop 20 hp at 1000 psi. The relief valve will be set to open at 1100 psi.

The loads on the motors may be unequal and normally of any value up to 10 hp, although occasionally under abnormal conditions they might be higher. Sometimes one or both of the loads may tend to drive, that is, overrun, its hydraulic motor. The piping will be such that with a load of 10 hp on each motor the pressure at the motor input will be essentially the full 1000 psi developed by the pump. To assure that the pressure drops will indeed be negligibly small, it is necessary as a preliminary to the piping design to know what pressures and flows may occur in each part of the system under various conditions.

22. Pilot Ejector

For emergency use in jet-propelled fighter aircraft, it is necessary to have a device to eject the pilot upward and out of the plane at a velocity sufficiently high that he will clear the tail structure. For a particular plane, the ejection velocity believed to be necessary for safe margin is 100 feet per second. It is desired that this velocity be attained within the very short vertical distance available, and thus extremely high acceleration is required. It is believed from physiological research that a person can withstand for a short time an acceleration not exceeding $20g$ ($g = 32.2$ ft sec^{-2}), but under these conditions the rate of change of acceleration (sometimes called the jerk) must not exceed $150g$ per second.

What is the minimum distance within which the ejection can be accomplished safely? It may be assumed that the pilot with the equipment attached to him weighs 300 pounds.

23. Chattering Ball

A vibration table is a machine for subjecting small devices to steady sinusoidal vibration of known amplitude and direction. The amplitude is adjustable by changes in lever ratios between the actuating cam and the table. The following scheme is proposed for checking the calibration of vertical amplitude: A small steel ball will be placed in a slight depression in the horizontal surface of the table. If the vertical amplitude is small the ball will follow the table but as the amplitude is increased a value will be reached beyond which the ball will break contact with the plate, that is, the ball will "chatter" on the plate.

For a particular frequency, say 1500 cycles per minute, work out an expression for the amplitude in inches above which chattering will occur.

24. Automobile Dynamometer for Acceleration

An automobile chassis dynamometer is an arrangement for laboratory testing the entire power plant of an automobile including engine, transmission, and rear axle. The front end of the car is fastened securely to the floor, and the rear wheels rest on a pair of large drums mounted in a pit so that their surfaces are tangent to the floor level. The rear tires then drive the drums so that in effect the road moves backward instead of the car moving forward. The drums are geared to an electrodynamometer of the usual cradle type, which serves as a means for both absorbing and measuring the power delivered by the rear wheels.

Commonly such an arrangement is used only for testing under conditions of constant speed. It is desired, however, to adapt it to acceleration tests. To simulate the part of the load on the engine which normally results from accelerating the car it is proposed to add suitable inertia to the rotating parts of the dynamometer, or to lighten these parts if there is already too much inertia. Analyze this proposal for provision of inertia loading, and, if you find it sound, show how to determine the proper inertia for testing a given automobile.

25. Reel Drive

In many industries material is produced in very long sheets which must be wound into coils. This problem concerns the design of the drive for the reel for winding the coils in an application where 0.025″ × 24″ steel sheet comes from the processing machinery at a speed of 2000 feet per minute and the reel must be driven so as to subject the strip to a tension of 1800 pounds whether the reel is empty or full. The diameter of the mandrel on which the coils are wound is 24 inches and the diameter of finished coils is 72 inches.

You are asked to recommend a suitable type of d-c driving motor (series, compound, or shunt) and its horsepower and speed rating. The motor is to be coupled positively to the reel either directly or through gears as may be indicated by costs.

For very precise control of such drives elaborate means for monitoring the tension are sometimes used. The output of the measuring device is then used to keep some electrical quantity, such as the motor field current, set at the proper value to maintain constant tension. In the present application, however, simplicity and reliability are more important than the closeness with which the tension is held. Consequently, the tension may be allowed to vary within 5 per cent of the desired value if this will permit simplification of the controlling device. Short of altogether eliminating the need for a regulator, a desirable simplification would be a device confined entirely to the motor supply circuits, that is, one not requiring any element to sense directly the tension in the strip. In addition to specifying the motor type and rating, you are to make a general recommendation as to the regulating device, if one is needed, that is, to specify what quantity the regulator should hold constant.

26. Inertia of a Gear Train

In some servo systems the load to be driven may be very small and may run at some lower speed than the motor. The speed reduction may be of the order of 10 to 1 or even much more. Under these circumstances the load itself introduces a negligible amount of inertia into the system, the primary sources of inertia being the motor armature and the gear train. Thus, since fast and accurate response of the system is associated with small inertia, it becomes important to make the inertia of the gear train itself as small as possible. To minimize the inertia of the gear train the question arises whether to make the whole speed reduction from motor to load in one step using a single pair of gears, or to make it in two or more steps. Investigate this question in the case of spur gears to see which way the advantage lies and if possible how much it may be in a specific case.

27. Gear-Box Reaction

The following problem arises in connection with the application of an electric drive to a machine tool. A 60-hp, 1200-rpm motor is to be coupled to the machine tool through a speed-reducing gear. The input and output shafts of the gear box are horizontal and parallel and protrude from opposite sides of the box. The shafts turn in opposite directions. The gear box stands on four feet which are bolted to the horizontal bed plate of the machine. Flexible couplings are provided between motor and gear box, and between gear box and tool.

The speed ratio of the gears is 6 to 1, so that the machine drive shaft turns at 200 rpm. The efficiency of the gears is 96 per cent.

With the motor delivering 60 hp at a steady speed of 1200 rpm (machine tool under steady load), determine the magnitude and direction of the torque reaction in foot-pounds on the feet supporting the gear box. This information is needed for determining deflections of parts of the machine relative to each other. Relate the directions of the reactions to the direction of rotation of the motor, and indicate clearly whether the reaction at each supporting foot is such as to add to or subtract from that due to the weight of the gear box.

28. Dynamometer Test

A small electric motor A which is under development was arranged as shown in figure P–28 so that the torque it developed could be determined accurately. The stator of the motor was in static balance and was supported by ball bearings B_2B_2 and was thus free to rotate about the shaft except for bearing friction, and except for the fact that there was an arm which pressed down against a scale pan so that measurements of torque could be obtained. Motor A was coupled through a flexible coupling C to a small d-c motor which could be used for driving or for loading.

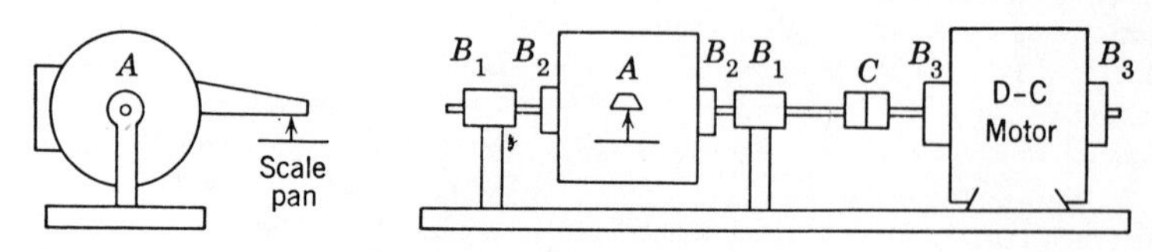

Fig. P-28.

With A developing driving torque under one condition at 1800 rpm, the scale pan indication was 1145 gram-centimeters. Under another condition, also with A developing driving torque at 1800 rpm, the scale-pan indication was 3 gram-centimeters. What torque was being developed internally in A under each of these conditions, and what must have been the direction of rotation in each case?

The following data from three other tests at 1800 rpm have been gleaned from the experimenter's laboratory notebook.

(*a*) A electrically disconnected. D-C motor driving A and the shaft in the positive direction.
Scale pan indicated a torque of 10 gram-centimeters.
(*b*) A driving the shaft and d-c motor, the d-c motor being electrically disconnected.
Scale pan indicated a torque of 60 gram-centimeters.
(*c*) Coupling C disconnected.
A driving the shaft.
Scale pan indicated 20 gram-centimeters.

29. Inertia Starter

An inertia starter for an internal-combustion engine is shown schematically in figure P–29. It consists of a flywheel geared to a shaft which can be connected to the engine crankshaft by means of a clutch and gearing not shown. With the clutch disengaged, the rotating parts of the starter are run up to high speed by a hand crank; then the hand crank is disconnected and the clutch is engaged so that the energy stored in the rotating gears and flywheel is dissipated in starting the engine.

Find the torque which is available at the starter shaft and the reaction torque at the mounting flange in terms of the deceleration of the starter shaft.

30. Landing-Wheel Slip

In evaluating a device to reduce airplane tire wear by bringing the landing wheels up to speed before they come into contact with the runway, an estimate is desired of the energy which must be dissipated as the tires slip against the runway when the prerotation device is not used.

Calculate the energy in literal terms, neglecting effects due to flexure of the tires and assuming that the brakes are not applied until after slipping has ceased.

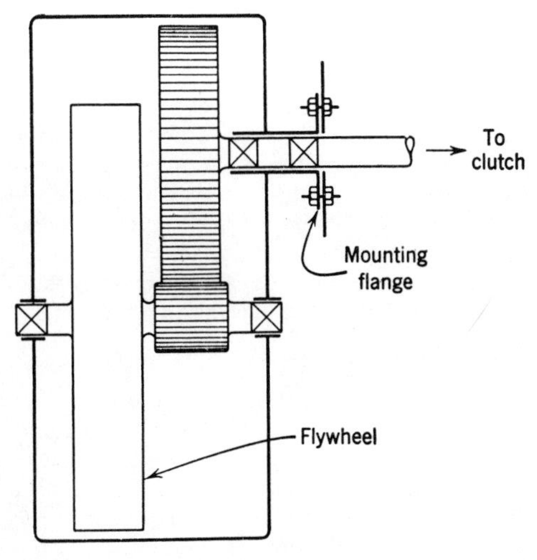

Fig. P-29.

31. Runaway Impeller

This problem concerns an accident which happened while a centrifugal compressor was being tested at high speed. The impeller weighed approximately 350 lb and was mounted on the end of its shaft as shown in figure P–31. While running at about 6000 rpm, the shaft broke and the impeller came out through the side of the casing, which was

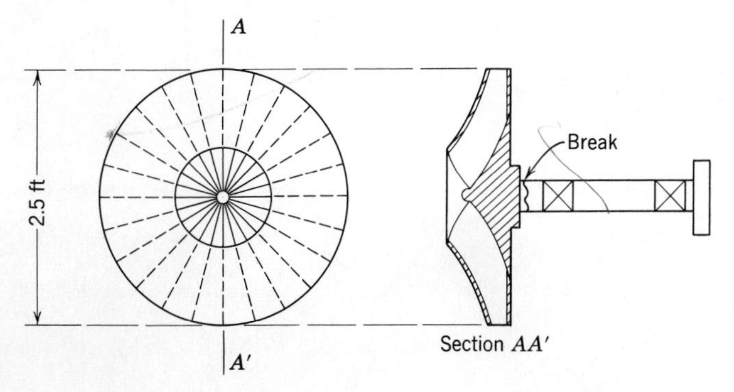

Fig. P-31.

of light construction. The impeller struck the concrete floor quite near the machine, as evidenced by deep gouges corresponding to the rim of the impeller. It then apparently bounced and rolled along the floor in a fairly straight line (again evidenced by marks on the ground) and out through an open door, finally crashing into the brick wall of another building some hundreds of yards away. In connection with subsequent damage suits it is desired to determine with what maximum velocity the impeller could have struck the wall.

32. Roller Conveyor

A long inclined roller conveyor of gentle slope forms part of the production line in a transformer manufacturing plant. The rollers are solid steel cylinders which spin very easily. Transformer tanks all alike go down the conveyor one at a time. It is observed that when tanks go down singly at sufficient intervals they apparently do not accelerate beyond a certain limiting velocity which they attain very quickly. An exception to this is that, when a tank is started along too soon after an identical one, the second usually overtakes the first, resulting in a collision which sometimes results in chipping of the enamel and consequent scrapping of one or both tanks.

To form a rational basis for remedying the trouble an analysis is desired which will show the conditions that may be expected to result in a collision.

33. Circuit to Reduce Inductive Kick

As is well known, the interruption of direct current flowing through a coil of large inductance can result in severe sparking at the switch and dangerously high voltages across parts of the circuit. In a particular instance your boss is faced with the problem of eliminating this difficulty and he remembers from an experience of many years ago that a rectifier somehow connected into the circuit was a very effective cure and that it did not interfere significantly with the normal steady-state operation of the circuit.

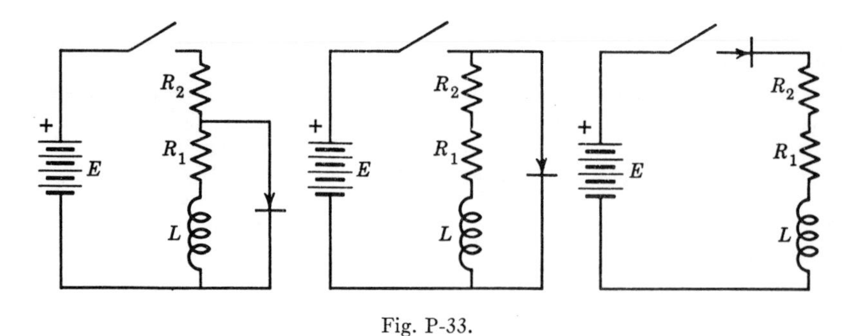

Fig. P-33.

Unfortunately, however, he cannot remember just how the connections were made but he thinks the circuits in figure P–33 may be possibilities. In these circuits:

E represents the d-c source (of the order of 300 volts).

R_1 and L are the resistance and inductance of the coil itself and are respectively of the order of 2 ohms and 10 henries.

R_2 is a resistor of about 4 ohms whose function is to limit the steady-state current to the desired value of about 50 amp.

The rectifier which your boss proposes to use has a resistance of about 4 ohms in the forward direction (current flow in the direction of the arrow), and about 4000 ohms in the opposite direction.

With this information you are expected to produce a circuit using a rectifier which from the standpoint of voltages following a very quick opening of the switch will be an improvement over a simple series connection with no rectifier. Your boss also wants to know how much the improvement will be so that he can decide if the extra complication is worthwhile.

34. Current in a Relay Coil

The d-c relay shown in figure P–34 is energized from a battery, and the current following closure of switch S is observed with an oscillograph. The circuit and a typical oscillogram are shown in the figure.

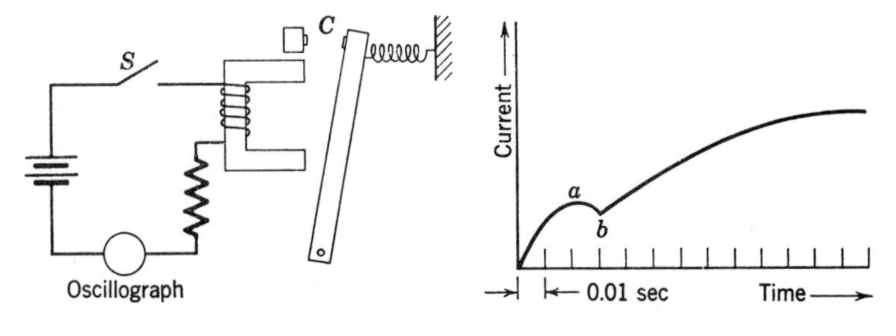

Fig. P-34.

Account for the dip in the current from a to b. It would be desirable to be able to determine from the oscillogram the time interval between closing of the switch S and the closure of the contacts C.

35. Insertion of Inductance in a Series Circuit

The circuit shown in figure P–35 is initially in the steady state with S closed. What is the current the instant after S is suddenly opened?

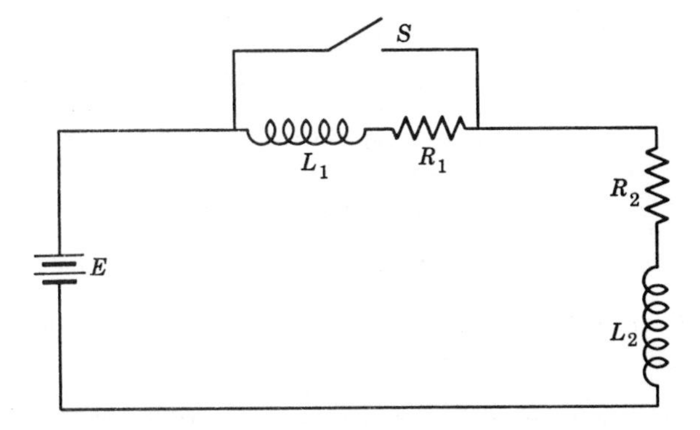

Fig. P-35.

Investigate the problem by assuming that as the switch opens the current through it decreases linearly with time, reaching zero in a time t_0. Find the current in the inductances at t_0 and then let t_0 approach zero. Compare the result with that using the principle of conservation of flux linkages. What if the current decreases in some fashion other than linearly?

36. Fluxmeter

The element of one kind of magnetic fluxmeter is similar to that of a d-c ammeter (D'Arsonval or moving-coil type) except there are no springs to provide restoring torque and thus the element is perfectly free to turn except for friction. Auxiliary means are provided to set the pointer to zero before a flux change is to be measured. The design is such that the flux linkage of the moving coil with the permanent magnet field increases directly with angle turned from the zero position, within the useful range of the instrument.

In use, the coil of the meter is connected in series with a distant search coil through which a flux change is to be measured. One use is for submarine detection; here search coils are located on the channel bottom and the fluxmeters, thousands of yards away in the harbor defense control post.

How is meter deflection related to change of flux through the search coil? Is it important to keep the circuit resistance as low as possible? Should mechanical damping of the meter element be minimized?

37. Short Circuit of an Alternator

A single-phase cylindrical-rotor alternator is running under open-circuit conditions with its field excited when it is short circuited at an instant of zero voltage. Determine expressions for the armature current, field current, and torque which will hold during the first few seconds after the short circuit. This may be done with good accuracy by entirely neglecting the effects of resistances. Also find the sustained short-circuit current, that is, the current that flows after transients have disappeared.

Data for such a machine with rating 2 poles, 5000 kva, 3600 rpm, 7200 volts are:

Armature inductance	0.025 henry
Field inductance	1.24 henries
Maximum value of mutual inductance	0.16 henry

38. Dollar Value of Blower Efficiency

A single-stage centrifugal blower is to be purchased for a manufacturing plant. Suppliers have been consulted and the choice has been narrowed down to two new units of modern design, both made by the same company and both having the same rated capacity and pressure. Both are driven at 3600 rpm by identical 200-hp electric motors.

One blower has a guaranteed efficiency of 72 per cent at full load and is offered installed for $5000. The other is more expensive because of aerodynamic refinement which gives it a guaranteed efficiency of 75 per cent at full load.

Except for these differences in efficiency and price the units are equally desirable in all respects such as durability, maintenance, ease of operation and quietness. In both cases plots of efficiency versus amount of air handled are flat in the vicinity of full rated load. The application is such that, whenever the blower is running, it will be at full load.

Develop a formula for calculating how much the manufacturer could afford to pay for the more efficient unit.

39. Purchasing a Motor

An electric motor is to be purchased for a manufacturing plant which is engaged in a very stable business. The motor is to drive a pump that requires 50 hp and that must run continuously day and night, seven days a week.

Two motors are being considered for the job, each rated at 50 hp. The motors are equal in every respect except guaranteed full-load efficiency. The efficiency of one is 80 per cent and of the other 88 per cent. Both motors are made by the same company. The difference is due solely to the use of core materials having different magnetic properties. The more efficient motor is offered installed for $1000, and the less efficient one for $900. Which one would you recommend for purchase?

You are to show clearly how you arrive at your decision. If numerical factors are needed, estimate them to the best of your ability and arrange your work so that it can be revised easily when you can obtain better values for the estimated factors.

40. Comparison of Distillation Methods

Figure P–40 shows a scheme for obtaining fresh water from sea water on board ship. In considering its adoption, we wish to compare the amount of fuel expenditure with that using a simple still comprising a separate evaporator and a condenser, and the greatest possible heating of the evaporator feed by the condensate.

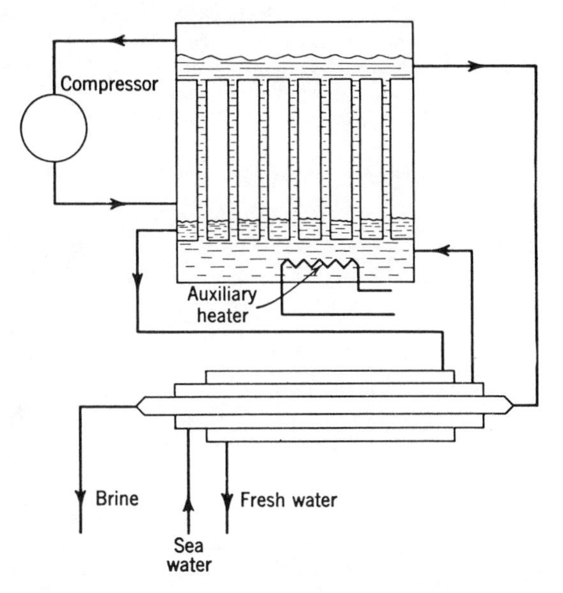

Fig. P-40.

The scheme consists of a shell-and-tube evaporator-condenser, a compressor, and a three-liquid heat exchanger. The compressor takes steam at about atmospheric pressure from the vapor space above the upper tube sheet, compresses it, and forces it into the shell space surrounding the tubes. The steam then condenses on the outsides of the tubes and is drawn off from the lower part of the shell space as the fresh-water output. The fresh water on the way out gives up some of its heat to the incoming sea water. Brine on the way out also gives up heat to the incoming water. An auxiliary heater is provided to start the process and maintain it as may be necessary.

Because of considerations depending on rates of corrosion, good practice with evaporators limits the brine concentration factor to five, that is, five times as much solid dissolved in the brine as in the sea water.

41. Water-Cooler Recuperator

Consider a drinking fountain with electrically powered refrigeration. Much of the cooled water goes down the drain instead of into the drinker. Some manufacturers of water coolers now use the waste water to precool the incoming water. The waste water may also be used to help cool the condenser, which usually is done by air circulated with a fan. Before embarking on the design of a new drinking fountain an analysis is desired of just what the extent of the commercial advantages may be in utilizing the drain water as a coolant. For a typical installation consider a drinking fountain in the hallway of a college engineering building.

42. Clothes Dryer

In the usual domestic clothes dryer, the clothes are tumbled in a drum heated either by electricity or gas, and the water in the clothes is driven off into the air. A disadvantage of this is that either outdoor venting must be provided or the humidity in the house is caused to become objectionably high, particularly if the house is small and tightly constructed. A proposal which would avoid this, and possibly have other advantages, would be to provide a closed system in which the air would be recirculated. The moisture would be removed with the aid of refrigerated coils in the air path and the clothes would be heated entirely or in part by heat derived from the condenser of the necessary refrigerating machine.

Investigate this proposal considering particularly whether or not it would be economically feasible.

43. Quick Freeze-Water Heater

A manufacturer of domestic refrigerators, quick freezers, gas and electric water heaters, and the like is looking for ways to increase the marketability of his product. A scheme that occurs to him is to introduce a line of refrigerators and quick freezers in which the usual air-cooled condenser would be replaced by one cooled by water. The water thus heated would be used to contribute to the domestic hot-water supply of the house and thus greatly reduce the consumer's total expenditure for gas or electricity.

Investigate the feasibility of the scheme.

44. Pan for Automatic Defrosting

A line of domestic refrigerators is being redesigned to incorporate automatic defrosting. Defrosting is to occur regularly once every day and is accomplished by pumping warm refrigerant through the evaporator so that melting of the frost is completed within a few minutes. To avoid the necessity for drain connections or removable pans the water is to be collected in a pan in the machinery compartment over which warm air from the condenser will pass. Thus the pan is to be emptied by evaporation into the room.

Recommend dimensions for such a pan to be used with a refrigerator having a 7 cubic foot box.

45. Rearward-Facing Seats

Rearward-facing seats are being adopted as a measure for increased safety of passengers in aircraft of the Military Air Transport Service.

Analyze the problem of adoption of such seats for commercial airliners. Assume that you are an engineer on the staff of the manager of an airline and that you must report on the *whole* problem; that is, you are expected not merely to consider the increased safety of the seats which has been firmly established by Air Force research, but the psychological and economic factors as well.

The end result desired is a recommendation whether or not to adopt the seats, and if they are to be adopted, a recommendation on how they should be introduced to the public.

46. Ventilating Duct

Cool air enters one end of a ventilating duct in a large electric generator, and during its passage through the duct absorbs heat from the walls. The duct is of uniform rectangular cross section. The flow of air is turbulent, which insures thorough mixing. The temperature of the duct walls may be assumed the same at all points, and only steady-state conditions are to be considered; that is, the temperatures of the walls and entering air do not change with time; likewise the air temperature at any fixed distance from the duct entrance is constant in time (but, of course, changes from point to point along the axis). The losses due to turbulent flow may be neglected.

Express the temperature of the air in the duct as a function of distance from the entrance in terms of the dimensions of the duct, surface heat-transfer coefficient, duct wall and entering air temperatures, velocity of flow, and any other quantities that may be significant.

47. Performance of an Automobile

For predicting the performance of a new-model automobile calculate a speed-time curve for smooth, level road conditions. The curve is to begin at a speed of 20 miles per hour and is to be at full throttle and direct drive throughout. The rear axle ratio is 3.93 to 1 and the tire size is 6.70 × 15. The car weight is 3200 pounds.

The torque of the engine at full throttle is as follows:

Engine Speed, rpm	*Torque, ft lb*
500	166
1000	182
1500	187
2000	185
2500	178
3000	167
3500	152
4000	132

From measurements of rear-wheel torque on cars of earlier model but of generally similar design as to body size and contour and weight distribution, the following prediction has been made of the road load for the car being designed. This is an estimate of the torque required at the rear wheels to drive the car at steady speed in still air and on a level road.

Wheel Speed, rpm	*Wheel Torque, ft lb*
0	130
200	155
400	195
600	270
700	325
800	390
900	470
1000	570
1100	700

48. Time to Reach Full Speed

It has been decided to use a 5:1 speed reduction in coupling the motor and mixer described in problem 16.

In order to coordinate the control of the whole system of which the mixer is a part, calculate the time for the mixer to come up to full speed after the motor is energized.

$$WR^2 \text{ of the motor} = 0.4 \text{ lb ft}^2$$

$$WR^2 \text{ of the load} = 75 \text{ lb ft}^2$$

49. Build-Down of a D-C Generator

In working out a voltage-regulating system for a d-c generator it is necessary to know how rapidly the voltage will change following a sudden increase or decrease in the resistance of the field circuit. Consider the following situation.

A d-c shunt generator is driven at 1800 rpm by a synchronous motor. With the field rheostat all cut out the field current is 6.0 amperes and the armature voltage is 300 volts, and thus the field-circuit resistance is 50 ohms. With the machine running under this condition and without load the field-circuit resistance is suddenly increased to 70 ohms by opening a pair of contacts which bridge across a 20-ohm resistor in series with the field. How will the armature voltage then vary as a function of time?

The no-load saturation curve is given below. From design data it has been calculated that the flux linkage of the field circuit equals 182 weber-turns when the voltage generated in the armature at 1800 rpm is 300 volts.

SATURATION CURVE
(Taken with separate excitation and at 1800 rpm)

Field Current, amperes	*Generator Voltage, volts*
0	5
0.5	38
1.0	71
1.5	104
2.0	137
2.5	168
3.0	197
3.5	222
4.0	242
4.5	260
5.0	275
5.5	288
6.0	300
6.5	308

50. Decay of Fields in Parallel

In designing electrical systems the consequences of such events as blowing fuses and tripping circuit breakers must sometimes be carefully investigated. For instance, consider the following situation.

Two identical 1800-rpm d-c generators shown in figure P–50 are operating on open circuit with their fields connected in parallel to an exciter bus which is maintained constant at 300 volts. The generators are exactly like the one described in problem 49 except that their fields are connected to the separate source. The field resistances are adjusted so the field current of generator 1 is 6.0 amperes and that of 2 is 1.0 ampere.

If the circuit breaker *CB* is tripped how will the terminal voltages of the two generators vary with time?

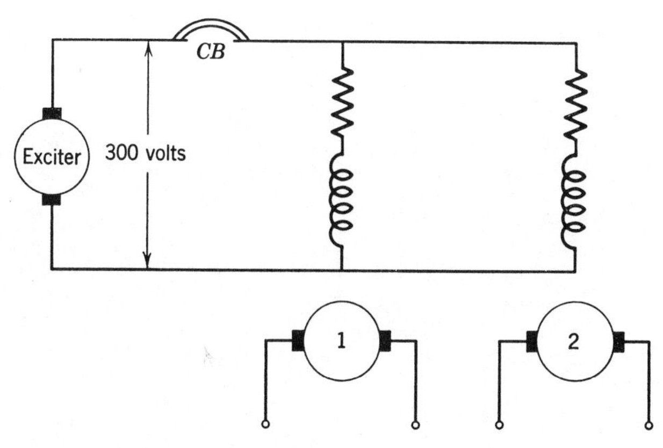

Fig. P-50.

51. Car and Buffer

A buffer is to be used to arrest the motion of a small car running on tracks. The buffer consists of a ram restrained by a compression spring and damping mechanism. The latter is a piston working in a cylinder of oil, there being a restricted bypass to permit flow around the piston. With the ram in the position shown in figure P–51 the

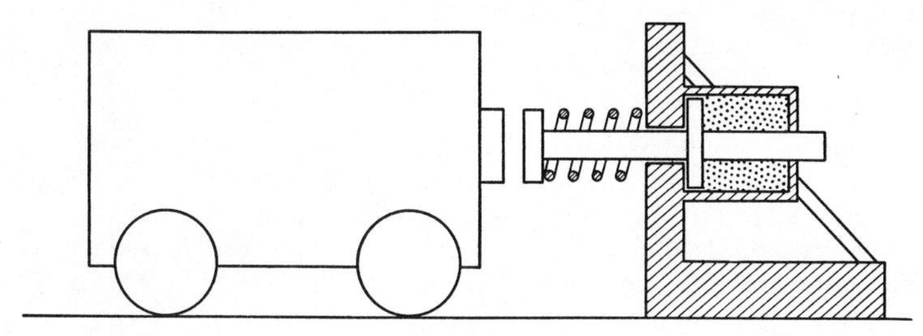

Fig. P-51.

spring is not stressed. In order to check the design it is necessary to calculate the performance. Determine the positions of the car and of the ram as functions of time.

Weight of car	= 258 lb.
Weight of the ram and piston	= 64 lb.
Damping constant	= 100 lb sec ft^{-1}.
Spring constant	= 90 lb ft^{-1}.
Initial velocity of car	= 5 ft sec^{-1}.

52. Thermocouple Ammeter

A problem in the design of thermocouple ammeters is proper compensation for variations in ambient temperature and variations in the temperatures of the meter terminals caused by heating in the external connections.

In one design, indicated at the top of figure P–52, the current is carried through a straight heater strip of uniform section connected between a pair of massive metal blocks. The hot junction of the thermocouple is attached to the center of the heater strip midway between the blocks. The cold junction is attached thermally to one of the blocks but insulated from it electrically. In another design there are two parallel strips between the blocks. One strip carries the current as in the first design while the other

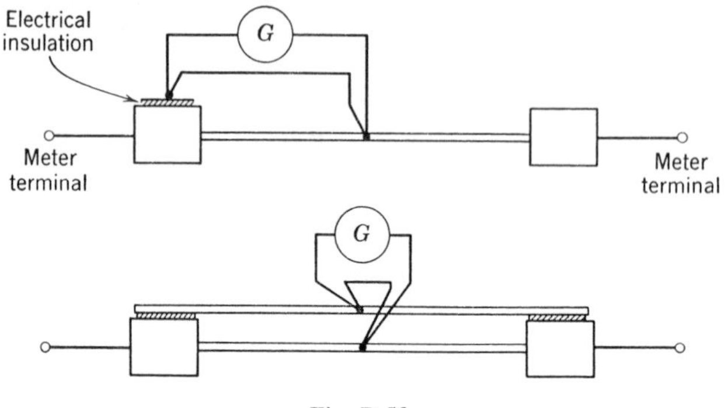

Fig. P-52.

makes good thermal contact with the blocks but electrically is insulated from them. The thermocouple junctions are connected to the midpoints of the two strips.

Investigate these two designs and decide on their relative merits.

53. Heating of a Cable near a Terminal

The current ratings of rubber-covered wire specified in a certain code are based on the maximum temperatures to which rubber may be subjected for long periods without deterioration. When the insulated wire is in still air, the code ratings result in the copper temperature rising 11°C above the ambient air when rated current is carried. This figure applies for points on wires far from connected equipment and is the same for all sizes.

Various apparatus standards permit maximum temperature rises of the connection terminals as high as 35°C above ambient. If rubber-covered wire is connected to such equipment there is an obvious inconsistency with the intent of the wire code, since the ends of the connecting wire will be subjected to excessive temperature. Study the problem with a view to determining how great a length of the connecting cable would be seriously affected. A typical case would be #0000 rubber-covered copper cable carrying its rated current of 225 amperes and connected to a terminal which runs at 35°C above ambient air.

54. Thermal Test of Circuit Breakers

Inconsistencies have developed in the acceptance tests of some small circuit breakers intended for the protection of wiring supplying lights and small motors.

The mechanism to trip the breaker incorporates a bimetal strip heated by the line current. This gives the desired characteristic of permitting moderate overloads for short periods but nearly instantaneous tripping on short circuits. At the rated current, 50 amperes in the case of interest, the breaker must remain closed indefinitely. To assure that the specifications for the tripping characteristic will be met it is necessary that the breaker run quite hot even at rated current when it is not supposed to trip. On the other hand it must not get too hot because the connecting wires are normally covered with rubber insulation which is subject to damage by high temperature. Therefore the specification includes the provision that the temperature rise of the circuit-breaker terminals must not exceed 40°C above ambient.

The engineer responsible for the design of these breakers had assured himself by actual tests that they would carry rated current continuously without tripping and that under these conditions the ultimate temperature rise at the terminals of the connecting wires was slightly below the allowed 40°C. However, when the breakers were tested by another agency to see if they met the specifications, it was found that they did not; they tripped out with normal current and while their terminal temperatures were still increasing.

Upon investigation it was found that the tests were equivalent in all respects but one. In his own work the engineer had tested breakers one at a time and with a considerable length of wire stretched out from each terminal. The testing agency, however, had tested a half-dozen breakers in series connected by wires a few inches long. In both tests the proper size of wire was used, rubber-covered #6, and in both cases the breakers and connections were spread out on a table top. Also, the ambient temperatures in both tests were within a degree or two of 25°C.

The engineer suspected that the difference in the tests was due to the difference in the path for heat transfer from the breakers to the surroundings by way of the connecting wires. In trying to evaluate this difference he tested a 10-foot piece of #6 wire carrying 50 amperes and found that in the region far from either end the temperature rise on the copper was 11°C and on the outside surface of the rubber insulation about 9.5°C. Also, he found by measuring the electric input that the heat developed in each breaker was of the order of 5 or 6 watts, the variation being due to differences in the condition of the contacts.

The engineer would like to know if a significant amount of heat is carried away from the breakers by the connecting wires, and if so what length of wire between breakers is necessary to make the test essentially independent of this length.

55. Spring and Dashpot Mechanism

In designing an automatic machine, it is desired to have a part which is initially restrained and then after release moves rapidly away from its initial position and finally

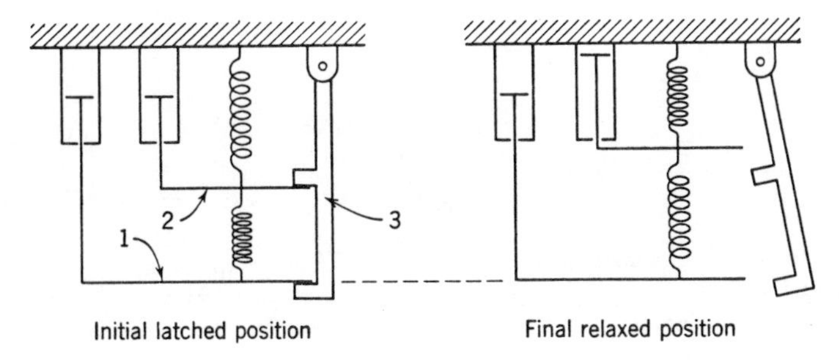

Fig. P-55.

returns to its initial position without overshooting. The means proposed for accomplishing this consists of two springs, two dashpots, and a latch arranged as indicated schematically in figure P–55.

Operation is intended to be as follows: At the left the mechanism is latched with the lower spring compressed and the upper spring stretched. Latch 3 is then swung to the right, releasing parts 1 and 2 simultaneously. Part 1 is supposed to move rapidly downward and then gradually back to its initial latched position without overshooting.

The design is to be such that parts 1 and 2 are constrained to straight-line vertical motion. All parts are made very light while the springs and dashpots are to be rather strong, so that neglect of all masses is justified, at least in this preliminary study.

Will part 1 move as desired, and how will its motion be related to the significant constants of the system?

56. Filling a Tank with Compressed Air

In planning a compressed-air system it is desired to have an estimate of the time it will take the compressor to pump the storage tank up to full pressure.

The compressor is double acting, has 8-inch bore and 8-inch stroke, and clearance equal to 5 per cent of the swept volume. The operating speed will be 200 rpm. The tank is 2 feet in diameter and 8 feet high. The volume of the piping connecting compressor and tank will be about 0.5 cubic foot.

Air will be taken in from the atmosphere, and the final tank pressure is to be 250 psig.

57. Pneumatic Controller Element

The mechanism shown in figure P–57 is being considered as part of a control device. Two identical brass bellows have their outer ends held stationary and their inner ends

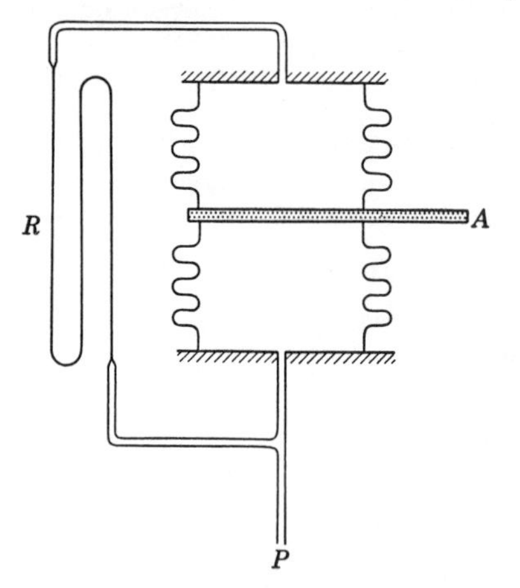

Fig. P-57.

fastened to a very light metal piece A which is free to move up and down in response to actions of the two bellows. The signal to this control device is a small change in air pressure P which is applied directly to the lower bellows and through a capillary resistance R to the upper bellows. The capillary resistance is a long tube with very small passage and is characterized by the fact that mass rate of flow through it varies in direct proportion to the pressure difference between its ends.

With the device initially in equilibrium it is desired to determine quantitatively how A will move as a function of time after a small (order of 1 psi) and sudden step increase in the pressure P which initially is 15 psig. Neglect all inertia effects.

58. Mass Suspended between Springs

A mass is suspended between two springs as shown in figure P–58. The constant of the upper spring is 150 pounds per foot and of the lower 25 pounds per foot. The mass weighs 25 pounds. The springs are such that they can act either in tension or compression. Each spring is 0.5 foot long when unstressed, but the total length of the space available for two springs is 1.5 feet.

The mass is initially pushed down to the point where the lower spring is neither stretched nor compressed, and then it is released. Determine the subsequent motion of the mass as a function of time neglecting air resistance and other damping forces.

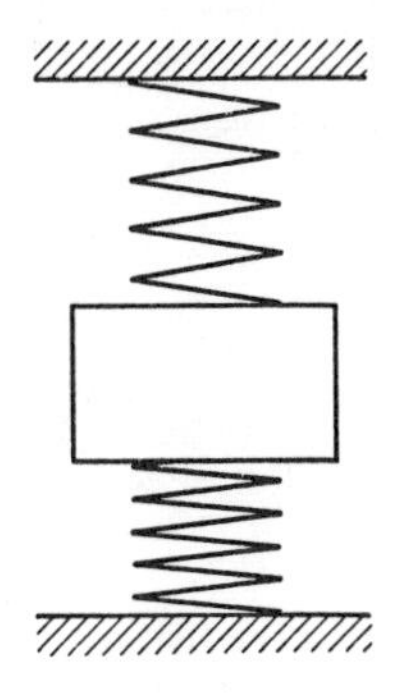

Fig. P-58.

59. A Spring and a Mass Partly Immersed

Determine the natural frequency of oscillation of a mass suspended by a spring, and partially immersed in water. The mass is a long cylinder with axis vertical.

60. Natural Frequency of Liquid-Level Accelerometer

Calculate the natural frequency of oscillations of the liquid in the accelerometer of section 1–2, neglecting effects of damping.

61. Radius of Gyration of an Armature

According to one edition of a certain test code for d-c machines, the radius of gyration R of a rotor may be determined experimentally by the following procedure:

"Support the armature in bearings having two to three times the diameter of the journals. If the journals are of different sizes, the smaller one must be bushed up to the same size as the larger. Displace the armature and allow it to roll or rock freely in the bearings, and determine the period or time in seconds (T) of a complete oscillation.

"Then:

$$R = R_2 \frac{0.815T^2}{R_1 - R_2} - 1$$

where R_1 = radius in feet of bearing.
R_2 = radius in feet of journal.
T = time in seconds of one period.
W = weight in pounds (must be given or determined by weighing)."

Before adopting this procedure it is desired to verify the formula by deriving it, particularly to see what limitations it may have, and also because of certain peculiarities in it.

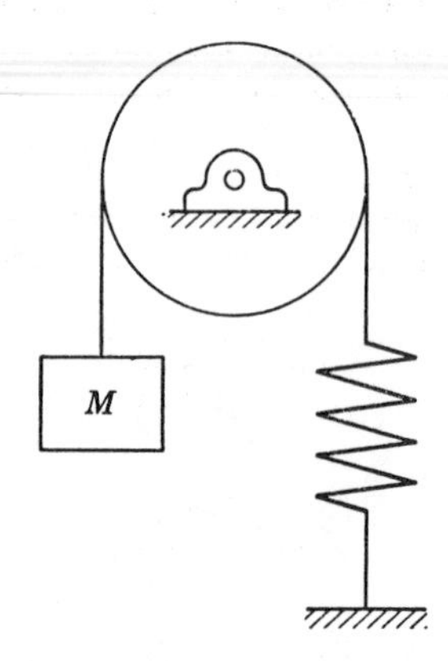

Fig. P-62.

62. A Mass, Spring, and Pulley System

Find the natural frequency of vertical oscillations of the mass M in the system shown in figure P–62. The mass and spring are connected by a light cord which does not slip with respect to the pulley. The pulley is massive and turns freely in its bearings.

63. Flywheel for a Punch Press

Some machine tools, for example punch presses, require driving torques which vary over a wide range in each cycle of operation. To smooth out the torque demand on the driving motor, and hence the current supplied, it is common practice in such cases to provide a flywheel and to use a motor with a drooping speed-torque characteristic.

To investigate this matter consider a machine which requires a driving torque varying cyclically with time, repeating again and again for long periods. The cycle consists of an interval during which the torque required is constant followed by a longer interval in which the torque required is zero. The motor is to have a speed-torque characteristic such that with increasing torque the speed decreases linearly from its no-load value, a condition approximated by induction motors and d-c shunt motors in their normal operating speed ranges. A flywheel is to be used to reduce the torque demand on the motor. Find how the maximum torque demand on the motor depends on the size of flywheel and other significant factors.

64. Spring and Mass with Dry Friction

A steel block is arranged to slide back and forth on a horizontal steel surface which is not lubricated (dry friction). The block is restrained by a horizontal spring which can act in either tension or compression. The block is pulled away from its position of equilibrium and then released. It is desired to describe as completely as possible the subsequent motion of the block.

65. Inductance, Capacitance, and Glow Tube

In figure P–65, L is an inductance which behaves linearly, C is a condenser, and G is a glow-discharge tube which does not conduct at all unless the voltage across its terminals is raised to a certain critical value called the breakdown voltage. Once G has broken down, it will conduct any amount of current with a voltage drop just equal to the breakdown voltage. These characteristics of G are similar in the two directions. The switch S is arranged so that C can be charged by means of an adjustable battery to any desired voltage and then can be connected to G and L in series. The resistances of L, C, and the connecting wires may be assumed negligibly small.

Investigate the behavior of this circuit.

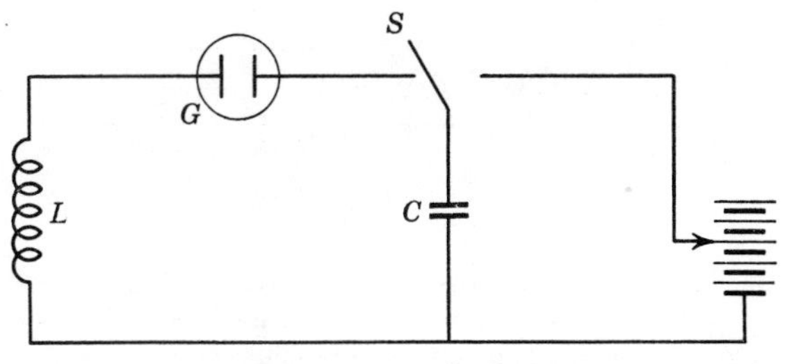

Fig. P-65.

66. Condenser to Reduce Sparking

In the circuit shown in figure P–66 the switch is closed long enough to establish a steady state and then is suddenly opened.

In order to reduce the voltage which appears across the switch it is proposed to connect the condenser as shown. Is this worth while? What voltage rating should the condenser have?

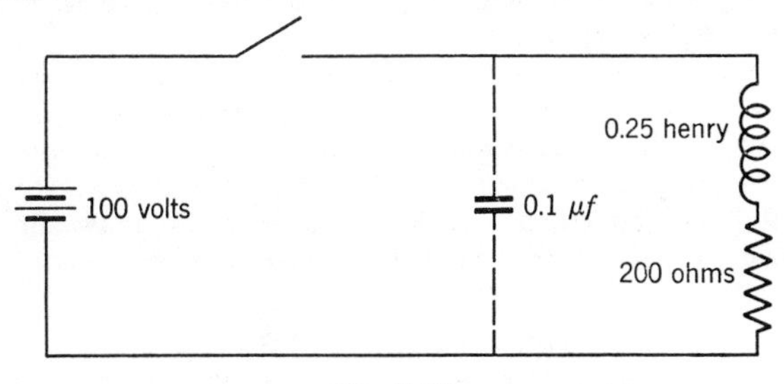

Fig. P-66.

67. Parallel L-C Circuit

Is the circuit shown in figure P–67 capable of oscillation on closing the switch, and if so, what is the influence of the value of R?

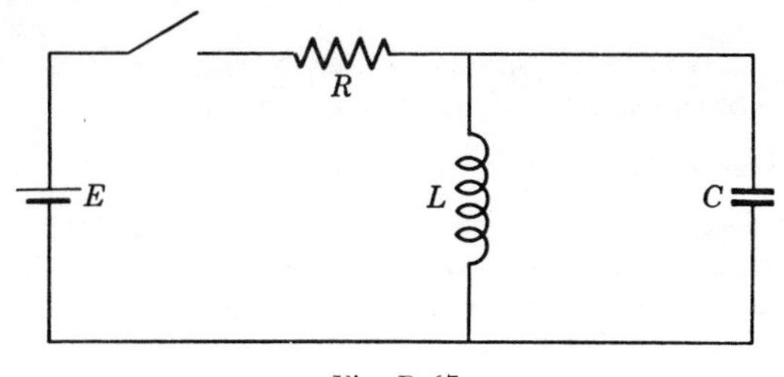

Fig. P-67.

68. Study of Chatter

This problem arises in an investigation of chatter, the sustained vibration which sometimes occurs, for example, with cutting tools and with brushes on commutators.

The mass in figure P–68 is set gently onto a horizontal surface which has constant velocity in its own plane. The mass is limited to linear motion parallel to the surface and is restrained by a spring fastened to a fixed point. In a preliminary study it was

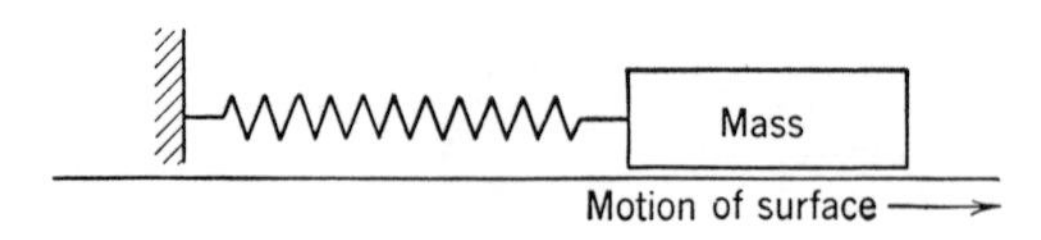

Fig. P-68.

found that if the friction between the mass and the surface is assumed independent of relative velocity, except for sign, that sustained vibrations cannot occur provided even a small amount of velocity damping is present. Find whether there can be sustained vibrations if the static friction is assumed greater than the dynamic friction and that the latter is independent of relative velocity.

69. Circuit Analog for Chatter

In order to pursue the study of chatter which was the subject of problem 68 it is proposed to construct an electric circuit analog. It is hoped that this can be done in such a way that by means of an oscilloscope records may be obtained which will represent to scale the time variations of mechanical quantities (forces, velocities, etc.) which are significant in the chatter phenomenon.

To represent the friction we hope to be able to use a certain glow-discharge tube which is available. The tube commences to conduct when the voltage applied across it is raised to 105 volts. The voltage required to maintain current flow through the tube is 75 volts for all currents up to about 30 milliamperes beyond which the characteristic ceases to be flat. The characteristic of the tube is similar in the two directions.

The mechanical system which it has been decided to study is characterized by the following quantities:

Weight of block	5 lb
Spring constant	100 lb ft^{-1}
Damping	Small
Surface velocity	6 ft sec^{-1}
Static friction force	70 lb
Dynamic friction force	50 lb

The mass does not simply rest on a surface but is squeezed between two parallel moving surfaces.

Try to devise a suitable circuit, specify the sizes of components to use, and tell how to convert between the mechanical and electrical quantities.

70. Ballistic Galvanometer with Short Period

In the conventional use of the undamped ballistic galvanometer the charge being measured is assumed to pass through the circuit in a time very small compared to the period of the galvanometer, that is, in effect instantaneously. To insure that this condition is fulfilled by employing a galvanometer of excessively long period is troublesome because the longer the period, the more difficult and more time consuming it is to make the measurements.

A situation has arisen where it is necessary to measure amounts of charge transferred by currents of the form $I\epsilon^{-\alpha t}$, and galvanometers of various periods are available. Determine in as general terms as possible what length of galvanometer period will suffice.

71. Effect of Inductance in a Ballistic Galvanometer

A voltage e of irregular form is applied to a shunted ballistic galvanometer. The resistance of the external circuit is R_e, the resistance of the shunt is R_s, and the resistance of the galvanometer coil is R_g. See figure P–71.

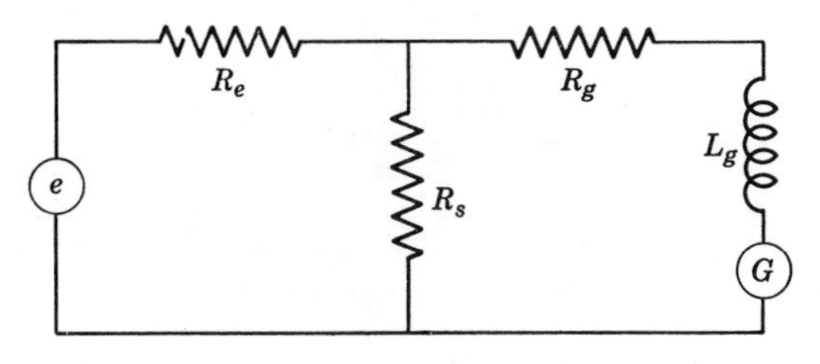

Fig. P-71.

Often in relating voltage impulse to galvanometer throw inductance L_g of the galvanometer coil is neglected. Does this make any difference as far as charge through the galvanometer coil is concerned?

72. Variable Inductor with Coils in Parallel

In a variable inductor the rotor and stator coils usually are connected in series and the inductance is varied by changing the coupling. Would it be feasible to connect the coils in parallel to reduce the range of inductances obtainable? It would be undesirable

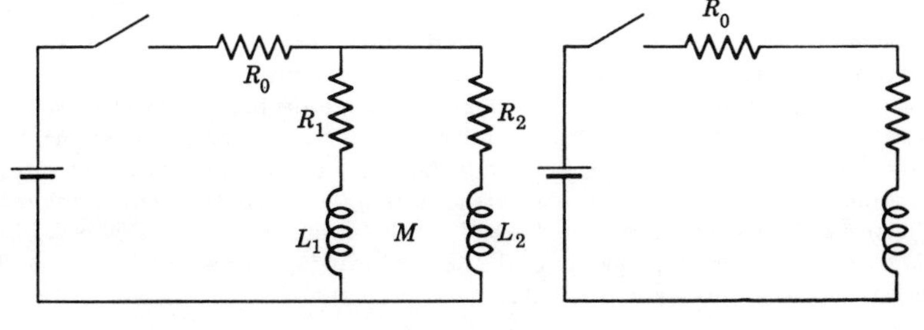

Fig. P-72.

if changing the setting effectively changed anything but the equivalent inductance between terminals. It is suggested that the question be investigated by comparing the transient behavior of the two circuits shown in figure P–72. The first is the variable inductor with coils 1 and 2 in parallel; the second is a proposed equivalent circuit. In each case there is a battery and an external resistance R_0.

73. Cooling a Gas-Turbine Rotor

An essential consideration in gas-turbine design is the removal of heat from the buckets so that their temperature will not be excessive. Consider a single turbine wheel shrunk on a hollow shaft through which cooling water is circulated. The wheel has a uniformly tapered section as indicated in figure P–73. It is desired to know the amount of heat that would be carried away from the buckets by conduction radially inward to the hub of the wheel neglecting any heat transfer to or from the lateral surfaces of the wheel.

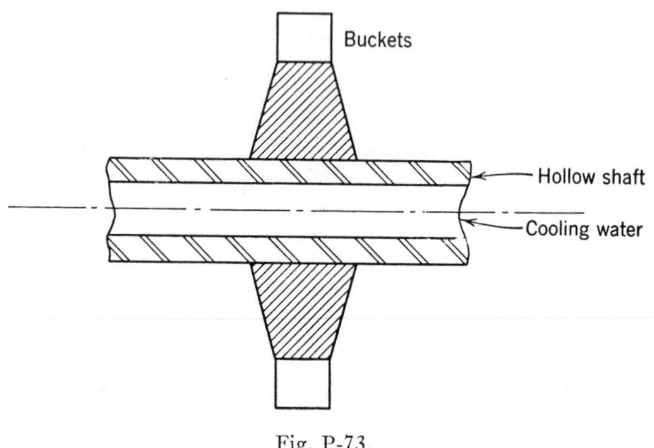

Fig. P-73.

74. Nuclear Reactor Element

Some tentative speculations about the design of a nuclear reactor for generating power envisage tubular elements packed tightly together with their axes parallel. Heat would be liberated by nuclear reactions uniformly throughout the material of which the tubes are constructed. The heat would be carried out of the reactor by a fluid pumped through the tubes. The interstices between the tubes would be simply dead space, with the result that no heat would flow radially through the outer surfaces of the tubes.

It is thought that a limitation of such a design might be imposed by stresses set up thermally within the material of the tubes. As a preliminary to estimating these stresses calculate how much power could be taken per foot of tubular element without exceeding a maximum temperature difference within the material of 500°F in the radial direction. The tubes contemplated are of 1-inch outside diameter and ⅜-inch inside diameter. The thermal properties of the material may be assumed similar to those for such things as porcelain and fused silica.

75. Friction Clutch for Starting a Load

For driving large fans for ventilating mines, synchronous motors are desirable because of their high efficiency and high power factor. On the other hand, the relatively poor starting characteristics of synchronous motors sometimes may make them unattractive for this application. A means for circumventing this difficulty is to couple the motor to the fan through a dry disc friction clutch. The motor is brought to synchronous speed with the clutch disengaged, and then the fan is started by engaging the clutch. The fan accelerates gradually to synchronous speed as the clutch slips.

Evidently the torque required to slip the clutch must be less than the torque at which the motor pulls out of synchronism, but more than that needed to drive the fan at synchronous speed. Recommend a suitable value for the clutch torque, keeping in mind that the energy to be dissipated as heat from the clutch may be dependent on the value of slipping torque selected.

76. Build-Up of Vibration

It is desired to have a study made of the build up of resonant vibrations in a mass-and-spring system with various amounts of linear (velocity) damping. More specifically, it is desired to know the displacement as a function of time after the application of a sinusoidal force whose frequency equals the undamped natural frequency of the system, and the effect of damping on this function.

77. Rapid-Reversal Motor

The following appeared in *Mechanical Engineering*, August 1950, p. 628:

"A method for reversing a small electric motor in three to four milliseconds has been developed by Jacob Rabinow at the National Bureau of Standards, Washington, D.C. Designed specifically to meet the need for high-speed reversal of magnetic tapes in the memories of electronic digital computing machines, the technique may prove useful in many other applications.

"In the Bureau's rapid-reversal motor, the kinetic energy of the rotor, instead of being dissipated as heat in a brake during deceleration, is converted into potential energy in a spring, which is then used to accelerate the rotor rapidly in the opposite direction.

"A small low-inertia two-phase motor operating at 3200 rpm was used. The reversal spring consists of a steel torsion bar approximately 31 in. long and $\frac{3}{16}$ in. in diam. Only one phase of the motor is connected to the a-c power supply; thus the motor will rotate in the starting direction, either clockwise or counterclockwise. The motor shaft is rigidly connected to one end of the torsion bar, which is equipped at the other end with two positive unidirectional clutches. One clutch prevents clockwise, and the other counterclockwise, rotation. If the motor is rotating in a clockwise direction and the proper clutch is engaged, the adjacent end of the torsion bar is thereby stopped; this brings the rotor to a stop in approximately 20 deg.

"The potential energy stored in the torsion bar is then returned to the rotor in the form of a counterclockwise impulse. The motor attains virtually full speed in the new direction within about two milliseconds.

"In the experimental model of the rapid-reversal mechanism built at the Bureau, the clutches are operated manually, but it is expected that in normal use the clutch mechanism will be operated by suitable electromagnetic controls. Although the studies thus far have used a small motor of about $\frac{1}{45}$ hp, it is anticipated that motors of all sizes could be reversed rapidly by this technique, the speed of reversal being limited only by the mechanical strength of the various parts . . .. ."

Analyze the mechanics to check the above statements, particularly the last sentence.

78. Air Mail Pick Up

A device is proposed to enable an airplane to pick up mail or express from small airports without stopping. The container holding the load would be set on the ground and attached to one end of an elastic cable laid out along the line of flight. The other end of the cable would terminate in a loop that could be engaged by a hook on the airplane.

For a given size of load to be picked up, it is, of course, desirable to subject the airplane to as little force as possible and still not have the cable stretch excessively. Investigate the problem with a view to recommending a suitable elastic constant for the cable. For the purposes of this analysis you may assume that the air resistance force on the load container varies as the first power of velocity.

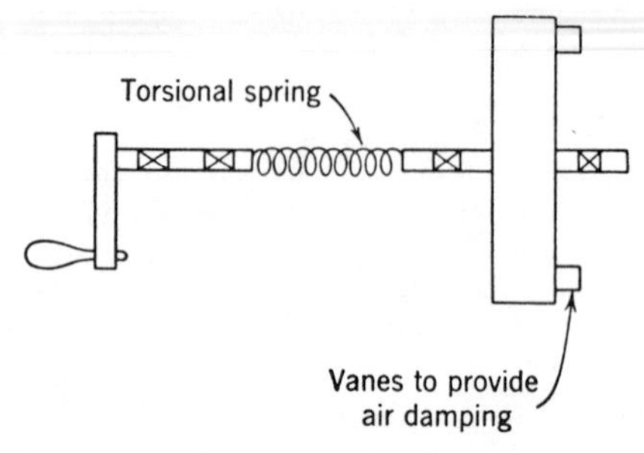

Fig. P-79.

79. Crank, Elastic Shaft, and Flywheel

The mechanical system shown in figure P–79 consists of a crank joined by an elastic shaft to a flywheel which has damping. The elasticity of the shaft is represented by the torsional spring. With the system initially at rest the crank is suddenly given a constant angular velocity.

Find the angular position of the flywheel as a function of time.

80. Instrument in a Rocket

A delicate, but massive, instrument is to be supported on springs within the body of a rocket used for exploring the outer atmosphere. In considering the design of the mounting it is desired to determine the motion of the instrument during the first part of the flight of the rocket. Assume for this analysis that the acceleration of the rocket reaches a constant value instantaneously at the start and continues indefinitely.

81. Spring-Suspended Packing Case

Delicate apparatus is sometimes packed for shipping in a container suspended by springs within the packing case.

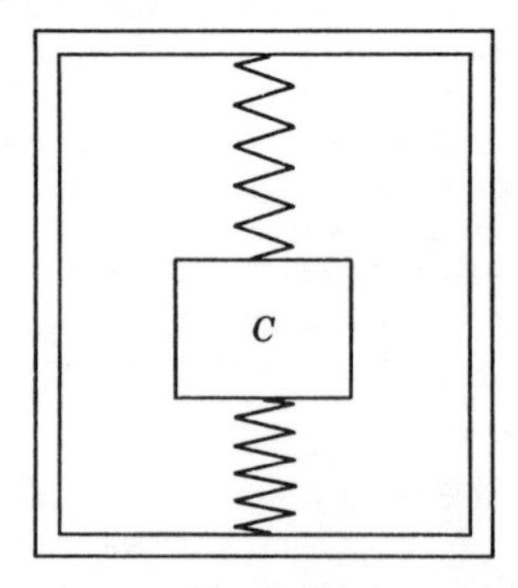

Fig. P-81.

For the purposes of design it is desired to analyze the simplified form of mounting shown in figure P–81. In the simplification the container C may be considered to be constrained to vertical motion. The two springs are identical. In an actual packing case, there would of course be additional sets of springs along the axes at right angles.

The stresses set up in internal parts of the apparatus being shipped depend upon the acceleration to which they are subjected.

Determine the maximum acceleration of the container C in terms of appropriate constants, when the framework shown is dropped vertically and strikes a rigid surface without bouncing.

82. Bucking Broncho

Two d-c machines are connected as shown in figure P–82. Machine 1 is a series generator whose armature is driven at constant speed by a motor not shown. Machine 2 is a motor whose field is supplied with a constant current by a separate source. The armature of machine 2 is free to turn and is not coupled to any load.

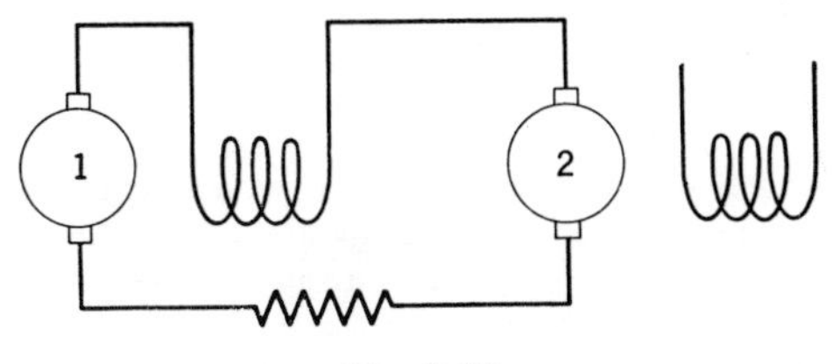

Fig. P-82.

Analyze the behavior of this system assuming that the series generator flux varies linearly with current, that is, that there is no magnetic saturation, and that the self-inductance of the series field is constant.

83. Series Generator and Condenser

The circuit in figure P–83 contains a d-c series generator and a condenser. In the approximately linear range of the generator the generated voltage is 200 when the current is 2.0 amperes and the speed is 1000 rpm. The capacitance is 200 microfarads. The inductance, consisting principally of the generator field, is 100 henries, and the resistance, including that of the generator, is 100 ohms. The condenser is charged to 100 volts by an external connection not shown and then discharged through the circuit.

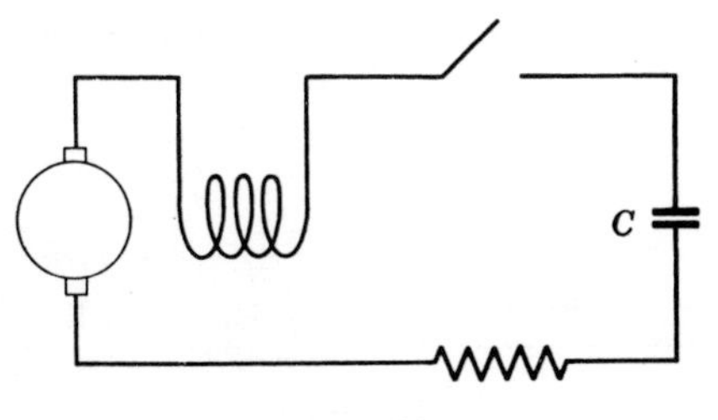

Fig. P-83.

Investigate the behavior of the circuit for three different constant speeds of the generator: 900, 1000, and 1100 rpm. Consider particularly the energy exchanges which may take place. What is the effect of reversing the connections to the armature?

84. Damping for Quick Response

A 150-volt d-c voltmeter of the D'Arsonval type has been designed except for the damping device. It is desired to have the damping such that when a constant voltage of 100 volts is suddenly applied the pointer will appear to reach 100 as quickly as possible and remain there.

Recommend a value of damping in terms of critical damping.

85. Effect of Inductance on Dynamic Braking

In section 6–5 the dynamic braking of a d-c motor was analyzed neglecting the effect of the self-inductance of the motor armature. Reexamine the problem, taking inductance into account. It may be assumed safely that the inductance will be such that the time constant of the armature circuit, that is, the ratio of inductance to total resistance, will be small compared to the time for the motor to stop.

86. Automatic Weighing

A device proposed for continuously balancing a varying force is shown in figure P–86. It consists of a beam supported at the left on a knife edge and carrying a counterpoise which is moved by means of a screw driven by a small d-c motor carried on the beam. The right-hand end of the beam is restrained through a pinned joint by a short leaf spring which is so stiff that motion of the beam is not appreciable. The upper and lower surfaces of the leaf spring carry resistance-wire strain gages. The strain gages are connected into a bridge circuit, the output of which is amplified and fed to the armature of the d-c motor. The field of the motor is separately excited. The characteristics of the strain-gage arrangement and associated circuits are such that a voltage is applied to the motor armature which is directly proportional at every instant to the force which the end of the beam exerts on the leaf spring. The polarity of the voltage is such as to cause the motor to turn in the direction to move the counterpoise toward the position where its weight will balance the force being measured. The value of the force then can be determined by reading the position of the counterpoise either against a scale on the beam or with the aid of a revolution counter on the motor shaft.

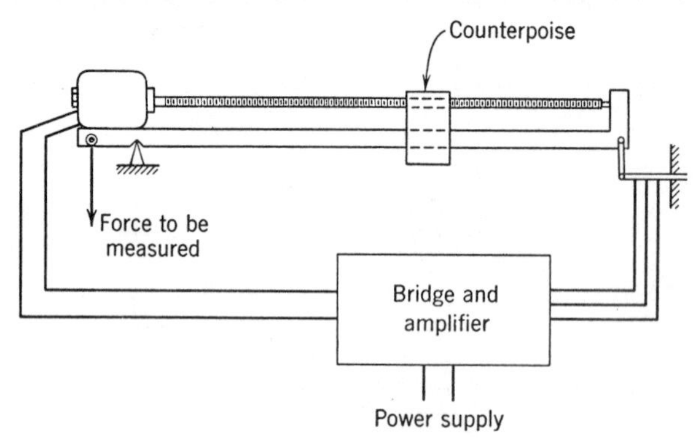

Fig. P-86.

Working in terms of appropriate system constants we wish to predict the performance of the proposed device, and to see what limitations there may be if the force being measured varies rapidly.

87. Positioning Mechanism

A massive radar antenna is to be positioned accurately in azimuth by remote control, using the system shown in figure P–87.

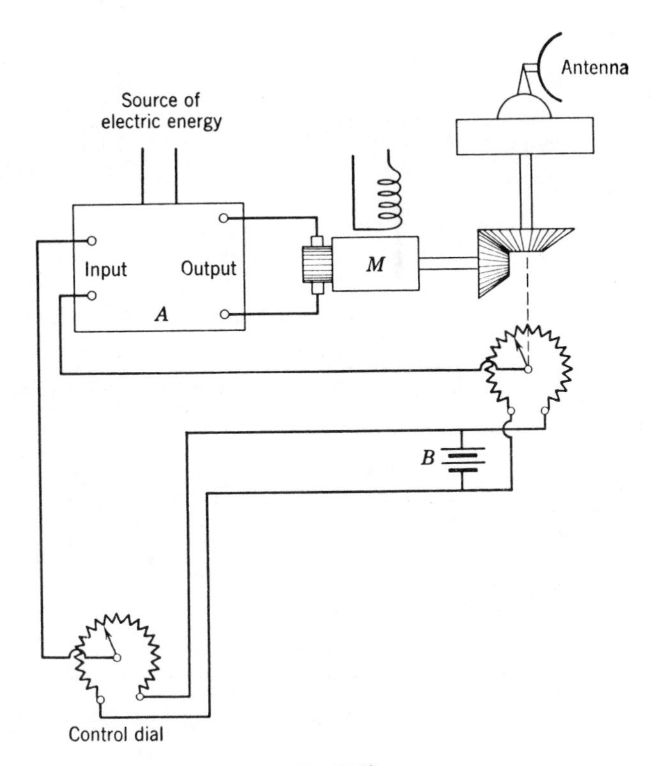

Fig. P-87.

The d-c driving motor M has separate and constant field excitation. The remote-control dial is fastened directly to a slider which can move around a uniform circular resistor, the ends of which are connected to the d-c voltage source B. An exactly similar resistor is connected to the same source and arranged so that its slider is driven directly by the shaft which turns the antenna. The two sliders are connected to the input of the amplifier A.

The amplifier produces a direct voltage output which is directly proportional to the voltage applied to its input terminals. The input terminals present such a high impedance that negligible current flows in the circuit through the sliders.

It is desired to have the antenna point accurately in the direction called for by the control dial and to have it assume this position quickly when a new setting is made. Moreover, if the control is turned at a uniform rate as in tracking, the position of the antenna should remain in correspondence. Analyze the system quantitatively with the above requirements in mind. The following magnitudes are known:

Voltage of B	60 volts
Angle subtended by the circular resistors	300°
Moment of inertia of the armature and load referred to the motor shaft	1000 lb ft sec^2
Viscous friction constant of motor and load referred to motor shaft	500 ft lb sec
Gear ratio (motor speed/antenna speed)	5:1
Motor constants:	
Generated voltage	40 volt sec per radian
Torque	30 lb ft per amp
Resistance of armature and amplifier output circuit	0.20 ohm
Ratio of amplifier output voltage to input voltage	250

88. Self-Balancing Potentiometer

A self-balancing potentiometer for continuously recording temperature as measured by a thermocouple is shown diagrammatically in figure P–88.

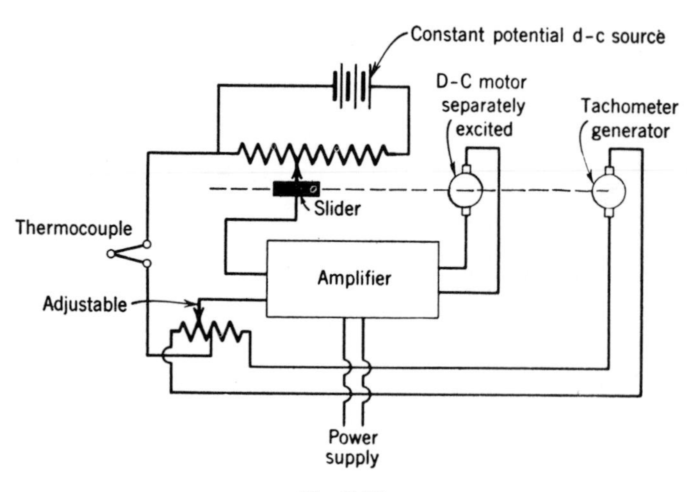

Fig. P-88.

When the temperature changes, the motor drives the slider along by means of gearing and a screw, thus tending to keep the voltage across the potentiometer in balance with the voltage developed by the thermocouple. The slider carries a pen, which draws a graph of temperature on a chart which moves continually at constant speed at right angles to the temperature axis.

The amplifier supplies current to the motor in direct proportion to the voltage across the amplifier input terminals. The internal resistance between these input terminals is so high that current flow in the circuit through the thermocouple is negligible.

Analyze the transient behavior of the device in terms of appropriate system constants and consider particularly the function of the tachometer which is driven by the motor shaft.

89. Exciter Overshoot

An alternator is excited by a d-c shunt generator (the exciter) whose armature is connected directly and permanently across the terminals of the alternator field. The alternator field is varied by opening and closing contacts which short circuit all or part of the exciter-field rheostat.

In most systems when the contacts are closed after they have been open for some time the alternator field current builds up gradually and smoothly, approaching its new value without overshooting. In some systems, however, the alternator field current builds up more rapidly, overshoots the new steady-state value, and oscillates before settling down. Investigate this phenomenon with a view to determining quantitatively the conditions under which it can occur. Consider only the case where the a-c armature terminals are open. It may be assumed that the speeds of both machines are held constant.

90. Decay of R-C Transient

An engineer has made the following statement:

"One cycle after an alternating voltage is applied to a circuit containing resistance and capacitance in series, the transient will be small compared to the maximum of the steady-state current."

Is the statement truc for all values of circuit parameters, any frequency, and any instant of closure?

91. Surge Generator

For applying voltage surges to test the insulation of electrical apparatus the circuit in figure P–91 may be used. The terminals at the right are connected across the insulation

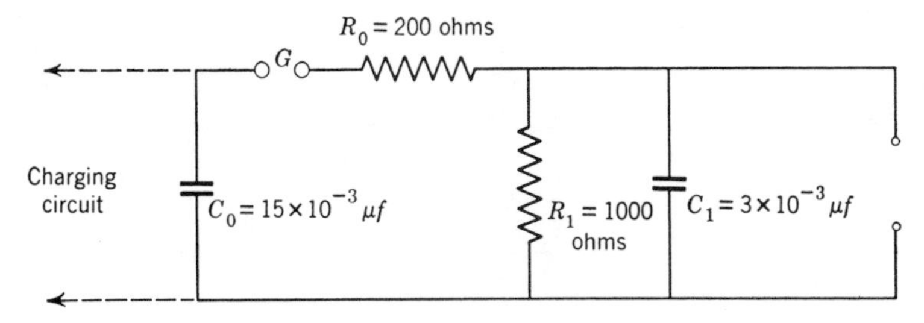

Fig. P-91.

to be tested; for example, one terminal might be connected to the windings of a motor and the other terminal to the frame. Capacitor C_0 is first charged from a d-c source through a very high resistance. The spark gap G breaks down when the voltage across it reaches a predetermined value, thus applying the test surge. Once G breaks down the voltage drop across it is negligible.

If it is desired to apply a surge at the test terminals with a crest value of 25 kilovolts, for what voltage should the gap be set to break down?

The severity of a voltage surge is usually described by specifying its crest value, the time for the crest to be reached, and the time for the voltage to drop to half the crest value. The shorter the former time and the longer the latter, the more severe the test. How will the severity of the test be affected in this case when a piece of apparatus is connected across the test terminals, supposing that its insulation does not break down and that it can be considered equivalent to a capacitance of 0.001 μf?

92. An Integrating Circuit

Figure P–92 shows a circuit proposed for integrating electrically. The quantity to be integrated is expressed as the voltage e, and it is desired that the voltage e_c across the condenser be proportional to $\int e\,dt$. For example, e may be the voltage induced in a magnetic exploring coil, proportional to dB/dt; then e_c should be proportional to B.

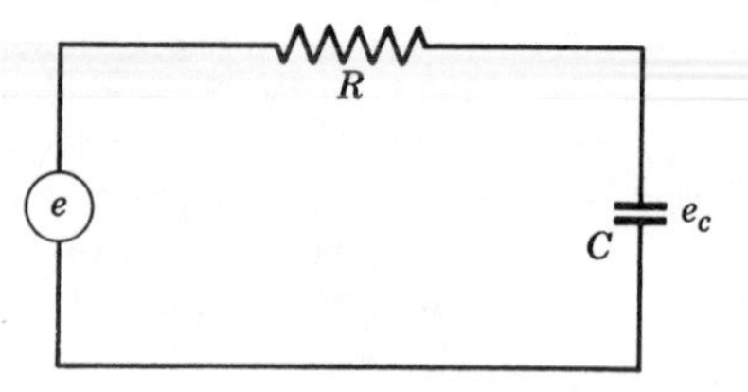

Fig. P-92.

Supposing that e is of the order of 100 volts maximum and that e_c as small as 1 microvolt can be measured, determine the performance of the circuit under various conditions and recommend the best values of R and C.

93. Electrical Differentiation

The circuit in figure P–93 is intended to differentiate electrically. The voltage across a small resistance is an approximate representation of the derivative of the input voltage, the approximation improving as the resistance is made smaller.

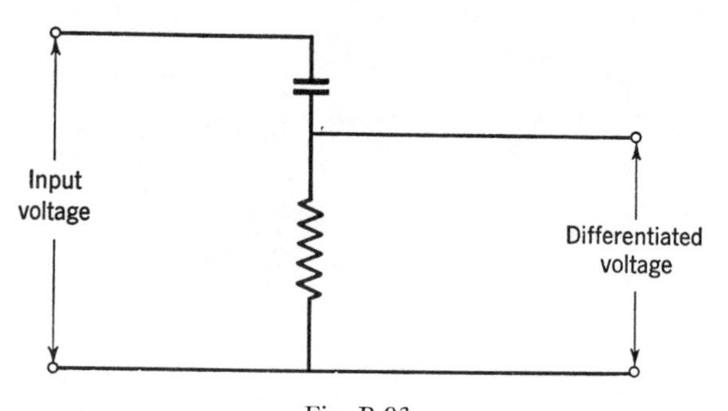

Fig. P-93.

Investigate the conditions under which the circuit will differentiate satisfactorily by trying various forms of input voltage.

94. Frequency Response of an Oscillograph

A magnetic oscillograph element is known to be critically damped and to have an undamped natural frequency of 1200 cycles. From this information it is desired to obtain if possible curves of frequency response and phase shift for the range 0 to 2000 cycles and also the response to a suddenly impressed direct current.

95. A-C Response from D-C Response

When direct current is suddenly applied to the element of a magnetic oscillograph the curve in figure P–95 is obtained. The frequency of the damped oscillation is 1500 cycles and the amplitudes of successive swings in the same direction are in the ratio of 2 to 1. From this it is desired to predict the response to alternating current and phase shift as functions of frequency.

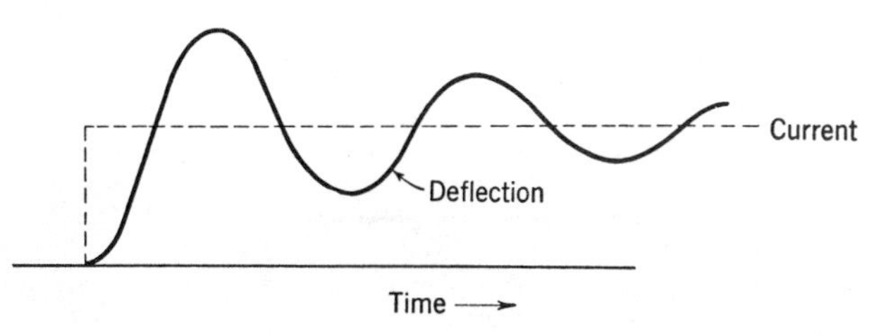

Fig. P-95.

96. D-C METER WITH TWO ALTERNATING CURRENTS

The period of a center-zero d-c ammeter is short enough so that instantaneous values of a current of ½-cycle frequency are indicated with reasonable accuracy. What will the meter indicate for superposed alternating currents of 119.5 and 120.0 cycles each of the same rms value?

97. A-C METER WITH SUPERPOSED CURRENTS

The current through an iron vane type of ammeter is made up of two superposed components of equal rms value but different frequencies. The frequency of one component is constant at 60 cycles, and the other is variable. Describe the behavior of the meter as the latter frequency is varied from 0 to 120 cycles.

98. REDUCTION OF HARMONIC VOLTAGES

It is found that a source of voltage to be used for testing purposes contains objectionably large fifth and seventh harmonics. To eliminate these harmonics, and thus produce a more nearly pure sine wave of voltage, the tuned circuits *FF* and *SS* shown in figure P–98 are proposed. Circuit *FF* is tuned so that its inductive and capacitive reactances are equal at the fifth harmonic frequency; *SS*, similarly for the seventh harmonic. The thought is that these circuits will effectively short circuit the unwanted harmonics.

Is the proposal feasible?

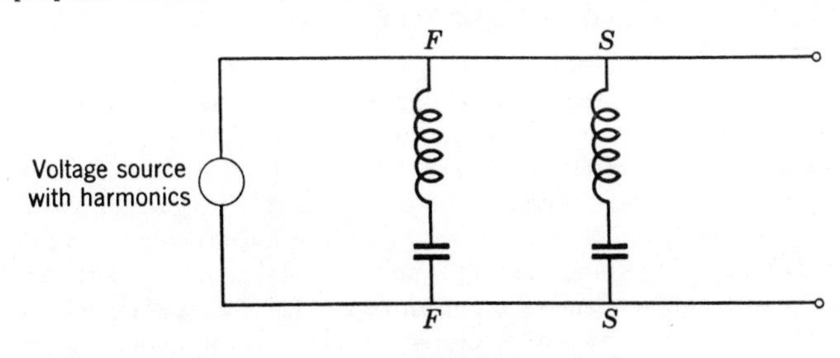

Fig. P-98.

99. CIRCUITS FOR A VIBROMETER

A certain vibration-measuring instrument (vibrometer) contains a coil which moves relative to a stationary permanent magnet. Thus there is developed in the coil a voltage which is proportional to the velocity of the part whose vibration is being measured. We wish to have a circuit to which the coil may be connected and which will give us as output a voltage proportional at every instant to the displacement of the part. We also want another circuit to be used alternatively whose output will be a voltage proportional to the instantaneous acceleration of the vibrating part. Steady-state vibrations only are of interest. Amplifiers are available so that even if the output voltage of the desired circuit is very small it may be magnified suitably for application to a meter or indicating device. Devise a circuit for each case and specify suitable proportions for its constants.

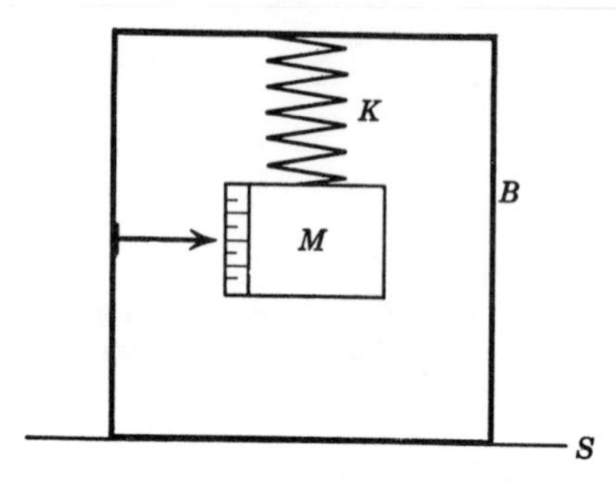

Fig. P-100.

100. Vibrometer

In figure P–100, B represents a vibrometer, which is to be used to measure the steady-state amplitude of the vertical displacement of a surface S assumed to be sinusoidal in time.

The instrument contains a mass M suspended by means of a spring of constant K. Position of the mass M with respect to the container B is measured by a dial indicator, electric strain gage, or other suitable means. How must M, K, and the frequency of the vibration of the surface be related so that the device will indicate directly the amplitude of the displacement of the surface over a wide range of frequency?

How should M, K, and the frequency be related if the indication is to be proportional to the amplitude of the acceleration of the surface over a wide range of frequencies?

101. Modification of Vibration Instrument

A certain apparatus for measuring vibration contains a piezo-electric crystal which gives a voltage output proportional to the force exerted through the crystal. One side of the crystal is fastened to a massive holder, and the other side has fastened to it a rigid probe whose mass is very small. In using the device, the end of the probe is fastened firmly to the surface whose vibrations are to be measured. Under these conditions the massive holder is forced to follow the vibrating surface since the probe and the crystal are essentially rigid. The voltage output from the crystal is thus proportional at every instant to the acceleration of the vibrating surface. This voltage is amplified without distortion and made to drive an oscillograph so that the time variation of the acceleration can be observed visually.

Often, however, it is more desirable to observe the form of the velocity or the displacement of the vibrating surface, and this can be done by incorporating in the amplifier either one or two integrating circuits, as the case may be.

As an alternative to integration by electrical means, it is desired if possible to modify the device mechanically so that its output voltage variation with time will represent the time variation of the velocity of the surface. It is also desired to modify the device mechanically in another way so that the voltage output from the crystal will be proportional at every instant to displacement of the vibrating surface.

Try to devise suitable mechanical modifications that will accomplish these two objectives, that is, voltage output from the crystal proportional to velocity of the surface, and voltage output proportional to displacement, retaining the same crystal throughout. Specify the ranges of frequency for which your proposals are good.

The vibrations to be observed are periodic and in a steady state.

102. Instrument to Measure Peak Value

A certain vibration-measuring device gives a voltage proportional at every instant to acceleration. It is desired to apply this voltage to an instrument that will give a reading proportional to the peak value of the acceleration when the wave form of the vibration is non-sinusoidal, but cyclic. The instrument should indicate the maximum acceleration occuring in the cycle independently of the wave form, provided only that the vibration is in the steady state. For this purpose the circuit in figure P–102 has been proposed. The ammeter A is of the D'Arsonval or d-c type; it indicates the average value of current through it. The rectifiers may be assumed ideal.

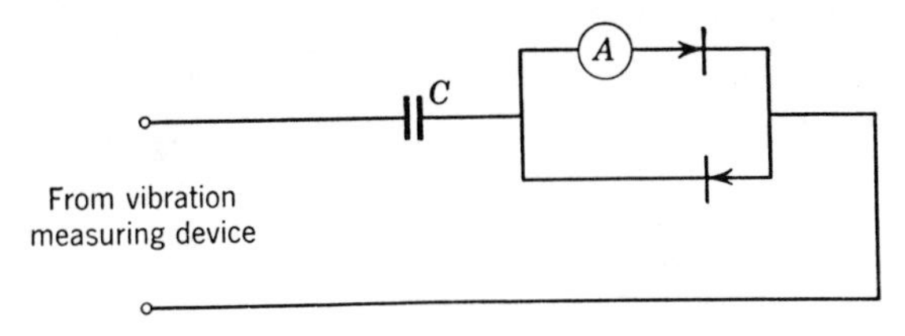

Fig. P-102.

Analyze the proposal, and, if it is at all suitable, consider how to choose the value of the capacitance C and the range of A. Perhaps some kinds of waves will result in correct readings and others not.

103. Vibrating Motor-Generator Set

A motor-generator set installed in a steel building produces objectionable noise and vibration. It is proposed, therefore, to provide rubber mounting pads, the rubber to have very little internal damping.

The set weighs 2000 lb, and its speed is essentially constant at 1200 rpm. Measurements of the disturbing frequencies have been made, and they are:

20 cycles per second, which corresponds to the rotation of 1200 rpm (this one tickles the feet).

120 cycles per second, which comes from magnetization of the iron in the motor by 60-cycle current (this is audible noise).

It is believed that the rubber mountings will result in a suitable cure if the 20-cycle and 120-cycle components of force transmitted to the building can be reduced to $\frac{1}{3}$ and $\frac{1}{10}$, respectively, of their values without the mounting. The static deflection of the set on the flexible mounting must not exceed 0.150 inches.

Recommend, if possible, a suitable stiffness constant for each of four rubber pads—one under each corner of the set.

104. Instrument Mounting

A delicate instrument is to be mounted on a support designed to minimize the vibrations due to tremors of the building. To this end it is proposed to set the instrument upon a small platform suspended by springs from the ceiling and without any damping device. Assume the ceiling to vibrate vertically in simple harmonic motion with a fixed amplitude.

To minimize the platform vibrations, should the mounting be extremely soft (large deflection under gravity), extremely stiff (small deflection under gravity), or have some intermediate value of stiffness?

105. Elastic Mounting on Non-Rigid Base

In elastically mounting a machine for isolation of vibration or shock it usually happens that the floor is not rigid but is itself a mass on elastic supports. Treating the floor as a mass supported on springs and assuming no damping, find how the natural frequency of oscillations is affected by the mass of the floor and the stiffness of its supports. Consider vertical motions only. It would be desirable, if possible, to have the results in terms of the two ratios: mass of floor to mass of machine, and spring constant of floor supports to spring constant of machine on floor. Ideally either or both these ratios would be infinite.

106. Vibration Absorber

Investigate the motion of a system consisting of a massive machine mounted on an elastic foundation, and a small mass attached elastically to the machine as indicated in figure P–106. Due to inherent unbalance of its rotating parts the machine is subjected to a vertical component of force which varies sinusoidally in time at constant frequency. All motions are constrained to the vertical.

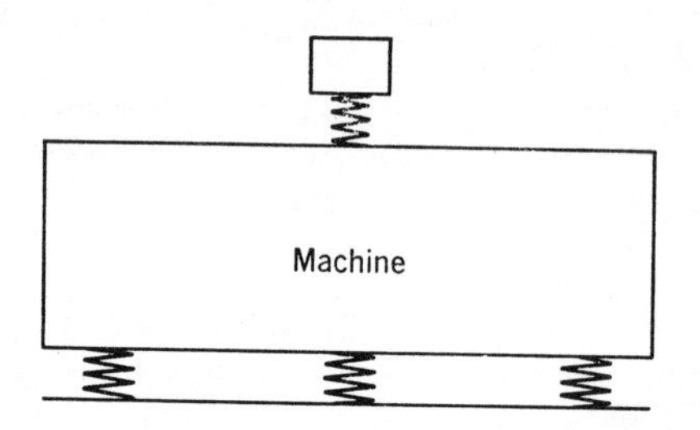

Fig. P-106.

Are there any conditions under which the presence of the small mass reduces the force transmitted through the elastic foundation of the machine?

107. Instrument for Investigating Turbulence

Figure P–107 shows an instrument for studying turbulence in an open channel, consisting of a sphere about $1\frac{1}{2}''$ in diameter fastened to a stiff rod about $18''$ long, and immersed in the stream of water. The rod is connected to a fixed support by a short leaf spring on which electric strain gages are placed. Motion of the rod is damped by an oil pot. The damping appears to be a little more than critical.

The output of the strain gages is known to be directly proportional at every instant to the moment acting across the leaf spring. This output is amplified and is used to actuate the pen on a recorder.

It is claimed that the output of the strain gages is also directly proportional at every instant to the force acting longitudinally on the sphere. This is certainly true if the force is constant, but what if it varies rapidly and irregularly as it does when the sphere is immersed in a highly turbulent stream? The form of a typical record is shown. During such a recording the sphere moves back and forth perhaps $\frac{1}{4}$ to $\frac{1}{2}$ inch.

An analysis of the dynamic behavior of the instrument is desired so as to know what errors may be inherent in it.

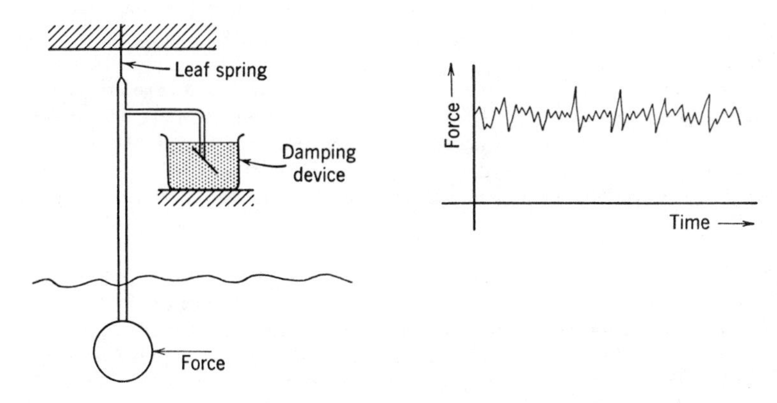

Fig. P-107.

108. Flywheel for a Motor-Driven Compressor

An air compressor driven by a synchronous motor is causing trouble because of severe pulsation in the line current to the motor. It is believed that this is due to mechanical oscillation of the motor rotor with respect to its average or synchronous position, brought about by the fluctuating torque of the compressor. It is thought that an addition to the flywheel might move the natural frequency of the system far enough away from the forcing frequency to cure the trouble. With this attack on the problem in mind, the following information has been gathered.

The compressor is 2-cylinder 2-stage with cranks at 90°. Indicator cards have been taken, and with proper consideration of the connecting-rod and crank geometry and the inertia of reciprocating masses the total crank effort has been calculated. A Fourier analysis of this shows that the fluctuating component of torque in pound-feet is given by

$$-542 \sin \omega t + 210 \cos \omega t + 1350 \sin 2\omega t - 263 \cos 2\omega t$$

where ω is the average shaft speed in radians per second.

The synchronous motor is directly connected to the crankshaft and is rated 28-pole, 210 hp, 170 kva, 257 rpm, 60-cycle, unity power factor, 2300 v. Its synchronous power has been calculated to be 623 kw per electrical radian of displacement. This means that if the rotor were displaced 1 electrical radian behind its no-load synchronous position a motor torque sufficient to give 623 kw at 257 rpm would be developed. The torque thus produced within the operating range may be assumed directly proportional to the electrical displacement.

The torque produced by the damper windings of the motor is estimated at 1800 lb-ft per mechanical radian per second. The inertia constant (WR^2) of the motor, compressor, and existing flywheel is 10,500 lb ft^2.

Experience has shown that the current pulsations will be within acceptable limits if the amplitude of the motor oscillation is less than about 3 to 5 electrical degrees.

Determine whether an addition to the flywheel is advisable and, if so, its size.

109. Machine on Flexible Support

A reciprocating machine bolted to a steel beam carried between masonry supports vibrates objectionably. The vibrations have a double amplitude (vertical) of 0.005 in. and a frequency of 40 cycles per second (determined with vibration-measuring instruments at a point on the beam where the machine is attached). The machine weighs 1000 lb, and this causes a static deflection of the beam of 0.015 in. measured at the same point as the vibration amplitude.

Suppose one of your subordinates, a young engineer whom you have just employed, has investigated the situation. He has decided that the supporting beam is too flexible, and he plans to provide a second and similar beam beside the original one and to move the machine slightly so that it is supported equally by the two beams. He comes to you for approval of the order for the material required for the change. What would you do?

110. Voltmeter for Very Low Frequency

An electrodynamometer type of a-c voltmeter is to be used to measure sinusoidal voltages in an experimental study of a servo system. If the frequency is not too low the instrument gives a substantially steady indication equal to the effective (rms) value of the voltage. On the other hand at much lower frequencies the pointer swings up and down the scale in step with the voltage so that the maximum indication is a nearly true measure of the instantaneous crest voltage (multiplied by a proper scale factor). Thus there are two possible ranges in which the voltmeter may be used: a high frequency range where the needle is steady enough to be read in the usual manner, and low frequency range where the maximum of the swing is proportional to crest voltage almost independently of frequency.

Recommend suitable limits for the two frequency ranges in the case of an instrument of the usual laboratory portable type with a 150-volt range. The manufacturer was unable to give the desired information but did supply the following about the instrument:

Frequency range intended	25 to 133 cycles per second
Moment of inertia of the moving element	13.8 gm-cm^2
Spring constant	0.14 gm-cm per radian
Damping	0.8 critical

111. Damping for a Vibrating Reed Tachometer

Vibrating reed tachometers and frequency meters contain a number of reeds R fastened to a block B as indicated in figure P–111. The block is subjected to a vibration whose frequency is to be determined. Each reed is loaded at its free end with a mass M which for the purposes of this analysis may be assumed large compared to the mass of its reed.

Fig. P-111.

The reeds in a given instrument are tuned to a series of equally spaced frequencies. When the block B is vibrated at a fixed frequency the mases at the ends of the reeds vibrate steadily to give a pattern such as that shown. The reed showing the largest amplitude indicates the value of the impressed frequency.

Investigate the influence of damping on the resolution of such an instrument.

112. Overrunning Clutch

Figure P–112 shows a one-way or overrunning clutch such as that incorporated in many automatic transmissions for automobiles. Such a clutch has been termed the highly civilized, toothless great-grandson of the pioneer pawl and ratchet. It is used to allow one shaft to rotate freely with respect to another for one direction of relative motion, and to prevent relative motion in the reverse direction.

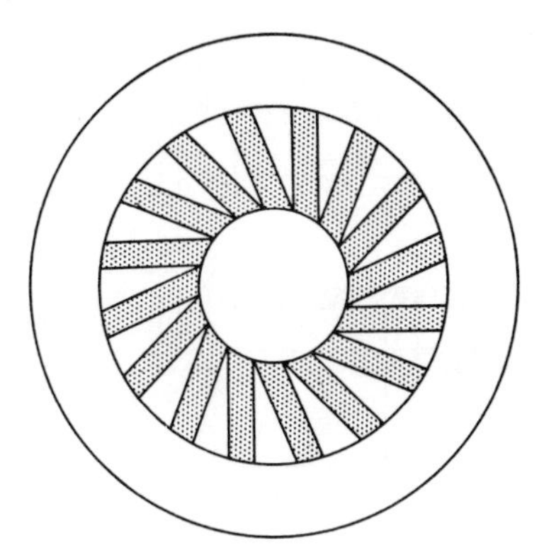

Fig. P-112.

The clutch shown is of the sprag type. The small pieces of roughly trapezoidal shape (the sprags) are prevented from falling out by retaining rings not shown and are otherwise unrestrained except at the surfaces which engage the inner and outer cylinders.

Work out the important basic design relationships (for instance, the ratio between length and width of the sprags) in terms of what you consider to be fundamental quantities.

113. Overloading a Transformer

It is desired to operate a small air-cooled transformer at double its rated output current at normal voltage for a short time without exceeding the winding temperature rise upon which the rating is based.

The transformer is rated 3 kva, 115/11.5 volts, 60 cycles. Its total weight is 95 pounds, and it may be assumed that the weights of the iron core and copper coils are respectively 60 and 30 pounds. The design is such that the iron almost completely surrounds the coils so that nearly all the heat generated in the latter must pass through the iron on the way out to the air.

The following test data pertain to conditions at rated full load output (3000 watts at 11.5 volts).

Iron loss	44 watts
Copper loss, cold	64 watts
Copper loss, hot	77 watts
Steady-state copper temperature rise	63.5°C
Steady-state temperature rise of outside iron surface	40°C

How long is it safe to operate the transformer at twice the rated load current? Is your estimate conservative?

114. Temperature Controller

In developing a plastic molding process the temperature of the plastic is to be maintained at fixed values which may be varied over a moderate range near 150°F. A scheme for doing this is to imbed a thermocouple in the plastic at some representative spot and use its output voltage as the input to an automatic controller, as indicated in figure P–114. The controller would regulate the heat to or from the plastic by controlling the flow of steam and cold water to the jacket spaces in the cast-iron mold.

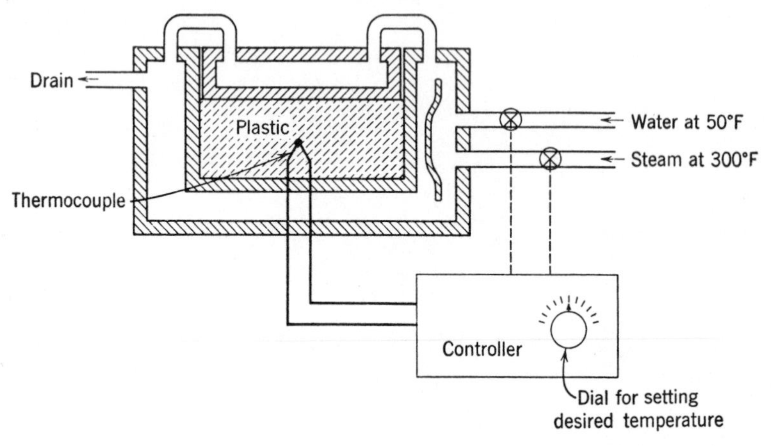

Fig. P-114.

In such a system, how may the temperature in the plastic be expected to vary with time after the controller is suddenly reset to call for a new temperature? In particular, might the temperature come gradually to the new value or might it oscillate about this value?

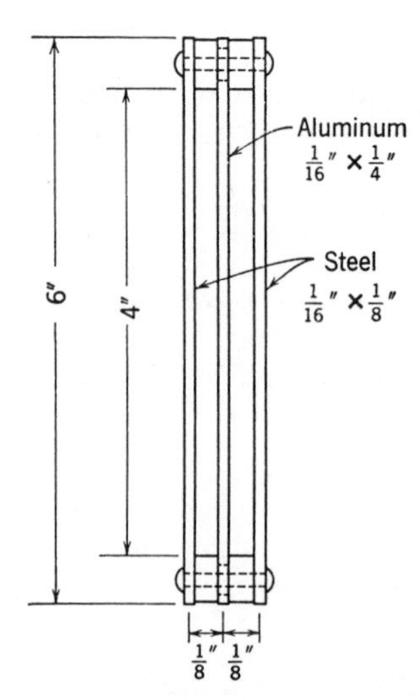

Fig. P-115.

115. Thermal Snap Switch

A thermal switch which is to close when the temperature increases about 90°C has been designed for incorporation in a low-cost product. The switch consists of three metal strips clamped together in an assembly of insulating pieces, as shown in figure P–115. At the actuating temperature the central strip is to snap aside, making contact with one or the other of the outer strips, which are electrically connected, and thus closes a circuit.

It is desired to modify the design so that operation will occur with a rise of 20°C. If possible, preserve the external dimensions so that the two switches will be interchangeable in the assembly of the product. It would also be desirable to vary only the thickness or material of the strips so that the same dies can be used to produce either design.

116. Heat Flow Meter

This problem concerns a heat flow meter described quite incompletely in a technical pamphlet issued by a competitor. The meter element is constructed as a flat woven

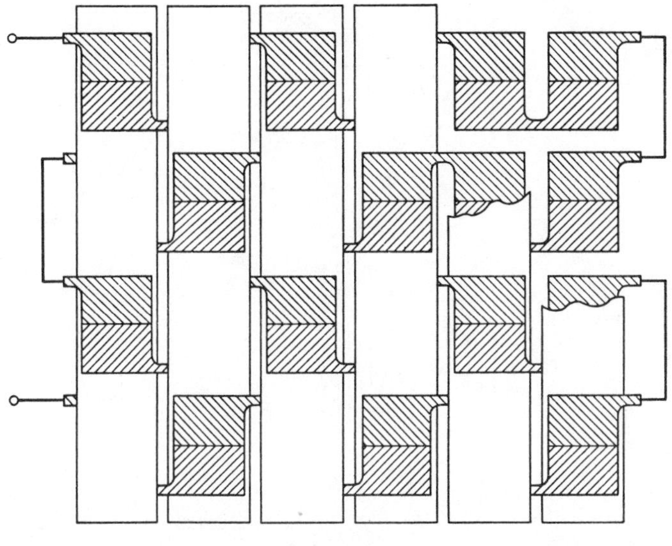

Fig. P-116.

mat of plastic and metal strips, as shown in figure P–116, which is scaled from photographs in the pamphlet. The plastic strips, which run vertically in the figure, are rectangles about 1 inch wide by 5½ inches long, and presumably quite thin. There are six such strips in the figure, the two at the right being torn off to show the metal strips more clearly.

The metal strips are made of "bimetallic foil" and are otherwise undescribed except for what can be deduced from the photographs. Apparently the two metals composing the foil are joined edge to edge along the center line of the strips, as indicated by the cross hatching in the figure. Notches are cut nearly through the metal strips alternately from either side at a spacing equal to the width of the plastic strips, that is, at intervals of about 1 inch. The metal strips are woven over and under the plastic strips as shown, and their ends are joined by what look in the photographs like thin strips of foil so as to form a series path through all four strips.

Although the pamphlet does not say so, photographs of the meters in use imply that the indication is electrical.

Photographs of meter elements in various stages of construction show that the woven assembly is covered on each face with a sheet of plastic. The assembled element is thus a thin flat rectangular pad which is cemented onto a solid surface such as a wall through which the heat flow is to be measured.

We need an explanation of the way the meter works and an estimate of its sensitivity.

117. Corona Voltage for a Bushing

A common form of bushing for insulating a conductor where it passes through the cover of a transformer or circuit-breaker tank consists of a tubular porcelain shell

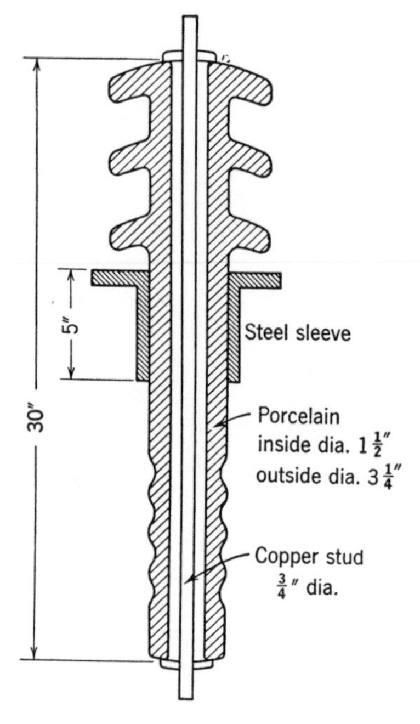

Fig. P-117.

grooved on its outside surface. At about the midlength the porcelain is cemented into a steel sleeve which in turn is bolted to the tank cover. The conductor is a straight bare copper rod which is held centrally in the hole through the bushing. To allow for the large manufacturing variations in the inside diameter of the porcelain it is necessary to leave a thick annular air space between the copper and porcelain. When such a bushing is operated above a certain voltage, corona forms in this air space. Corona is objectionable on account of the radio interference which it induces and the deteriorating effect it has on organic insulation.

For the bushing in figure P–117 determine the corona starting voltage (rms line-to-line 3-phase).

To raise the corona voltage it has been proposed to wrap the stud with varnished cambric tape before putting it into the porcelain, leaving about 1/16- or 1/32-inch clearance between the wrapped stud and the porcelain. Give an engineering opinion on this proposal.

The dielectric constant of porcelain is 5.5 and that of varnished cambric tape is 3.0.

Corona commences when the crest value of the voltage gradient in the air reaches a critical value. The critical value is dependent somewhat on the geometry of the surfaces which bound the field. For a space bounded by concentric cylinders the corona starts at the inner surface at a value of gradient which for normal atmospheric conditions is as follows:

Inner Radius, cm	*Corona Starting Gradient, kv (max) per cm*
0.8	40
1.0	39
1.5	37
2.0	36

118. High-Voltage Circuit Breaker

The advantage of the multibreak circuit breaker over the single-break type depends on its ability to divide the circuit voltage more or less equally between the several gaps. The division of voltage depends almost entirely upon the electrostatic capacitances between the various elements of the breaker and between them and ground.

When one end of a certain horizontal four-break circuit breaker is grounded and 146 kv (60 cycles) is applied to the other end, a test gives the voltages across the separate breaks shown in figure P–118.

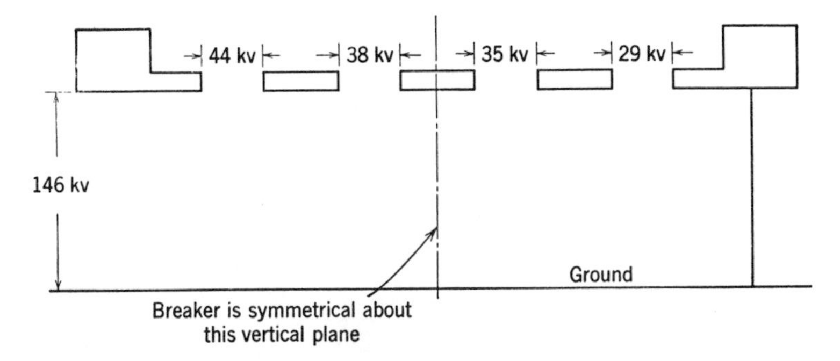

Fig. P-118.

Find the voltage distribution with $+E$ applied at one end of the breaker and $-\frac{1}{2}E$ at the other. This is the condition which applies to the first pole to clear when a three-pole breaker interrupts a symmetrical three-phase ungrounded short circuit on a system whose line voltage is $\sqrt{3}\,E$.

119. Single-Phase Motor Mounting

Single-phase a-c motors develop a torque which pulsates between zero and twice the average value at double the supply frequency, that is, at 120 cycles for a 60-cycle motor. When the stator is rigidly mounted the result is transmission of the pulsating torque through the base with consequent generation of objectionable noise.

To overcome the difficulty it is common to support the stator by rubber rings clamped around the bearing housings. This gives the desired torsional springiness and at the same time assures reasonable lateral stiffness as is necessary for coupling to the driven device.

To avoid the use of rubber in an application where there is a high concentration of ozone an all-metal mounting is desired. The proposed design is indicated in figure P–119. Steel straps bent as shown and fastened rigidly to the bearing housings, which may need modification, and to the base plate will provide torsional flexibility and lateral stiffness.

Specify suitable straps for a ¼-horsepower, 60-cycle, 1750-rpm motor.

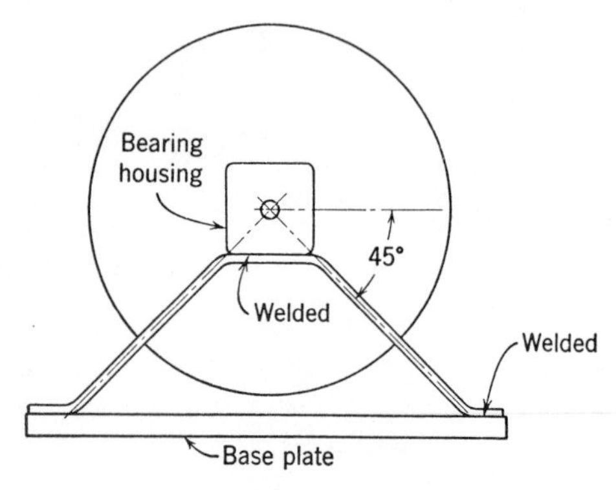

Fig. P-119.

120. Dynamic Braking of Two Motors

In a tire-fabric plant the fabric is passed through two sets of rolls driven by separate d-c motors, as indicated in figure P–120. The motors are shunt wound and are adjusted so that the fabric between rolls is under tension in normal running.

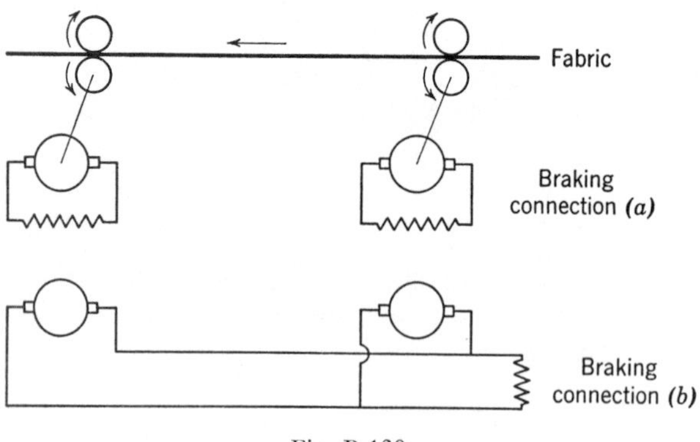

Fig. P-120.

For quick stopping, dynamic braking is to be used on both motors. While stopping, the fabric between rolls should neither pile up nor be subjected to excessive pull. The motors are nominally alike, but the inertias of the rolls and the retarding torques may be different.

Choose between the following proposals for braking:

(*a*) A separate braking resistor for each motor, the connections to be made simultaneously after the power supply circuit to each motor is opened.

(*b*) A single braking resistor to be connected with the armatures left in parallel as they are during running.

121. Capacitors for Electrostatic Precipitator

The high-voltage supply circuit for an electrostatic precipitator for cleaning the air in a house is shown in figure P–121.

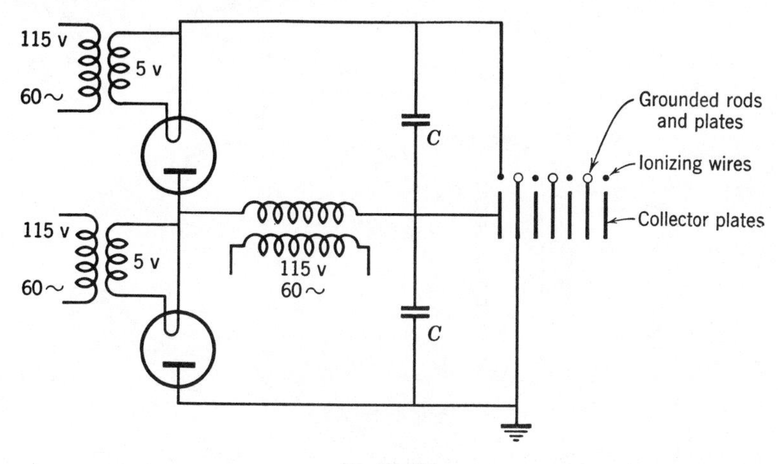

Fig. P-121.

The ionizing wires are to be maintained at 12,000 volts positive with respect to the grounded rods. This voltage should remain substantially constant, but a small decrease during the cycle can be tolerated. At this potential of 12,000 volts the total discharge current between the ionizing wires and the grounded rods and plates is 1.0 milliampere. Current to the collector plates is negligibly small. Capacitances between the precipitator elements are negligible.

For economic reasons the condensers, which are of equal size, should be as small as possible. It is therefore desired to have a relation between their size and the voltage fluctuation on the ionizing wires, and a recommendation as to a suitable size of condenser.

122. Manometer System

Because of space limitations, the pressure-measuring systems used in wind-tunnel testing frequently employ capillary tubes of considerable length between the manometer and the point where the unknown pressure exists. Sometimes the capillary has to be so fine and long that serious time lag may be introduced in the measurement, particularly in testing at very low pressures.

In attempting to work out more satisfactory arrangements the experimental set-up shown in figure P–122 has been made, in which the performance of the measuring system can be studied under conditions more easily controlled than in a wind tunnel. The experimental procedure is to pump the tank down to some low pressure representative of the wind-tunnel condition and maintain it there. Then the air inlet valve is opened momentarily to establish a pressure difference across the manometer. With the air valve closed again the left-hand column of the manometer gradually rises toward equilibrium and the scale readings are observed as functions of time. Thus the response of the measuring system to a step-function of pressure is determined.

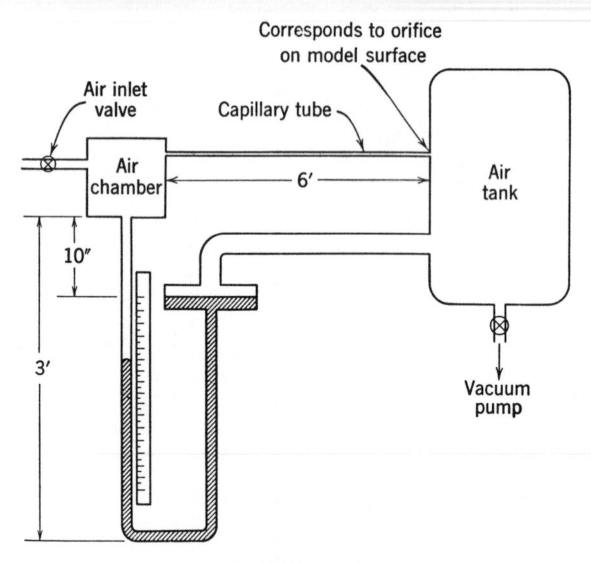

Fig. P-122.

However, the experimenter does not have a clear conception of what factors are important and his investigation is not proceeding smoothly. Consequently, it is desired to work out analytically a theory from which the response of the manometer system can be calculated. It is hoped that by so doing the need for further experimentation can be eliminated or at least that the program can be placed on a rational basis.

The following data pertain to one of the experiments.

Pressure in the air tank (constant), 1/50 atmosphere

Difference in Pressure as Read using the Manometer Scale, lb ft^{-2}	*Time (seconds)*
136.4	0
80	18.5
50	40
25	78
4	204

The initial pressure difference of 136.4 lb ft^{-2} corresponds to an initial depression below equilibrium of exactly 15 in. The manometer fluid has a specific gravity of 1.749 at 72°F, the temperature of the test.

The internal diameter of the manometer tube is 0.160 in. The length from the bottom of the air chamber to the bottom of the U is 3 ft, and that to the equilibrium level is 10.0 in.

The capillary is made of Monel metal 6 ft long and 0.039 in. inside diameter. The volume of the air chamber including the pipe to the air inlet valve is 5.70 $in.^3$

Index